ENTERPRISE SECURITY

Robert C. Newman

Georgia Southern University

Upper Saddle River, New Jersey
Columbus, Ohio

Library of Congress Cataloging-in-Publication Data
Newman, Robert C.
　　Enterprise security / Robert C. Newman.—1st ed.
　　　p. cm.
　　Includes bibliographical references and index.
　　ISBN 0-13-047458-4 (alk. paper)
　　1. Electronic commerce—Security measures. 2. Business enterprises—Computer networks—Security measures. 3. Computer security. 4. Data protection. I. Title.

HF5548.32.N49　2003
658.4'78—dc21　　　　　　　　　　　　　　　　　　　　　　2002074826

Editor in Chief: Stephen Helba
Assistant Vice President and Publisher: Charles E. Stewart, Jr.
Production Editor: Alexandrina Benedicto Wolf
Production Coordination: Preparé, Inc.
Design Coordinator: Diane Ernsberger
Cover Designer: Ali Mohrman
Cover art: Stockmarket
Production Manager: Matthew Ottenweller
Marketing Manager: Ben Leonard

This book was set in Palatino and Helvetica by Preparé, Inc. It was printed and bound by R. R. Donnelley & Sons Company. The cover was printed by Phoenix Color Corp.

Pearson Education Ltd.
Pearson Education Australia Pty. Limited
Pearson Education Singapore Pte. Ltd.
Pearson Education North Asia Ltd.
Pearson Education Canada, Ltd.
Pearson Educación de Mexico, S.A. de C.V.
Pearson Education—Japan
Pearson Education Malaysia Pte. Ltd.
Pearson Education, *Upper Saddle River, New Jersey*

Copyright © 2003 by Pearson Education, Inc., Upper Saddle River, New Jersey 07458. All rights reserved. Printed in the United States of America. This publication is protected by Copyright and permission should be obtained from the publisher prior to any prohibited reproduction, storage in a retrieval system, or transmission in any form or by any means, electronic, mechanical, photocopying, recording, or likewise. For information regarding permission(s), write to: Rights and Permissions Department.

10 9 8 7 6 5 4 3 2 1
ISBN 0-13-047458-4

PREFACE

WHO SHOULD READ THIS BOOK?

The current Enterprise networking market has been growing at an unbelievable rate, but security preparedness has not kept pace with this growth. This is in a large part due to the expansion of Internet access to almost all segments of society. Everyone who has a computer can now have access to the Web. Children are introduced to the Web at an early age, both at school and at home. The opportunity for expansion of services to the entire population is astronomical. This book is designed to provide a broad working knowledge of the security issues that permeate today's Enterprise networks and to provide the serious student with the tools to address the numerous opportunities that abound in this field.

This book is oriented toward both the business and education communities, especially toward the undergraduate students who aspire to succeed in the telecommunications and data communications industry. It is also directed at the marketing and sales professionals and the education providers that support these organizations. Some prior knowledge of the telecommunications and computer industry, as well as of basic data-communications functions and client/server processes, is beneficial. Database and computer-processing knowledge is also a plus, as these two topics are closely interrelated. *Broadband Communications* (Newman, 2002) provides a basic understanding of the current broadband communications technologies.

This book covers the entire field of Enterprise security, including physical locations, computer and client/server systems, communication networks, application and operating system software, and WAN, MAN, and LAN topologies. A general understanding is developed for each component of the Enterprise environment that can be impacted by security issues. Wiring and cabling issues are addressed in detail, because this is a common point of attack across the network. Chapters 12 and 13 are devoted to virtual networks and distributed networks, which address the security issues of VPNs and e-commerce systems. Chapter 14 is devoted entirely to wireless networking security, which addresses issues prevalent in the cellular, microwave, satellite, and mobile radio environments.

Although this book is not highly technical, just enough detail is provided to allow the reader to apply the information provided to be successful in addressing the security issues prevalent in the Enterprise environment. Anyone who is part of the strategic, tactical, or operations structure will find this book interesting and useful. This may lead the reader to look at the more technical aspects of this field.

ORGANIZATION OF THIS BOOK

This book is divided into two major parts. Part One provides a general understanding of security and integrity issues of computer and networking systems that are part of the Enterprise environment. Part Two provides an in-depth look at specific applications of these areas.

Part One includes chapters on Enterprise Security Basics; Enterprise Access Control; Cryptosystem Techniques; Security Systems; Enterprise Security Management; Computer and Network Systems; Software and Database Systems; Attacks, Threats, and Viruses; and lastly, Building, Campus, and Facility Security. Because many principles are developed and terms explained in this part, these chapters should be read and understood before moving to Part Two. Some prior exposure to programming or data processing would be a plus for understanding the subjects presented in Chapter 7.

Part Two includes chapters on the Local and Metropolitan Area Network Security Environment; Wide Area Network Security; Virtual Networks Security; Distributed Systems Security; and Wireless Networking Security. Cross-references are presented, when appropriate, to refer back to information found in Part One. Some knowledge of data communications would be beneficial for a complete understanding of the material presented in Chapters 10 and 11.

The Enterprise Systems Security Review is discussed throughout the book and should be reviewed when reading Chapter 1. Appendix A should be included as a reading assignment with Chapter 7.

More than 600 key terms are provided throughout the book and defined in the glossary. The glossary terms and definitions were gathered from sources listed in the References and Other Resource section, including *Newton's Telecom Dictionary* (Newton, 1999) and the *Dictionary of Communications Technology* (Held, 1996). A number of LAN, MAN, and WAN concepts presented in this book were extracted from *Broadband Communications* (Newman, 2002).

HOW TO USE THIS BOOK

Each chapter presents a set of exercises, varying in complexity, which can be used to support the material that has been presented. These exercises require external resources, such as the Web, to develop the solutions. Some of these exercises may qualify as semester projects as some intense research is required to develop complete solutions.

Many Internet and Web resources were used in developing this book. A representative sample of URLs is presented at the end of each chapter. The amount of information and data on the security subject is staggering and cannot possibly be digested during one semester.

More than 600 questions are presented throughout the book, which provide complete coverage of the material. The answers to these questions can be found in the Instructor's Manual. Also included in the Instructor's Manual are answers to the true/false questions and exercises that are presented at the end of each chapter. A CD-ROM containing PowerPoint slides is packaged with the Instructor's Manual.

ACKNOWLEDGMENTS

I would like to thank the following reviewers for their invaluable feedback: Bill Hessmiller; K. R. Kirkendall, Montana State University; Bill Liu, DeVry University, California; Fred Seals, Blinn College, Texas; and Costas Vassiliadis, Ohio University. Comments and suggestions concerning this book can be sent to me at newmanrc@GaSou.edu.

Robert C. Newman

This book is dedicated to all the public service personnel involved in the protection of the nation's population and resources, especially those who are members of the numerous fire responders and law enforcement agencies.

CONTENTS

PART ONE 2

CHAPTER 1 Enterprise Security Basics 4
　　　　　　Objectives 4
　　　　　　Introduction 5
　　1.1　A Historical Perspective 5
　　1.2　The Current Environment 8
　　1.3　Trusted and Untrusted Networks 10
　　1.4　Computer and Networking Issues 10
　　1.5　Corporate Security Policy 11
　　1.6　Site Security Policy 12
　　1.7　Network Security 13
　　1.8　Hardware, Software, and Network Environment 13
　　1.9　Threats 14
　　1.10　Security Goals and Objectives 17
　　1.11　Threats To Network, Hardware, and Software Components 18
　　1.12　Data and Database Security 22
　　1.13　Electronic Commerce and Business-to-Business 23
　　1.14　Protecting the Enterprise Assets 24
　　1.15　Intrusion Detection 26
　　1.16　Physical Plant Security and Control 26
　　1.17　Continuity Plan 27
　　1.18　Security Consulting and Outsourcing 28
　　　　　　Summary 29
　　　　　　Key Terms 30
　　　　　　Review Exercises 31

CHAPTER 2 Enterprise Access Control 34
　　　　　　Objectives 34
　　　　　　Introduction 35
　　2.1　Network Resource Access 35

2.2 Passwords 38
2.3 Authorization 40
2.4 Access Control 42
2.5 Privileges and Roles 44
2.6 Digital Certificates and Signatures 44
2.7 Certificates and Certification Systems 45
2.8 Public Keys and Private Keys 46
2.9 Data Security and Encryption 49
2.10 Key Management 53
2.11 Access Security of Computer and Network Resources 54
2.12 Biometric Access Control 57
2.13 Security Cost Justification 58
2.14 Internet and Web Resources 58
Summary 59
Key Terms 60
Review Exercises 61

CHAPTER 3 Cryptosystem Techniques 64
Objectives 64
Introduction 65
3.1 Cryptosystems 65
3.2 Cryptography 66
3.3 Encryption Algorithms 66
3.4 Public Key Cryptography 74
3.5 Public Key Infrastructure (PKI) 77
3.6 Private Key Cryptography 78
3.7 Cryptanalysis 78
3.8 Steganography 80
3.9 Security Models 81
3.10 Internet and Web Resources 83
Summary 84
Key Terms 84
Review Exercises 85

CHAPTER 4 Security Systems 88
Objectives 88
Introduction 89
4.1 Security Systems Design 89
4.2 Trusted Systems 90
4.3 Biometric Systems 91
4.4 Secure Protocols 94
4.5 Secure Hypertext Transfer Protocol (S-HTTP) 96
4.6 Secure Electronic Transaction (SET) 97
4.7 Secure Multipurpose Internet Mail Extension (S-MIME) 101
4.8 Kerberos 102

4.9 Point-to-Point Protocol (PPP)   104
4.10 Instrusion Detection   105
4.11 Security Evaluation   107
4.12 Security Specifications   108
4.13 ISO 17799 Standard   114
4.14 Encryption Exporting   114
4.15 Internet and Web Resources   114
Summary   116
Key Terms   116
Review Exercises   117

CHAPTER 5 Enterprise Security Management   120
Objectives   120
Introduction   121
5.1 Administration   122
5.2 Corporate Planning   122
5.3 Security Requirements Assessment   123
5.4 Maintaining Network Integrity   124
5.5 Network Management System   132
5.6 Network Management Architecture   133
5.7 SNMP Management Functions   136
5.8 Common Management Information Protocol (CMIP)   138
5.9 Network Management Standards   142
5.10 Network Management Support Systems   143
5.11 Network Troubleshooting   154
5.12 Test Equipment and Resources   156
5.13 Security Reform Act   159
5.14 Internet and Web Resources   160
Summary   160
Key Terms   161
Review Exercises   161

CHAPTER 6 Computer and Network Systems   164
Objectives   164
Introduction   165
6.1 Computer Systems   166
6.2 Mainframe Components   167
6.3 Client/Server Components   171
6.4 Securing Mechanized Transactions   174
6.5 Network Components   175
6.6 VPN Hardware and Software   185
6.7 Voice Communication Systems   188
6.8 Internet and Web Resources   191
Summary   192
Key Terms   192

Review Exercises   193

CHAPTER 7   Software and Database Systems   196
　　　　　　　Objectives   196
　　　　　　　Introduction   197
　　7.1　Protecting the Data and Database Asset   197
　　7.2　Database Management System   201
　　7.3　Operating System   202
　　7.4　Directories   204
　　7.5　Access Rights   205
　　7.6　Application Programming Interface   206
　　7.7　Internet Search Tools   211
　　7.8　Standards and Protocols   214
　　7.9　Wireless Technologies   219
　　7.10　Enterprise Network Management   222
　　7.11　Management Information Base (MIB)   223
　　7.12　Virtual Private Network Software   229
　　7.13　Electronic Mail   236
　　7.14　Remote Monitoring   237
　　7.15　IPv4 and IPv6 Security   238
　　7.16　CERN   239
　　7.17　Internet and Web Resources   240
　　　　　　　Summary   240
　　　　　　　Key Terms   241
　　　　　　　Review Exercises   242

CHAPTER 8   Attacks, Threats, and Viruses   246
　　　　　　　Objectives   246
　　　　　　　Introduction   247
　　8.1　Enterprise Threats   248
　　8.2　Virus Threats   249
　　8.3　Countering the Virus Threat   250
　　8.4　Malicious Attacks   254
　　8.5　Other Security Breaches   259
　　8.6　Software Threats   262
　　8.7　Social Engineering   262
　　8.8　Threat Targets   263
　　8.9　Responding to Hacker Attacks   264
　　8.10　Threat Assessment   264
　　8.11　Gap Analysis   267
　　8.12　Ethics   268
　　8.13　Cybercrimes   269
　　8.14　Security Tools   271
　　8.15　Internet and Web Resources   273
　　　　　　　Summary   273

Key Terms 274
Review Exercises 275

CHAPTER 9 Building, Campus, and Facility Security 278
Objectives 278
Introduction 279
9.1 Physical Security Overview 280
9.2 Physical Security Categories 280
9.3 Preventing Damage to Physical Assets 285
9.4 Physical Security Controls 287
9.5 Security Forensics 291
9.6 Disaster Recovery 293
9.7 Auditing 294
9.8 Disaster Planning 295
Summary 295
Key Terms 296
Review Exercises 297

PART TWO 300

CHAPTER 10 Local and Metropolitan Area Network Security Environment 302
Objectives 302
Introduction 303
10.1 Local Area Network (LAN) 304
10.2 Wired LAN and MAN Connectivity 305
10.3 LAN Components and Concepts 308
10.4 LAN Media 317
10.5 Wired LAN Security 318
10.6 Wireless LAN Overview 319
10.7 Extended Wireless LANs 321
10.8 Wireless LAN Technologies 322
10.9 Wireless LAN Security 326
10.10 Virtual Local Area Network (VLAN) 329
10.11 Metropolitan Area Network (MAN) 331
10.12 Intranet 332
10.13 Extranet 334
10.14 Extranet Security Issues 335
10.15 Content Management 336
10.16 Electronic Data Interchange (EDI) 337
10.17 Computer Telephone Integration (CTI) 338
10.18 LAN, MAN, and WAN Comparisons 338
10.19 Standards 339
10.20 Internet and Web Resources 341
Summary 341

Key Terms 342
Review Exercises 343

CHAPTER 11 Wide Area Network Security 346
Objectives 346
Introduction 347
11.1 Wide Area Network Environment 348
11.2 Telephone Network Structure 349
11.3 Telecommunications Network Functions 352
11.4 Telecommunications Network Components 353
11.5 Elements of a Data Communications System 354
11.6 Internetworking 357
11.7 Network Configurations and Design 358
11.8 Access Circuit Types 359
11.9 Interfaces 360
11.10 WAN Switches 365
11.11 The Internet and the Web 370
11.12 Internet Network Components 375
11.13 Access and Transport 376
11.14 IP Addressing 378
11.15 World Wide Web (WWW) 381
11.16 Security in the WAN 382
11.17 Internet and Web Resources 384
Summary 385
Key Terms 386
Review Exercises 387

CHAPTER 12 Virtual Networks Security 390
Objectives 390
Introduction 391
12.1 The New Workforce 393
12.2 Virtual Networks 393
12.3 Virtual Networks Overview 395
12.4 VPN Types 396
12.5 The Internet VPN 397
12.6 Virtual Networks Connectivity 398
12.7 VPN Connectivity and Design 403
12.8 VPN Hardware and Software 404
12.9 VPN Implementation 408
12.10 VPN Standards 412
12.11 VPN Applications 413
12.12 VPN Issues and Considerations 414

12.13 Security Issues   415
12.14 Voice over Internet Protocol (VoIP)   418
12.15 Internet and Web Resources   421
Summary   422
Key Terms   423
Review Exercises   423

CHAPTER 13   Distributed Systems Security   426
Objectives   426
Introduction   427
13.1 The Distributed Computing Environment   428
13.2 E-commerce Overview   429
13.3 E-commerce Infrastructure   430
13.4 E-commerce Security   433
13.5 Financial Transactions   436
13.6 Implementation Issues   438
13.7 E-security versus E-thieves   439
13.8 Security Protocols   440
13.9 E-commerce System Design Concerns   441
13.10 Distributed Security Recommendations   443
13.11 Internet and Web Resources   444
Summary   444
Key Terms   445
Review Exercises   446

CHAPTER 14   Wireless Networking Security   448
Objectives   448
Introduction   449
14.1 Wireless Transmission   450
14.2 Wireless Network Configurations   451
14.3 Mobile Telephony   453
14.4 Microwave Technologies   454
14.5 Satellite Transmission   456
14.6 Cellular Telephony   458
14.7 Two-way Messaging   467
14.8 Wireless Advances in Technology   469
14.9 Wireless Standards   470
14.10 Security and Privacy in Wireless Systems   471
14.11 Internet and Web Resources   473
Summary   473
Key Terms   474
Review Exercises   474

APPENDICES 477

Appendix A: OSI Model   479
Appendix B: Enterprise Systems Security Review   484
Appendix C: Request for Comments   487
Appendix D: ISO 17799 Security Standard   490
Appendix E: Certification Programs   492

References and Other Resources   494

Acronyms   498

Glossary   500

Index   509

ENTERPRISE SECURITY

PART ONE

CHAPTER 1
Enterprise Security Basics

This chapter sets the stage for the remainder of the book by presenting basic concepts and numerous definitions associated with the computer and networking environment. Historical perspectives along with the current environment are presented. Corporate and site policies are discussed, as are the numerous threats that must be addressed in the Enterprise environment. Security goals and objectives are discussed as they apply to the network, computer hardware and software components, and databases. An overview is presented on the basics of e-commerce, which is thoroughly covered in Chapter 13. The chapter concludes with a general discussion on the physical plant security and control. All of the basics presented in Chapter 1 are discussed further in subsequent chapters.

CHAPTER 2
Enterprise Access Control

Chapter 2 is concerned with the various methods and techniques that are utilized to access computer and network resources. Considerable information is presented that concerns authentication, authorization, and accounting, or AAA. This includes password issues and the devices that can be deployed to provide access control. Subjects include digital certificates and signatures, public and private keys, and the various encryption/decryption techniques. The important process associated with key management is presented. Physical security of computer and network assets is discussed along with the issues of resource access. The chapter concludes with a discussion on the various methods of access control for internal and external people.

CHAPTER 3
CryptoSystem Techniques

The various cryptographic methods and techniques are discussed that are part of a cryptosystem. Various ciphers are presented, which include monoalphabetic, transposition, and polyalphabetic processes. Details are provided on the Data Encryption Standard (DES) and Triple DES. The Rivest, Shamir, Adelman (RSA) public-key encryption method is discussed as is the important public key infrastructure (PKI). The chapter concludes with a discussion on the activities of a cryptanalysis and governmental encryption policies.

CHAPTER 4
Security Systems

This chapter begins with the important topic of biometrics and discusses all of the various attributes of this emerging security access technology. The various topics presented include finger scanning, iris imaging, retina recognition, and signature verification. The attributes of a brute force attack are discussed, along with possible countermeasures. Considerable information is presented on the Secure Electronic Transaction (SET) specification and its relation to e-commerce. Also discussed in detail are S-HTTP, SSL, PGP, and S/MIME. The chapter ends with a detailed discussion of Kerberos authentication.

CHAPTER 5
Enterprise Security Management

This chapter is concerned with the management of the Enterprise resources. Discussed are organization planning and polices and development of system baselines. These are related to security threats and detection techniques. A considerable amount of detail is presented concerning network management systems and supporting protocols such as SNMP and CMIP. Data management issues are discussed including data backup, archiving, data warehousing, and those methods that are deployed to protect the database asset. Network troubleshooting techniques and the devices used to monitor and detect intrusions are discussed. The chapter concludes with information concerning the Security Reform Act.

CHAPTER 6
Computer and Network Systems

This chapter covers the topics of hardware, the components utilized, their functions, and how they play in the security of the Enterprise computer and networking environment. Considerable emphasis is placed on the security provisions of routers, bastion hosts, gateways, firewalls, and proxy servers. Also addressed is the Fibre Channel storage technology and Virtual Private Network (VPN) components. The chapter ends with a discussion of the security issues with voice systems, such as Private Branch Exchanges (PBXs), and Automatic Call Distributors (ACDs).

CHAPTER 7
Software and Database Systems

Software is the vehicle for protecting software and database systems. Various techniques are presented for protecting the Database Management System (DBMS) and the other data resources. Many of the operating system software components are discussed, including Windows NT and UNIX. Sections are provided on directories and database access rights, and the role they play in resource security. Numerous Internet search tools are examined, as are many of the protocols utilized in the Enterprise to Internet environment. The chapter concludes with a discussion on the various software standards utilized in today's Enterprise computer network.

CHAPTER 8
Attacks, Threats, and Viruses

This chapter presents information concerning the numerous threats that can occur in the Enterprise environment. The primary focus of this chapter is on the computer virus and the variations, including logic bombs, and worms. Numerous countermeasures are presented to counter these attacks. A section addresses threat assessment and the various techniques of identifying potential threats. This is followed by an assessment on the issue of ethics for both internal and external users. The chapter concludes with a discussion on the various security tools that can be deployed to counter these threats and attacks.

CHAPTER 9
Building, Campus, and Facility Security

This chapter focuses on the physical security of the Enterprise computer and networking assets. Suggestions for a proactive stance that can mitigate possible threats, attacks, and natural disasters are presented. Discussions are directed at reducing the impact of natural disasters and man-made threats and attacks that do occur. Processes for disaster recovery are presented along with disaster planning methods.

CHAPTER 1

Enterprise Security Basics

OBJECTIVES

Material included in this chapter should enable the reader to

- Become familiar with the historical perspectives of the Enterprise network security environment.
- Recognize the various issues associated with computer and network security.
- Learn the language and terms associated with security.
- Identify the security problems and shortcomings that occur in most organizations.
- Become familiar with the various network topologies that are part of the security issue.
- Identify the various threats and risks that may occur in the computer and network structures.
- Look at techniques and solutions to counter the various security vulnerabilities.
- See how these security issues impact e-commerce and B2B operations.
- Evaluate the usefulness of contracting and outsourcing security.

■ INTRODUCTION

Chapter 1 sets the stage for an in-depth discussion of the various aspects of computer and networking security. It includes a historical perspective starting with the early computer system environments and proceeds to the current state of affairs. Discussions include the historical and present-day structures of the Enterprise network and the impact of the Internet, intranet, and extranet configurations on security issues.

Sections are provided that delve into security basics, security technologies, and security goals. Corporate security policy and site security policy issues are discussed in detail. The concept of trusted and untrusted networks is presented. An overview of the issues involved with physical site security includes various alternatives to counter malicious activities and threats.

Considerable emphasis is placed on security threats and vulnerabilities to network, hardware, and software systems. Details are provided on data and database security, integrity, availability, and confidentiality. The concepts and issues introduced in this chapter will be explored in detail in subsequent chapters.

Security and integrity relating to e-commerce and business-to-business enterprises are discussed. Lastly, a brief overview is presented concerning the issue of outsourcing the task of security to a contractor.

1.1 A HISTORICAL PERSPECTIVE

In the past, security was not an issue as all computer resources were contained in a central location. In today's environment, security lapses can bankrupt a company and destroy a wealth of information. What are these new threats and what can be done to negate their impact? This book explores the computer and network environment and provides a number of possible solutions to counter a multitude of pervasive threats.

In the sixties and seventies the computer industry was in its infancy. Hardware and software resources were only available to those users who were located at the central host location. There was not any network or computer communications access from remote locations. At universities, students created their programs on punched cards and had them loaded in locally attached card readers. As the industry developed, remote access was made available so that the programs could be loaded from remote locations on the main campus. Usually cables were run from the main host computer to the on-campus locations.

During the mid seventies came the capability of students to access the host site using a dial-up, slow speed connection. Development of this remote access appears to have gained momentum in the educational environment. As students graduated and secured jobs in industry, they carried with them this knowledge of a new method of accessing computer resources. Starting with a very slow speed of a 300-baud link to current megabit speeds, many of the networking developments pervasive in industry today started in the educational environment.

Security was not much of an issue when the computer resource was contained in a protected central location. Figure 1–1 depicts a common mainframe environment. It was fairly easy to build protection into the physical facility and a guard could be employed to watch this physical asset. As technology progressed, and the capability to access the computer resources from distant locations became feasible, different issues and challenges surfaced.

As more and more members of the general public became computer literate and computing devices became available to most of the population, issues relating to computer security became a major issue. A number of horror stories made the computer industry aware that something must be done to quell this threat to the computer resources. The physical controls that were used to protect centralized facilities had become obsolete (Amoroso 1994).

In today's networking and Internet environment, remote access to computer systems and resources can be achieved quickly and easily with very little expense to the user. Advances in usability and accessibility have been countered by losses in security of the resources. It is still necessary to have the physical site security of locked doors and alarms, however additional controls are necessary to provide security for remote access. The **client/server** explosion has provided the impetus to develop additional

client/server

Figure 1–1
Mainframe Environment

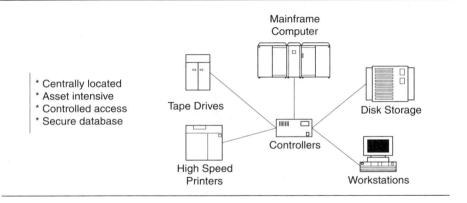

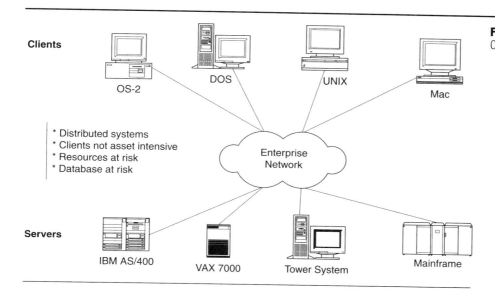

Figure 1-2
Client/Server Environment

security programs. Why is this environment a candidate for security initiatives? The client/server model is defined as a desktop computing device or program "served" by another networked computing device. Computers are integrated over the network by an application, which provides a single system view. The server can be a number of computer processor devices with attached storage devices. A client can be served by a number of servers (Newton 1999). Figure 1–2 depicts a client/server configuration. Different operating systems can be resident on a number of different server systems.

There are two major issues that must be addressed when discussing network and computer security—accidental and malicious events. An accidental situation can cause as much grief as someone trying to be destructive. It is, therefore, necessary to protect resources from both of these situations. There are a multitude of methods and procedures available that can address these issues.

Industry analysts estimate that in-house security breaches account for a major number of corporate computer network attacks. Many of these intrusions may go undetected. A disgruntled employee may seek revenge by deleting or altering files or applications. Another may participate in corporate espionage given a promise of large rewards. Still another may be looking for insider information for a stock purchase. Most organizations can successfully discourage insider attacks by assigning specific access rights that provide restrictions to this type of information.

Chapter 1 is devoted to general topics that establish a framework for the rest of the book. Discussing issues relating to the current security atmosphere that permeates the Enterprise networking environment sets the stage for the chapter. Efforts will be directed at presenting the three "Ws" (who, what, and where) concerning security issues.

1.2 THE CURRENT ENVIRONMENT

There are many individuals in today's society whose goal is to "get something for nothing," or are willing to wreak havoc for havoc's sake. The law enforcement establishment can usually handle the task of guarding against these individuals. However in today's computer and networking environment, these individuals can appear anywhere in the world at any time, to cause distress to the corporate, government, educational, health care, and public resources. This means that the legitimate users must initiate activities and actions that can minimize or eliminate these threats. Security boundaries can be set up around groups, subnets, and data centers within an organization. Such boundaries make it difficult for internal personnel to access restricted systems, and also make it more difficult for outside attackers to penetrate into an organization's network.

There are three major grouping descriptors that reflect the current state of the computer and networking environment. These include the value of the resource, the portability and size of the resource, and personal contact required.

Value of the Resource

Access to the Enterprise resources can improve productivity by making applications, processing power, and data available to employees, customers, and business partners. Networks, however, make organizational data and information more vulnerable to abuse, misuse, and external attack.

The computer and network resource might consist of a palmtop, cellular phone, laptop computer, client/server, or a mainframe computer. The cost of the physical device can be small or in the millions of dollars. The cost of the hardware, however, is only a small part of the resource. The value of the data and information stored in a computer system can be substantial. (Note that data are any material that are represented in a formalized manner that can be stored, manipulated, and transmitted by machine, whereas information is the meaning assigned to data by people). Computer systems contain confidential information about the population's taxes, investments, medical histories, education, criminal histories, military records, and so on. Even more sensitive information may be stored concerning corporate operations, which may include new product lines, marketing strategy, sales figures, and other strategic and tactical information. Governmental computers may store information on military targets, troop movements, weapons, intelligence, and other state secrets. The value of these assets is enormous and the repercussions of losing them can be quite devastating to the country, its organizations, and the general public.

Portability and Size

The first computers filled up rooms with hardware. The use of software was cumbersome and difficult, and was part of the technical realm. The palmtop that can be carried in a pocket has more computing power than the original mainframe computers. This device size has significant implications due to its portability. It is now possible to access the Internet from anywhere in the world using a hand-held computer. The other element associated with device size is the cost. An attaché case can hold thousands of dollars worth of computer and networking equipment. The information stored on these portable devices will probably be a subset of that resident on the client/server and mainframe systems. It is now possible to lose or misplace these portable devices, which could compromise the corporate or governmental resource (Figure 1–3). Implications for security are immense.

Personal Contact

In the near past, most transactions were one-on-one and up-front and personal. Either cash or personal check was the common medium of exchange. The time is probably at hand where money, in the traditional sense, will not be necessary or viable. Most citizens and organizations use some form of credit or debit card to conduct financial transactions. They not only are easier to use, but also provide an electronic audit trail of the transactions. Electronic funds transfers account for most transfers of money between banks and other financial institutions. Many organizations pay employees by direct deposit to the employees' bank account. Mortgage companies, credit unions, utilities, insurance companies, and other institutions can automatically process deductions against their client's bank accounts. It is now possible to make banking transactions and stock market transactions from portable and remotely deployed computer and networking devices. There are many opportunities inherent in this environment for fraud and loss.

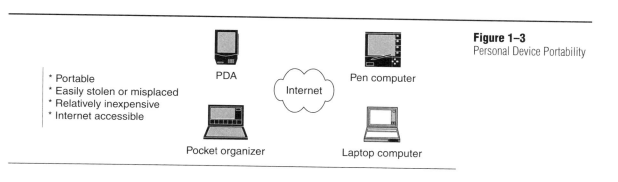

Figure 1–3
Personal Device Portability

1.3 TRUSTED AND UNTRUSTED NETWORKS

trust

Trust can be defined as confidence in the integrity, ability, and truth of a person or thing. In computer networking it refers to the ability of an application to perform actions with integrity, to keep confidential information private, and to perform its functions on a continuing basis. The components of trust in a network computing system consist of availability, performance, and security.

trusted networks

untrusted networks

A network or security administrator must be concerned with the topology of both trusted and untrusted computer networks. **Trusted networks** are inside the security perimeter, whereas known **untrusted networks**, such as the Internet, are outside the security perimeter. An unknown network is also considered an untrusted network (Internetworking Technologies 2001).

Someone in an organization administers the computers that comprise the trusted networks, and the organization controls security measures. The internal trusted network consists of a number of devices including firewalls, routers, gateways, and secure servers.

An untrusted network is outside of the organization's control, and there is no control over the administration or security policies for these sites. They consist of private, shared networks with which the trusted network often communicates. The trusted network security devices must be configured to explicitly accept communications from these untrusted networks and simultaneously protect the organization's assets. Much of the information presented in this book addresses the issue of untrusted network traffic.

1.4 COMPUTER AND NETWORKING ISSUES

One of the primary concerns of most organizations is protecting its computer and networking assets. Of particular interest are the security and integrity of the data and database resources. Companies can lose competitive position, and even fail, if information gets into the wrong hands. Securing data against illegal access and alteration is even more of an issue on networks, since there are many opportunities for snooping and interception when transmitting data between computers and between LANs.

Many techniques are available for computer and network protection, from user passwords to biometrics, and from firewalls to intrusion detection technologies. The challenge is to employ appropriate security measures that will provide sufficient protection of the organization's assets without impeding productivity.

It is obvious that security in the computing and networking environment has a high priority with corporate entities. The issues must be addressed by security professionals, which have a background in the

Enterprise environment. The purpose of this book is to develop a broad and general understanding of security issues that relate to the computing and communications industries. Various areas of organizations will be visited to identify candidates for security solutions. Vulnerabilities will be identified and countermeasures or controls will be suggested. In addition, a number of risks and their related costs will be identified. Corporate and site computer policies are addressed, followed by topics on basic threats and attacks and vulnerabilities of the networking asset.

1.5 CORPORATE SECURITY POLICY

Official corporate security policies and guidelines have been in place for many years. In the past, these guidelines usually covered the rules associated with the security of confidential information, intellectual property, and corporate secrets. There were also rules and procedures for physical security, such as access to computer assets. With the advent of networking and the accessibility of the Internet, security has moved to a position of prominence. The ramifications of not having a corporate security policy are significant, due to the costs that can be incurred. Corporations have responded to this need by establishing a code of conduct that employees are required to sign.

The corporate code of conduct should clearly set forth the standards of conduct that are expected from all employees. Rules must be established that identify the corporate elements that must be protected. Finally, it is necessary to state the repercussions of failing to adhere to these standards. A corporate code of conduct should accomplish the following (Kaeo 1999):

- Provide the scope of coverage—The scope of the security policy should state the audience impacted and the components of the organization covered under the rules. It should also provide some detail as to the specific objects covered under the policy.
- Identify the professional responsibility—Employees have a responsibility to support the goals and mission of the corporation. Employees are expected to perform in an acceptable professional manner.
- Set forth a confidentiality policy—Confidentiality of privileged information and communications is set forth. Levels of security must be set forth as they apply to each specific entity. Employees are expected to comply with and advise others of the policies.
- Address conflict of interests—Corporate interests and personal interests must be kept separate. This means that private interests, obligations, and transactions must not be in conflict with the corporate mission. Employees must perform duties and responsibilities in a fair and just manner in the best interests of the corporation.

- Present the legal ramifications—This is a statement of penalties resulting from failure to comply with the corporate security policy. Employees are expected to refrain from unlawful and dishonest conduct.

Employees in an organization and other authorized users are the source of most breaches of network security. Measures that minimize these breaches can include passwords and other means to identify authorized users. Networks can be monitored for abnormal patterns, and employees can be educated in security practices.

One major challenge to the organization is to balance the cost of security against the risks being addressed. The cost of providing solutions and controls should be in proportion to the value of the assets being protected. It makes no sense to spend more money to protect an asset than the asset is worth. An analysis of the situation must prove the feasibility and viability of the proposed security controls. This analysis needs to look at the criticality of the systems to the business (Dempsey 1997). What is the exposure of the systems to fraud and illegal disclosure of information? The risk to assets from both internal and external sources must be explored. One of the first steps toward securing the organization's assets is to develop a site security policy.

1.6 SITE SECURITY POLICY

security policy

A **security policy** is a formal statement of rules by which users who have access to an organization's technology and database assets must abide. This policy must be formulated by key organizational individuals who have authority to enforce the policies developed and who understand the ramifications of these policies. A site security policy is developed for the following reasons (Kaeo 1999):

- It enables security implementation and enforcement.
- A basis is created for legal action.
- A process is developed for conducting security audits.
- Formalized step-by-step procedures are developed.

There are a number of key characteristics of an acceptable security policy. Primarily, it must be capable of being implemented both organizationally and technically. It must be flexible to allow for changing conditions. The policy must clearly define the areas of responsibility for the users, both inside and outside the organization, and corporate and site management. Lastly, it must be enforceable with a security management system or set of security tools. It is important to note that a security policy should not determine how a business operates. A policy chosen should fit the current organization operation and structure. This policy should meld with the hardware, software, and network environment.

1.7 NETWORK SECURITY

Network security includes the development and implementation of methods and techniques for protecting an organization's networking assets. A primary goal is to ensure the network is as secure as possible against unauthorized access or use of network resources. The majority of network intrusions that are responsible for lost or stolen resources are perpetrated or facilitated by internal personnel.

The first step in establishing network security is the development of written policies covering all aspects of network security in the organization. These policies must include both acceptable and nonacceptable uses of computer and networking resources. Included would be policies relating to hardware and software components and employee time. The policy would also delineate the consequences for violating user policies.

To be effective, the organization's management at all levels must accept these network security policies and make a serious commitment to them. Security policies must meet the organization's concerns without conflicting with other policies or limiting user's access to necessary resources.

Most important, rules and procedures can be followed only if the users are aware and understand them. They cannot be effectively implemented without some type of notification and explanation. While security-licensing concerns are burning issues to the Information Technology (IT) department, most employees are not too concerned (Larson 2001).

1.8 HARDWARE, SOFTWARE, AND NETWORK ENVIRONMENT

Security efforts must be directed toward network, hardware, and software issues. Protection of hardware is possibly the easiest to accomplish. Equipment located at a secure central location is fairly easy to protect and maintain. Software is another issue. Software used to be fairly simple, however in today's environment it is exceedingly complex. If software is complex, then the network is much more difficult to manage. Because network users can access the computer facilities from anywhere in the world (even from space), this task is the most complex of all from a security standpoint.

A computer system includes many components. These include the central processor, often called a mainframe, a communications processor, called a Front End Processor (FEP), and considerable database and file storage devices. These collective devices are called **hardware**. In addition to these hardware devices, there are operating systems and application programs, collectively called **software**. The combination of network, hardware, and software will be called the Enterprise resource throughout this book.

hardware

software

14 CHAPTER 1

A good place to start this discussion is with the various types of threats that are posed and to continue with various attacks that can be expected. These threats and attacks will become apparent through numerous vulnerabilities.

1.9 THREATS

threat

A **threat** is generally defined as an expression to inflict some evil, pain, or punishment against some entity. This threat can come from an individual, a group of individuals, or an organization. It is regarded as a possible danger or menace and can be very expensive if not countered by some form of protection. A threat to a computer system is defined as any potential occurrence, either accidental or malicious, that can have an undesirable effect on the assets and resources of the organization. The assets might be software, databases, files, or the physical network itself. A threat is significant from a security viewpoint, because the computer security goal is to provide insights, methodologies, and techniques that can be employed to mitigate threats. These can be achieved by recommendations that provide guidance to computer system administrators, designers, developers, and users toward the avoidance of undesirable system charac-

vulnerabilities

teristics called **vulnerabilities** (Amoroso 1994).

Vulnerabilities
A characteristic of a computer system or a network that makes it possible for a threat to occur is called vulnerability. A vulnerability presence provides an opportunity for disasters to happen. To avoid these bad situations, threats to a computer system or network can be mitigated by identifying and eliminating these vulnerabilities. Solutions and suggestions are presented throughout this book. Figure 1–4 provides an advance look at the issues that will be presented in subsequent chapters.

Figure 1–4
Enterprise Computer Network Vulnerabilities

Hardware

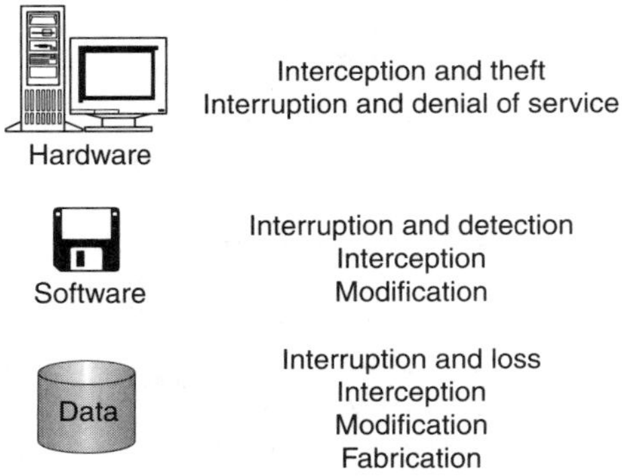

Interception and theft
Interruption and denial of service

Software

Interruption and detection
Interception
Modification

Data

Interruption and loss
Interception
Modification
Fabrication

From the outside, the most vulnerable routes for break-ins are through remote access facilities and from the Internet. To prevent these break-ins from damaging the computer and network resources, networks can employ firewalls to block unsafe services and sources. Detection techniques and vulnerability testing can identify and block intrusion attempts. Software can be used to block virus attacks. There are professional consulting and outsourcing services available for security planning and deployment.

It should be noted that no security measures or products can completely secure a network. Therefore, organizations must strike a balance between user accessibility and the level of protection required for maintaining a safe and secure environment. Higher levels of security might mean inconvenience for users, which could reduce productivity and increase job dissatisfaction. The likely result will be circumvention of security measures by the users. Realistic network protection strategies use measures that meet security needs while causing minimal interference with employee job performance and satisfaction.

Attacks

An **attack** on a computer system or network involves the exploitation of the vulnerabilities, which can result in a threat against the resource. An attack is often heuristic, which means the attacker has some knowledge of the vulnerabilities of the resource. This book is not concerned with general programming and operational errors that can occur at any time in this environment. The objective, therefore, is to reduce the possibilities that would allow an attacker into the network resource. Four general categories of attacks are fabrications, interceptions, interruptions, and modifications. Identifying and responding to threats, vulnerabilities, and attacks are complicated and nontrivial tasks, and may be too expensive and time consuming to eliminate. Efforts must be made to reduce the occurrences as much as possible.

    attack

Attacks are classified as either active or passive. An **active attack** involves some modification of the data stream or the creation of a false data stream. A **passive attack** would include monitoring and eavesdropping on a transmission. Both can be detrimental to the Enterprise organization. Chapter 8 provides methods that can be implemented to mitigate the effects of attacks and threats. Table 1–1 shows the various active and passive threats (Stallings 2000).

    active attack

    passive attack

Active Threats	Passive Threats
Denial of service	Release of message contents
Masquerade	Traffic analysis
Modification of message contents	
Replay	

Table 1–1 Active and Passive Threats

Additional discussion is directed toward specific types of threats, which include integrity threat, denial of service threat, and disclosure threat. These are explained further in Chapter 8.

Integrity Threat

integrity

Integrity is defined as rigid adherence to a code or standard of values; the quality or condition of being unimpaired and sound. An integrity threat is any unauthorized change to data stored on a network resource or in transit between this resource and another entity. A system is compromised when the integrity of the data has been maliciously or otherwise altered. Note that the integrity of the network resource can be compromised by both authorized and unauthorized modifications. This book is concerned with the unauthorized integrity threats.

The integrity threat can cause both minor and major consequences. Advance preparation can reduce the severity of the integrity compromise. If there is a backup copy of the data, then the impact of a breach will probably be less severe than a situation where a backup is not available. Loss of stored critical information can be disastrous, particularly when health or national security issues are involved.

A system fabrication or modification can compromise the integrity of a network resource. Forgeries or counterfeit objects may be fabricated and placed on a system. A modification can occur when an unauthorized party tampers with an asset. Changes can be made to database files, operating systems, and even hardware devices. It is, therefore, essential that assets be modified by authorized parties, or only in authorized ways. Modifications include creating, changing, deleting, and writing information to a network resource (Pfleeger 1997).

Denial of Service Threat

denial of service

Denial of service is defined as an attack that attempts to deny Enterprise network resources to legitimate users. This situation arises when there is an intentional blockage as a result of some malicious action by a user. This means when a legitimate user requires access to a resource, another user prevents this access by some malicious activity. This situation can be either a temporary or permanent blockage for the legitimate user. An interruption of service can occur when an asset becomes unavailable or unusable. This situation is called a denial of service.

Common examples of denial of service attacks involve users "hogging" shared resources such as processors or printers so that other users cannot access them. This type of attack may be bothersome or benign, depending on the criticality of the access. If this network resource was part of a mission critical system, the impact could be severe. Department of Defense systems would be candidates for efforts to avoid denial of service attacks.

It may be difficult for the user to identify poor response time with denial of service. It today's networking environment, over-subscription of network facilities can make the situation look like a denial of service when in fact it is a provider issue. The inability to reach some network resource may be caused by the provider or an error in the user's actions. The provider may have put all interfaces in an inactive state to accomplish a system update or a Web site may be undergoing modifications. The user will probably not know the difference.

Disclosure Threats
The last threat discussed in this section involves that of **disclosure**, where dissemination of information has occurred to someone other than the intended recipient. The disclosure threat occurs whenever some secret element that is stored on a network resource or in transit between a network resource is compromised to someone not cleared to receive such information. This is sometimes called a "leak" when it has been purposely disseminated without proper authorization.

    A significant amount of data and information are stored on computer systems that is not intended for general distribution. This could take the form of personal information on a user's PC or IRS tax records. Dissemination of a user's personal information could cause some embarrassment, however public disclosure of tax records could cause sever repercussions. This could be even more problematic if military secrets are disclosed. An interception by an unauthorized party who copies programs or files, and who does so without any trace, adds to the disclosure problem.

    Considerable resources have been devoted to the disclosure issue. Research and development in security has focused on the disclosure threat and how to counter it. One reason for this emphasis is the impetus given by the federal government.

disclosure

1.10 SECURITY GOALS AND OBJECTIVES

This section is devoted to the goals and objectives of secure computing. The two main goals discussed are availability and confidentiality. These components are subsets of resource integrity.

Availability
Availability means that the network resources are accessible to authorized parties. These resources are also available to illegitimate users. An authorized party should not be prevented from accessing a resource. It is possible to make the resource so secure that it is difficult even for the authorized user to access. Availability is sometimes known by the term *denial of service*, which is actually the opposite of availability. The goals of availability include the following:

- Controlled concurrency
- Fair allocation

availability

- Fault tolerance
- Timely response
- Usability

Confidentiality

Confidential means secure. This means that only authorized people can see protected data or resources. The main issue here is to decide what information is confidential and who has the right to access it. Confidentiality is probably the best understood of the security properties. When the statement "This is confidential" is made, it is usually understood that access is limited to a select few. Confidentiality may impact availability. Confidentiality asks the following questions:

- What resources are confidential?
- Who can access the resources?
- How often can the resources be accessed?
- Who can modify or delete the resources?
- Can the confidentiality status be changed?

1.11 THREATS TO NETWORK, HARDWARE, AND SOFTWARE COMPONENTS

There are different threats that can impact network, hardware, and software components. Each has different identities and repercussions.

Threats to Network Components

What exactly is a **network**? It may be a computer network, a telecommunications network, a **Wide Area Network (WAN)**, a **Local Area Network (LAN)**, or a **Metropolitan Area Network (MAN)**. A computer network is a data communications system that interconnects computer systems at various sites. A network may include any combination of LANs, WANs, and MANs. A LAN is typically a communications network within a building or a campus that is used to link together computers, servers, and printers. A MAN is a high-speed data network that links multiple locations, usually LANs, within a city, campus, or metropolitan area. The WAN uses carrier facilities to link dispersed network locations, typically LANs. Figure 1–5 shows the relationship between LANs, MANs, and WANs. Chapters 10 and 11 address the many issues associated with these network topologies.

Historically, computer systems were self-contained entities located at a central location. A network did not exist, which meant security from remote access was not a major issue. Protection of the system was straightforward—controls were added on the centrally located hardware and software. With the proliferation of distributed computer systems, which are accessible by various types of communications networks, the security task has become much more complex (Dempsey 1997).

Enterprise Security Basics 19

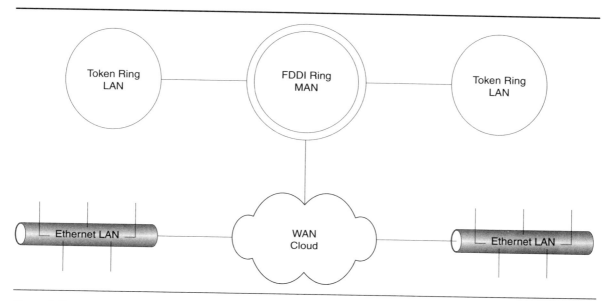

Figure 1–5
LAN, MAN, and WAN Environments

Networks are now being implemented to support remote access, which was historically outside the corporation. Access to the **Internet** has gained wide acceptability. With the advent of the **extranet**, contractors and suppliers can now have access to the local computer system resources. Corporate users communicate over an **intranet**. The differences between intranets and extranets are subtle, however these differences are significant from a security standpoint. An intranet is an internal network that implements Internet and Web technologies. Contrast this with an extranet, which is an intranet extended outside the organization to a business partner, with transmissions moving over the Internet or across private facilities (Sheldon 1998). Table 1–2 compares attributes of the Internet with those of an intranet and extranet. The interconnection of networks may be one of the most important factors for future business successes, however it has the potential for the greatest security risks. Intranets and extranets are

Internet
extranet

intranet

Attribute	Internet	Intranet	Extranet
Geographics	Worldwide	Within an organization	Between an organization and a supplier
Accessibility	The entire population	Corporate users	Corporate users and suppliers
Security Level	Many risks from everywhere	Internal Risks	Risks from suppliers

Table 1–2
Attribute Comparisons of the Internet, Intranet, and Extranet

being used as network vehicles to complete electronic commerce transactions. There are serious ramifications for financial applications using networks that interface with the Internet.

The networking environment has two new candidates that will require sophisticated security protection systems—wireless and virtual networks. Wireless networks are being integrated into all aspects of the Enterprise environment, including LANs and WANs. Virtual Private Networks and other tunneling networks are being implemented at a fast pace in all sectors of the economy. Chapter 12 contains a wealth of information concerning virtual network security.

Threats to Hardware Components

Because hardware devices such as computers and routers are physical, they are supposedly easy targets. Since they are the most visible of the three components, it would seem that security efforts should be increased for computer systems, however reasonable safeguards are usually already in place. Security programs for computer systems have been in place since the early seventies, during the Vietnam War era. Motion detectors, cameras, locks on doors, pass codes, access restrictions, physical location, and a number of other security precautions are commonplace.

There are a number of other situations that can cause havoc with computer and network hardware devices. These devices have been subjected to water, fire, gas, and electrical surges, which can be as devastating as sabotage. Either humidity or a lack of moisture can impact the electrical components of these devices. Dust and dirt and cleaning materials are equally harmful. Abuses by employees and visitors from spilled food and drink have caused considerable damage (Pfleeger 1997). Smoking used to be a problem before it was restricted in most work places. Rough handling by employees and delivery personnel can cause damage to hardware components. Circuit cards can become dislodged and connector pins damaged through handling abuses.

Since there are so many opportunities for hardware components to become damaged, how can one determine if a specific situation is a vicious act? Both employees and nonemployees can be candidates for security violations. Physical security programs are usually directed at nonemployees, however internal attacks may be more insidious and harder to detect. It is fairly easy to sabotage a computer system by dislodging circuit cards or disconnecting power sources, however these actions are easy to detect. Management must be aware of any situations where a disgruntled employee or service personnel might cause damage to the system.

Most of the comments so far refer to centrally located computer and network control systems. What can be done to ensure the integrity of office and desktop computing devices? Often personal computers, calculators, palmtops, scanners, and printers are left unattended on office desktops. These assets account for thousands of dollars of a company's property. Service personnel, maintenance personnel, and even employees can damage or pilfer these devices at will. Management must be vigilant in enforcing procedures for the protection of computer and network assets. Chapter 9 addresses the aspects of physical security.

Threat to Software Components

Software components include operating systems, utility programs, network software, and application programs. Additions, modifications, and deletions are possible to software by both authorized and unauthorized people. How can one limit these activities to authorized personnel? If hardware failures are fairly easy to detect and correct, the same cannot be said for a software attack. Leaving a personal computer logged on and unattended increases the opportunity to create software attacks. It makes sense that someone wishing to cause damage through a software attack would not use a personal login. This situation could be especially disastrous in the case of an administrator's workstation, which had root directory privileges.

A hardware device might show some evidence it has been altered; however, software may not display any outward evidences of tampering. Subtle modifications to software might not be evident if the primary function is retained. A number of software threats contain code components that are set to execute at some future date and time. A particularly difficult software threat to counter is called a **logic bomb**. A software modification can be made that causes some unintended action based on a particular incident or activity. A program may work correctly for years and fail based on some trivial piece of data, such as the temperature on a certain date.

logic bomb

Whereas modifications and additions to software may be difficult to spot, deletions are usually obvious very quickly. Almost everyone has accidentally deleted software and data files. Some people even delete their backup files. It is imperative the users keep backup copies of development software and databases for their own protection. If software and databases are deleted due to some illicit activity, the organization must quickly restore the deleted elements. A good change management system for production software that backs up and restores systems automatically will reduce the chances of permanent deletion of software and databases (Pfleeger 1997).

Intelligent software can be used to monitor access across internal security boundaries and the use of applications and protocols throughout the network. Such monitoring allows network administrators to identify users who are accessing or trying to access unauthorized software and applications.

Special categories of software modifications include **information leaks**, **trapdoors**, **Trojan horses**, and **viruses**. An information leak makes information from a program accessible to unauthorized people. A trapdoor is a hidden process that allows system-protection mechanisms to be circumvented. A Trojan horse is a computer program that appears to function normally, but performs unauthorized actions. A virus is a type of

information leak
trapdoor
Trojan horse
virus

Trojan horse program that infects a computer, causing the operating system to lose integrity. These software modifications will be discussed in more detail in Chapters 7 and 8.

Another issue of importance is the unauthorized copying of software. Software developers and distributors do not receive compensation for their products when this activity is allowed. Legal action has been pursued in a number of instances. It is essential that management have a plan in place to ensure only legal copies of software are available to employees (Pfleeger 1997).

1.12 DATA AND DATABASE SECURITY

Data has been defined as any material that is represented in a formalized manner, so that it can be stored, manipulated, and transmitted by machine. It follows that a **database** is a collection of data stored electronically in a predefined format and according to an established set of rules (Held 1996). There are many potential threats to data that are located in databases. The data threat is probably more widespread than either a hardware or software attack. Data items appear to have a greater value than hardware or software, because more people know how to use and interpret data. Data becomes **information** when people assign it meaning. This means, essentially, data has no intrinsic value unless it can be converted to information. The cost of data and by extension, the database, can be measured by the cost of creation, and ultimately, the cost incurred if it is lost. It may be expensive or impossible to recreate the data. Confidential data that comes into the possession of a competitor may cause a business to fail. There might be a liability if certain data are released to the public. Data often loses its importance over time. To summarize, data attributes include the following:

- Data, per se, has little intrinsic value.
- The cost of data is difficult to measure.
- Reconstruction of data may be expensive or impossible.
- Data must remain confidential.
- The useful time period of data is short.
- A data attack is more widespread than a hardware or software attack.

Data and database security can be described by the following three qualities:

- Availability—Preventing denial of authorized access
- Confidentiality—Preventing unauthorized disclosure
- Integrity—Preventing unauthorized modification

Availability of data and the database require a number of expectations be met. Security and accessibility will involve a balancing act.

Authorized users will demand fast and easy access, however sufficient controls must be in place to ensure both confidentiality and integrity of the data. It will be essential that data is available in a usable form and that there is enough capacity to meet the user's requirements. Access to the database must be timely and response time must be within a reasonable range. An overlay of security procedures can slow down database access. Software audits that might be applied to the database and the data can also create a negative impact on the performance of the access. As is often the case with controls, security techniques may cause negative performance issues. Because availability applies to both data access and service, it is essential both be given a thorough consideration before initiating security controls.

1.13 ELECTRONIC COMMERCE AND BUSINESS TO BUSINESS

Electronic commerce (e-commerce) uses electronic information technologies to conduct business between trading partners, sometimes using electronic data interchange (EDI) and the Internet. EDI is discussed in Chapter 10. **Business to business (B2B)** relates to the concept of different and similar organization types conducting business electronically over some type of network, such as the Internet.

electronic commerce (e-commerce)

business to business (B2B)

To support Internet- and Web-based commerce, organizations must make use of extranet, intranet, and security technologies to build a networking infrastructure to support the business. These subjects are discussed in Chapter 10. Figure 1–6 depicts a logical representation of the computer and networking structure that could be deployed in this environment.

The Internet is being used as the communication mechanism to support corporate-to-corporate communications for such activities as joint partnership projects or to support industry organizations. Many organizations maintain Web sites for information related to their products and

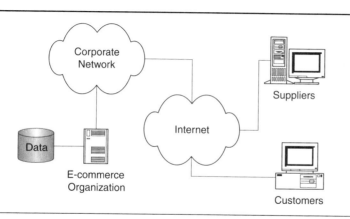

Figure 1–6
Logical E-commerce Structure

services. A natural extension to the business use of the Internet is to include commercial transactions on this untrusted network. Although this is an exciting prospect, it is also a serious security concern, because commerce on the network may refer to the transmission of financial or financial-related transactions. Commercial applications require the parties involved to authenticate one another and transact business confidentially. An important security element of these commercial transactions is nonrepudiation, which prevents a user from denying a legitimate transaction. The security of the client relationship when using Web browsers and other Internet services must be similar to those when using an automated teller machine (ATM).

When using the Internet to carry out transactions using the established payment methods of credit and debit cards, there are usually four entities that are involved:

- The issuer of the credit/debit card
- The merchant selling the product or service
- The acquirer of the transaction
- Security of the transaction by a security authority

The growing interest in these types of transactions promotes the need for a way of providing proof of identity and confidentiality over a network. Before electronic commerce, there was adequate legal protection against defective goods, fraudulent payments, and failure to deliver. New protection mechanisms are required when engaging in this new type of commerce. When purchasing from the Web, it is possible to use cryptographic techniques and digital signatures to provide some protection of the transaction. There are a number of other issues that might arise from electronic transactions. How does the user return damaged goods for credit? How does the user prove non-receipt of an item? What is the recourse if the item arrives damaged or arrives too late? Chapter 13 provides an in-depth discussion of the e-commerce environment.

The availability of a common security authority will help promote commercial transactions on the Internet. Security protocols are being developed and promoted to address the specific problem of transmitting financial transactions on the Internet. Two major credit card companies, along with other partners, have developed the secure electronic transaction (SET) protocol to address these security issues.

1.14 PROTECTING THE ENTERPRISE ASSETS

The primary mission of this book is to present options and alternatives that can be used to protect an organization's computer and networking assets. Chapter 2 presents the various techniques for providing access con-

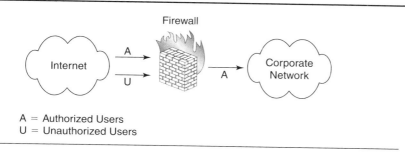

Figure 1–7
Firewall Barrier to Untrusted Internet

trol using the Triple-A technique of authentication, authorization, and accounting. A major topic of this book is the issue of firewall implementations in the computer-networking environment. Essentially, a firewall is hardware or software that uses various filtering and screening methods to determine if a user will be allowed access to the resources of an organization (Figure 1–7). It protects the organization's assets from the unsecure and untrusted Internet and ensures that only legitimate users are allowed access.

The fundamental purpose of Triple-A mechanisms is to maintain security by preventing unauthorized use of the Enterprise network. A number of authentication and access control mechanisms are employed to accomplish this task.

Authentication Mechanisms
- Biometric techniques
- Certificates
- Challenge-Response handshakes
- Kerberos authentication
- One-time passwords
- Passwords and PINs
- RADIUS
- Security tokens

Access Control Mechanisms
- Access control by Authentication Servers
- Access Control Lists
- Intrusion detection
- Physical access control
- Policy filters
- Traffic filters

These authentication and access control mechanisms are discussed in Chapter 2.

1.15 INTRUSION DETECTION

intrusion detection

Another method for protecting the Enterprise assets is **intrusion detection**. Intrusion detection systems are network based, host based, or hybrid based. An intrusion system is a conglomeration of capabilities to detect and respond to threats. Intrusion detection systems are available that provide the following functions (Proctor 2001):

- Event log analysis for insider threat detection
- File integrity checking
- Network traffic analysis for perimeter threat detection
- Security configuration management

Intrusion methods and systems are discussed in Chapter 4.

1.16 PHYSICAL PLANT SECURITY AND CONTROL

Efforts to secure the computer, database, and network resources will all be in vain if the physical plant that houses them is not protected. This section provides an overview of the issues involved and presents various alternatives to counter these situations and threats. Chapter 9 provides an in-depth analysis of building, campus, and facility security. Every organization should assume that it will be the target of some kind of threat, and as many barriers as feasible should be erected to counter these threats. Although sometimes expensive and troublesome, it is fairly easy to erect barriers against physical access. Equipment rooms and network **demarcation** closets can be kept locked. A demarcation is the physical point where a regulated service is provided to the user. Access can be restricted to only those with a specific need to be there.

demarcation

As concentration points for large numbers of circuits, equipment rooms and telecommunications closets are particularly vulnerable. They should be kept locked, and the keys should be given only to authorized personnel. Circuit records and wiring schematics and other communication master records should not be kept in these rooms. As cabling and wiring comes closer to the workstation or PC, it becomes increasingly easy to determine which one to tap. It is essential that these connections that attach to sensitive computer and network devices not show conspicuous tags. Most organizations are also vulnerable to malicious activities in these locations, which means that access to these areas be monitored and restricted (Green 1996). Physical access to a network link would allow someone to tap into that link, jam it, or inject network traffic into it. An important concept that applies to the network world is to avoid the "single point of failure." It would be possible for a saboteur to disrupt an entire organization's network by destroying or incapacitating its network demarcation point.

In addition to accessing networks through computers and terminals, it is possible for intruders to physically break into network cabling systems. Fiber-optic cable is considerably more difficult to tap into than copper wire. This is because it does not radiate signals that can be collected through surreptitious means. An intruder might resort to tapping copper connectors in a wiring closet or at cable outlets. Physical security measures at plants and offices are the primary means to prevent such break-ins.

Physical access to a building or data center can be controlled via locks and access-control mechanisms. Enterprise physical security can be based on security guards, closed circuit television, and card-key entry systems. Data centers and limited-access buildings may employ the double door system, which requires that anyone entering the facility must pass through two sets of security checkpoint doors, one door at a time. With these security measures in place, organizations can feel confident that their assets are protected and high user productivity is maintained.

1.17 CONTINUITY PLAN

Many organizations develop a business **continuity plan** or **disaster recovery plan** in the event of some natural or man-made catastrophe. Whether a natural disaster such as a storm or flood or a man-made virus occurs, no organization is immune to the costs and disruptions that result. A business continuity plan is a strategic instrument aimed at mitigating the threats to critical functions and assuring continuous operation.

<small>continuity plan
disaster recovery plan</small>

In a business environment that is increasingly dominated by electronic commerce, reliable network and computer connectivity is a must. The cost for system downtime for most companies is staggering. Cost categories include the following:

- Lost customers
- Lost revenue
- Lost market share
- Increased expense
- Tarnished reputation

The scope of the plan will depend on a number of characteristics, including company size, product, number of employees, and site locations. Objectives that must be considered are as follows:

- Guarantee safety of all employees, customers, and partners.
- Coordinate the activities of recovery personnel.

- Recover critical functions quickly.
- Limit any disaster-related damages.
- Mitigate financial losses and legal liabilities.
- Minimize cost of recovery operations.

Note that a function is defined as critical if it effects a high-profile client, generates significant revenue, or satisfies a legal or regulatory requirement.

1.18 SECURITY CONSULTING AND OUTSOURCING

Organizations spend large sums annually on security hardware, software, and services without any real objectives or plans. Protecting a computer network against damaging attacks and other risks is a complex task that involves many aspects of the asset structure and operation. Specialists, who are well versed in computer and network security issues and solutions, may possess the qualifications needed to undertake this task.

Independent and vendor-supplied professional **security consulting** services are available. Consulting services can help network managers assess network vulnerability. They can also suggest security products and services and assist in developing effective security management practices. These firms can provide managed-security solutions, which consist of access, monitoring, and problem resolutions. Personnel from these firms have continual exposure to security issues, challenges, and attack methods, and their experience can supplement a limited in-house security staff (Falk 2001).

Many organizations believe they can solve security issues by buying a solution. This might be a false assumption, because security should be embedded in the core operations of the organization and not applied to the surface. Enterprise system design should incorporate security as it is being developed and not included as an afterthought.

The proper location for implementing asset security is the Enterprise itself, starting with the employees who need access to computer and networking resources to perform their functions. The most powerful security measure is to obtain and maintain the loyalty of these employees. Security measures should be designed and implemented so they protect the organization's assets without unduly interfering with legitimate uses of the network.

Security elements that should be part of the Enterprise are depicted in Figure 1–8. These elements are discussed in detail throughout this book.

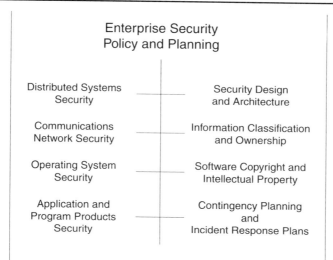

Figure 1–8
Enterprise Security Elements

■ SUMMARY

Requirements for security have increased in importance because computer assets are now accessible from remote users. Historically, a centrally located database could only be accessed onsite by local personnel. With the advent of the Internet and the derivations of intranets and extranets, many more people can now communicate with a centrally located data center. Both site security policies and corporate security polices have been developed to ensure the corporate resource will be protected from external threats. These are especially relevant with the advent of e-commerce and B2B organizations, which could use untrusted networks to conduct business activities.

It is essential that efforts be directed toward network, hardware, and software assets security. The network might contain elements that interact with WANs, LANs, and MANs. The hardware components consist of the computer, the data and databases, and the peripherals necessary to support the operation. Software includes both application and operating-system (OS) components.

Threats can come from all corners. It is essential that vulnerabilities are identified and methods developed to counter the threats identified. Specific attacks can threaten the integrity of the data, database, and the system. Specific attacks can include the denial of service, which limits access to legitimate users, and the disclosure threat, which allows the dissemination of secret information to unauthorized parties.

Security goals and objectives include availability and confidentiality. The system must be accessible in a timely manner, usable, and accurate. It must also be secure from unauthorized people and organizations. Software threats include information leaks, logic bombs, trapdoors, Trojan horses, and viruses. Both the data and the database must be protected from these threats, as they are the foundation and lifeline of the organization.

Security programs and policies will be incomplete if physical security is not part of the plan. Efforts must be made to protect equipment rooms and communications closets from malicious attacks. Finally, the organization must evaluate the cost of the security measures versus the risk of threats avoided. Security consulting and outsourcing companies can assist in this endeavor.

Key Terms

Active attack
Attack
Availability
Business-to-business (B2B)
Client/server
Confidential
Continuity plan
Data
Database
Demarcation
Denial of service
Disaster recovery plan
Disclosure
Electronic commerce (e-commerce)
Extranet
Hardware
Information
Information leak

Integrity
Internet
Intranet
Intrusion detection
Local Area Network (LAN)
Logic bomb
Metropolitan Area Network (MAN)
Network
Passive attack
Security consulting
Security policy
Software
Threat
Trapdoor
Trojan horse
Trust
Trusted network

Untrusted network	**Vulnerabilities**
Virus	**Wide Area Network (WAN)**

REVIEW EXERCISES

Questions and Analysis

Provide a short answer or in-depth analysis as required.

1. Describe the past, present, and future security environments.
2. Why does it matter from a security perspective if present data centers are not centrally located?
3. How has remote access capabilities impacted corporate data center security?
4. Describe a site security policy.
5. Why does an organization need a corporate security policy?
6. Describe a code of conduct.
7. Why is a site security policy developed?
8. Describe the cost versus risk issue.
9. Describe the network resource and comment on its components.
10. What happens when the system experiences a disclosure threat?
11. Discuss the subject of security goals and objectives.
12. What issues must be addressed with confidentiality?
13. What is the interplay among a LAN and the Internet, an intranet, and an extranet?
14. Why are security issues different between an intranet and an extranet?
15. What are the security issues caused by the Internet?
16. Describe threats to hardware components.
17. Describe threats to software systems.
18. Why are viruses dangerous to the corporate network?
19. What are the attributes of data?
20. Data and database security can be described by what three qualities?
21. Why is system availability important to the corporation?
22. Why are data and database integrity candidates for corporate security?
23. Why is security a concern to e-commerce and business-to-business activities?
24. Who are the entities involved in credit card transactions?
25. What are the authentication mechanisms for access control?
26. What are the access control mechanisms that can be employed to limit Enterprise access to legitimate users?
27. Why is it necessary to provide a security plan for a telecommunications closet or network demarcation point?
28. Describe a continuity and disaster plan including its objectives and components.
29. Why would an Enterprise organization want to employ a security firm for consulting and outsourcing?
30. What are the Enterprise security elements?

Definitions and Terms

Provide the most correct answer from the list of key terms.

1. A method of breaking the integrity of a cipher.

2. The amount of time that a computing system is available for use by the user.

3. An attack that attempts to deny corporate computing resources to legitimate users.

4. The interchange of business documents via an electronic medium, such as the Internet.

5. The process of ensuring that the data has not been altered except by the people who are explicitly intended to do so.

6. Logic embedded in a computer program that checks for a certain set of conditions to be present before executing some action.

7. Any circumstance or event with the potential to cause harm to a networked system.

8. Secret undocumented entry point into a program, used to grant access without normal authentication methods.

9. A type of computer program that performs an ostensibly useful function, but contains a hidden function that compromises the system's security.

10. The composite of security, availability, and performance.

True/False Statements

Indicate whether the statement is true or false.

1. A distributed client/server network is more secure than a centralized mainframe network.

2. Attributes for personal device portability include Internet accessibility, low cost, easily stolen or misplaced, and they do not pose a compromise threat to the Enterprise.

3. Trust is defined as a confidence in the integrity in a network.

4. The Internet is a trusted network.

5. Employees in the organization and other authorized users are the source of most breaches of network security.

6. A corporate code of conduct sets forth the standards expected in the organization.

7. Managers in the marketing department have the responsibility of developing a site security policy.

8. It is essential that security efforts be directed toward computer software, however it is not necessary to include computer hardware in the security program.

9. A threat is defined as any potential occurrence, either accidental or malicious, that can have an undesirable effect on the assets of the Enterprise.

10. Attacks are classified as active, delayed, passive, or punitive.

11. An active threat could include a denial of service, masquerade, or a replay attack.

12. A denial of service attack is the same as an oversubscription of facilities.

13. A firewall uses both hardware and software components to filter and screen Enterprise network traffic.

14. An intrusion detection system can be either host based or network based.

15. A demarcation point is not a security issue with the systems administrator.

Activities and Research

1. Use the Internet to access Web sites that provide an overview of network security. This can be accomplished by typing "network security" into a search engine such as Yahoo. Pick a subject and develop a one-page overview of that subject.

2. Select a term from the Key Terms list and develop a five minute oral presentation.

3. Develop a spreadsheet that compares the different types of threats.

4. Arrange for a security manager or network administrator to give a presentation on the subject of data center security.

5. Arrange a field trip to an organization that implements physical security.

6. Identify e-commerce and B2B organizations that use the Internet. Provide a brief overview of their product and service offerings. Address potential security issues.

7. Produce a list of security consulting and outsourcing companies, along with their capabilities.

8. Develop a security policy for a hypothetical organization.

CHAPTER 2

Enterprise Access Control

OBJECTIVES

Material included in this chapter should enable the reader to

- Learn how authentication, authorization, and identification techniques are used to protect the Enterprise network's resources.
- Look at how the issue of access control relates to the mission of the core organization.
- Recognize the various issues associated with computer and network access.
- Look at options for controlling ingress into facilities and buildings.
- Learn the language and terms associated with security access.
- Identify the security problems and shortcomings that occur in the area of Enterprise resource access.
- Identify the various methods, techniques, and solutions that can be used to combat unauthorized access into the Enterprise systems.

■ INTRODUCTION

Chapter 2 is oriented primarily toward the various techniques associated with computer and network access issues. Enterprise assets must be protected from unauthorized access, therefore discussion begins with the issues of identification, authentication, and authorization processes. The various methods, procedures, and vehicles that address the various access control issues include biometrics, passwords, firewalls, filters, and access control lists. A section is devoted to the implementation of digital signatures and certificates. Public and private keys are looked at in detail along with their relationship to asymmetrical and symmetrical security systems. The issues of trust and trust relationships, which are closely aligned with authentication and authorization, are discussed.

In distributed systems, the communications traffic between and among clients and servers is a point of attack for intruders. Vulnerabilities introduced by insecure communications can be counteracted by mechanisms and services, which are part of communications security.

Data security and encryption information presented show how the various encryption/decryption methods are implemented in the Enterprise network environment. Included in this section is a comparison of hash functions to asymmetric and symmetric methods. A section is devoted to a discussion of key management for encryption systems.

Access into computer and network facilities, and buildings, has become an issue with most organizations, especially those with large asset bases. It is essential that only authorized personnel be allowed, unattended, into a facility. Not only is ingress an issue, egress can be just as important, as hardware, software, documentation, and other items can "walk" if not protected. The chapter concludes with a discussion of access control for personnel, contractors, and visitors.

2.1 NETWORK RESOURCE ACCESS

A basic requirement of a security system is to identify the user that is trying to gain access to the network. The owner of the network must determine whether this user is a friend or foe. It is essential that networks, computers, and information be protected from unauthorized disclosure. This concept is called **confidentiality**. In a computer system or network there must be a method of identifying anyone that is allowed into the

confidentiality

environment. This takes the form of identification, authentication, authorization, and accounting, collectively called **AAA** or **Triple-A**.

Identification

This section examines techniques for identifying those active entities that are responsible for initiating specific actions on a computer system or network. This class of techniques is called identification. **Identification** is defined as consisting of procedures and mechanisms that allow an entity external to a system resource to notify that system of its identity. The requirement to perform an identification technique occurs when it is necessary to associate an action with some user or entity that causes each action. Computer systems and network devices can determine who invoked an operation by examining the reported identity of the entity that initiated the session in which the operation was invoked. This identity is usually established via a login sequence.

A system administrator may have assigned this login. The login is the simplest type of mechanism that will comprise identification. Once users are logged in, they are allowed to access various resources based on the rights and privileges assigned to their user accounts or the objects they possess. Identification is usually combined with another procedure, called authentication, which allows a system to determine if the identification sequence was correct (Amoroso 1994).

Authentication

Authentication is the process whereby a user proves they are who they claim to be. Additionally, to authenticate is to establish that a transmission attempt is authorized and valid. There are several techniques for authenticating a user. The first is to enable network logon, which ensures that only valid users are allowed to access the network. The next step is to provide user authentication on servers. Authentication at the server level operates independently of the network logon. Authentication is also required for message exchange to verify that a particular message has not been fabricated or altered in transit.

A username/password is the most common form of identification for network security. This authentication method can be used as either a manual or dynamic entry:

- A manual username/password allows users to choose their own passwords. The password remains the same until the user changes it.
- When using the dynamic username/password, the password continually changes. A device such as a security card can be used to implement the dynamic technique.

Remote Authentication

A **Remote Access Server (RAS)** gives remote users access to all servers on the LAN, which means authentication is required on each device. The two major remote-access server systems are RADIUS and TACACS.

RADIUS

A policy-based authentication method used on RASs is called **Remote Authentication Dial-In User Service (RADIUS)**. When a user logs in to the RAS, it does not process authentication by itself; rather it passes the user's login information on to the RADIUS server. The RADIUS server then either authenticates the user or not, and passes this information back to the RAS serving the user, where the RAS then accepts or rejects the connection. Figure 2–1 depicts this activity (Panko 2000).

Remote Authentication Dial-In User Service (RADIUS)

TACACS

The **Terminal Access Controller Access-Control System (TACACS)** offers Triple-A features (authentication, authorization, and accounting). Unlike the peer relations designed into Password Authentication Protocol and Challenge-Handshake Authentication Protocol, TACACS is designed to function as a client/server system, which provides more flexibility, especially in security management. Central to both RADIUS and TACACS is an authentication server.

Terminal Access Controller Access-Control System (TACACS)

One advantage of TACACS is the ability to act as a proxy server (address translator) to other authentication systems, such as Windows NT security domain, NetWare Directory Services (NDS), and Unix-based NIS maps. But, transmitting authentication packets between the parent server and the proxy server across a public network poses a security risk.

Both RADIUS and TACACS encryption is based on static keys, which could be compromised. This is because user names, passwords, and authentication server information are contained in a single packet, making them easy to use if intercepted.

Accounting

Accounting services are provided by some network operating systems to track the users who access the resources. Network resources that are commonly tracked are as follows:
- Disk space utilized
- User logons and logoffs
- Applications started

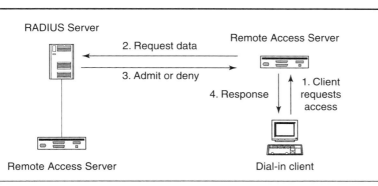

Figure 2–1
RADIUS Authentication

- Files accessed
- Traffic volume
- Resources engaged

In many cases, it is possible for the administrator set limits and boundaries on the user and the amount and types of resources used. If this information is captured, it can be useful in security forensics.

2.2 PASSWORDS

password

The first line of defense in network security is the password. A **password** is defined as a word or string of characters recognized by automatic means, permitting user access to a location, database, or file storage. It is a secret code required to login to a secure computer system. A successful password system depends on the passwords being kept secret. Security is often compromised because users tend to violate security procedures by posting the password on the computer monitor or committing other violations of security policies. Guidelines for password management include the following:

- Memorize the password.
- Keep the password secret.
- Choose a password that is different from a user name or an account name.
- Use an invalid word—not in the dictionary.
- Use a combination of alphanumeric characters.
- Change the password when required.
- Change the entire password—and not just one character.
- Use the longest allowable string of characters.

Table 2–1 shows samples of valid and invalid passwords. Note that the valid passwords are case sensitive, which allows for a larger number of combinations.

Passwords are the most widely used network security feature and almost all organizations employ some password scheme. Office applications and databases are routinely secured with login passwords. They provide a minimum level of security and may be used to restrict user

Table 2–1
Valid and Invalid Passwords

Valid Passwords		Invalid Passwords
Q1wT1p5D	(random, alphanumeric, upper/lowercase)	Account
iAm40YoT	(I am 40 yrs old today, upper/lowercase)	Robert
WysiWyg	(what you see is what you get, upper/lowercase)	12345678

access to specific network facilities during designated shifts or working hours.

An area of vulnerability is the database that contains the password file. This database must be protected from those that would attempt to illegally gain access. Many systems are susceptible to unanticipated break-ins. Once the attacker has gained access, it is possible to obtain a collection of passwords to use different accounts for different logon sessions, which could decrease the chances of detection. It is also possible that a password file could become readable, which could compromise all accounts.

There are several actions that would be prudent when managing passwords:

- When an employee leaves the organization, this password must be disabled.
- Login attempts must be limited to three before access is denied.
- An audit trail needs to be created that tracks logins and denials.
- An alert needs to be activated when password violations occur.
- The last login date stamp can be displayed when a user logs in. This will flag the user that someone else has been trying the user's password.
- Don't allow multiple users know the password for a single user ID.

Some users have accounts on multiple computers in other protection domains, and they might use the same password. Thus, if duplicate passwords are used and are compromised, multiple computers may be affected.

The use of a password is commonly referred to as first-function authentication. A second-function authentication is based on something the user possesses. This could include a disk, token, or unique card. A more personal form of authentication could include a thumbprint, retina scan, or a voiceprint. A second-function authentication might be used to control the access to individual offices within a floor, whereas a password might allow access to common areas on a building.

The password process can be compromised by the following activities:

- Post-it notes containing the password on the front of the display.
- Discovery of the password because it is too simple.
- The password is sent in plaintext across the network and is monitored.

The most common security problem comes from having passwords written down and kept in close proximity to a terminal. An off-hours maintenance or cleaning person could easily discover and copy the information or logon to a system and gain access to a database. Do not assume that these employees are not knowledgeable. They could be computer science students working their way through college.

One-Time Passwords

A method to prevent the unauthorized use of an intercepted password is to prevent it from being reused. **One-time password** systems require a new password for each session. These systems relieve the user of the difficulty of always choosing a new password for the next session by automatically generating a list of acceptable passwords for the user. **S/Key** was developed by Bellcore to provide one-time passwords, and is documented in RFC 2289.

S/key uses a secret pass-phrase, generated by the user, to generate a sequence of one-time passwords. The user's secret pass-phrase never travels beyond this local computer and does not travel on the network, which precludes it from replay attacks (discussed in Chapter 8.) Since a different one-time password is generated for each session, an intercepted password cannot be used again, and there is no information provided relating to the next password.

One-time password systems require that the server software be modified to perform the required calculations to generate the password, and each remote computer must have a copy of the client software. These systems may not be highly scalable, because it is difficult to administer the password lists for a large number of users.

Password Maintenance

The cost of maintaining passwords includes the many calls to the help desk wanting to know a forgotten password. It has been estimated that this activity costs organizations approximately thirty dollars for each occurrence. This includes the technician's time and the downtime of the user. In addition, it is necessary to maintain a database of legitimate users and their associated passwords. This password database must be afforded a high level of security or an intruder could compromise the organization's network resources through this single database.

To summarize, any integrity check would be meaningless if the identity of the sending or receiving party is not properly established. Authentication accomplishes this task. Choosing the proper authentication technology largely depends on the location of the entities being authenticated and the degree of trust placed in the particular facets of the network or computer system. Authentication is tightly coupled with authorization in most resource access requirements.

2.3 AUTHORIZATION

Authorization refers to securing the network by specifying which area of the network—whether it is an application, a device, or a system—a user is allowed to access. The authorization level varies from user to user. For example, payroll personnel have access to the payroll systems and engineers have access to design systems, which means that access is based on the need to know. This also means that access to system resources and services to users or processes, previously authenticated is selectively granted. It is possible to explicitly grant or deny authorizations. By explicitly denying access, only access to confidential information is restricted. This compares with explicitly granting access, which does not allow access to any information without specifically granting an access. Authorization can be defined system-wide or limited to specific data elements. Authorization can be based on the type of access, which includes read, write, delete, or execute. There are different authorization processes for routers, gateways, servers, and switches. Figure 2–2 shows the authentication and authorization flow of a login session.

Routers, gateways, and switches require authorization so that unauthorized personnel do not change configurations. Access is limited to a server because directory integrity must be preserved. Often a user is allowed a maximum of three attempts to enter the correct authorization password before the access attempt is terminated.

A concept closely related to authentication and authorization is trust. **Trust** is the composite of security, availability, and performance. It includes the ability to execute processes with integrity, keep confidential information private, and perform the required functions without interference or interruption. It follows that a **trusted system** is one in which a computer, network, or software can be verified to implement a given security policy. A framework for an organizational secure systems policy that addresses resource authorization is called a **trust model**.

A **trust relationship** is the link between two entities on a network (e.g., two servers) that enables a user with an account in one domain to have access to resources in another domain.

Trust is essential among organizations that engage in business transactions. Internet users are often required to trust someone they have never met. Likewise, commercial transactions between diverse Enterprise

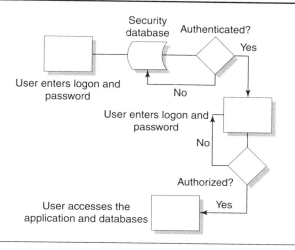

Figure 2–2
Authentication and Authorization Flow

computer systems often contain sensitive and proprietary information. Trust must be established in both cases before secure transactions can occur.

Trust relationships are established between systems so those systems can exchange information without the need for an administrator to actively monitor and authorize these exchanges. Security and system administrators establish relationships for trust in a cooperative way. Access control techniques that support trust systems are presented in subsequent sections. Also presented are technical solutions that can provide support to Enterprise trust systems.

2.4 ACCESS CONTROL

Network security within the infrastructure can be augmented with the addition of access control. This additional protection can take the form of a **filter**, **access list**, or a **firewall**.

Filters, also called access lists, can be configured on routers. Filters enable the router to either accept or reject access based on the contents of the access list. Figure 2–3 shows a user attempting to access an application or database on a server, through a router that controls the access with an access control list. A filter is also called a firewall. It blocks users who are not authorized to access segments of the network and allows access only to authorized users. Implementing filters on the routers, therefore, creates another barrier against unauthorized users, thereby preventing access to parts of the network that are off limits to those users.

Another access control technique used within the network is called **route authentication**. The routers verify among themselves that they are valid routers and that they provide valid paths to the other segments of the network. Route authentication is used to verify that the routing paths available on the network are real routing paths, not paths to an unauthorized device.

Security cards keep the network secure by preventing network access to users who don't possess one. This card requires a permanent username and a temporary password. The username remains the same; however, the password is temporary. This password changes after a specific amount

Figure 2–3
User Identity Flow

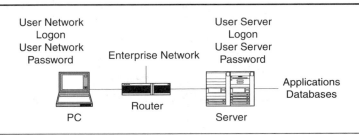

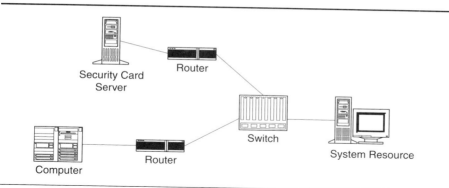

Figure 2–4
Security Card Environment

of time, often 60 seconds. The passwords belonging to a specific security card are synchronized with a security card server. Figure 2–4 depicts the security card environment. The steps involved with using this dynamic username/password pairing is as follows:

- The user turns on the card and types in the username.
- The card displays a password for the user to enter for network access.
- The user types in this dynamic password.
- This password is validated through a central security card server.
- If the password is correct, access is allowed to the resource; if not, the user must restart the process.

Smart card systems allow users to access data by inserting a card into a card reader. The card system provides access for authorized users to any reader-equipped workstation or computer. Automatic teller machines (ATMs) use this type of security combined with a **personal identification number (PIN)**. Smart cards containing microprocessor chips can perform such tasks as on-line encryption and user activity logging, as well as user identification. Some smart card setups require that two individuals perform an entry procedure before admittance is granted. Smart cards are more difficult to administer than passwords and more inconvenient for end users.

smart card

personal identification number (PIN)

Many Enterprise networks restrict user access by establishing **closed user groups (CUGs)** (Beyda 2000). Users who need to communicate only with each other and do not want other users to access their computers can form a CUG. As an example, an organization may allow its branch offices to connect to its central computer system for pricing information, but at the same time ensure that competitors and customers cannot gain access to this confidential information. The organization can create a CUG to ensure that only its branch offices have this access. CUGs are typically employed with packet switching networks (Chapter 11).

closed user groups (CUGs)

2.5 PRIVILEGES AND ROLES

privilege

In the Enterprise environment, a **privilege** is defined as a collection of related computer system operations that can be performed by users of that system. These include the access rights to a directory, file, or program, which typically includes read, write, delete, create, and execute. This is closely related to the role or job-related function that the user is trying to accomplish. These **roles** are defined as a collection of related privileges. Least privilege is a principle that may be used to guide the design and administration of a computer system or Enterprise network. This suggests that users only be granted the privilege for some activity if there is a justifiable need for its associated authorizations (Amoroso 1994).

roles

There is a relationship between roles and privileges. A role might contain only one privilege or many. This means that a role might be defined to include only one privilege for one activity or it might be defined to include many different privileges to perform a number of system operations. Privileges can also exist in multiple roles. This implies a number of privileges (e.g., 2 to n) can be shared by multiple different roles (e.g., 2 to n). Some privileges are more powerful than others, due in part to the level of access required to perform the role. "Normal" users who are assigned non-administrative roles typically include less powerful privileges. The security administrator must decide the appropriate privilege for each role.

Anonymous Access

anonymous

Many computer systems and network servers provide **anonymous** or guest access. An account is established with the name anonymous or guest, which does not require a password. Multiple users can login to the account simultaneously just by entering a username. This account usually has very restricted access to the system and is allowed access only to public files. These accounts can be used for FTP and Web servers on the Internet.

2.6 DIGITAL CERTIFICATES AND SIGNATURES

certificate

A **certificate** is a unique collection of information, transformed into unforgeable form, used for authentication of users. A certificate can verify the authenticity of a user who is logging on to a secure server. It follows that a Web server can have a certificate to prove its authenticity to users that access it (Sheldon 1998). Users on these servers must be assured that malicious people are not collecting personal information or distributing infected documents. A **certificate authority (CA)** is responsible for issuing these certificates. This CA is a trusted organization that verifies the credentials of people and puts a stamp of approval on these credentials.

certificate authority (CA)

digital certificate

Both certificates and signatures are mechanisms that provide an identity of an individual or user. A **digital certificate** is a password-protected file that contains identification information about its holder. This file includes a public key and a unique private key. The electronic exchange of keys and certificates allow two parties to verify their identities before communicating.

A **digital signature** is an authentication mechanism that enables the creator of a message to attach a code that acts as a signature. The signature guarantees the source and integrity of the transmitted message. A message can be digitally signed by a system by including a header, a body, and a signature as part of the message. These components are described as follows:

- The header describes the identity of the sender.
- The body contains the message to be sent.
- The signature contains a computed checksum of the message contents, encrypted with the secret key of the sender.

The receiver can decrypt the checksum using the sender's public key and ensure the checksum matches a computed checksum of the transmitted message (Amoroso 1994). The **checksum** process provides a sum for a group of items in the message, which is used for error checking.

The risk of message forgery is an important reason to use a digital signature. Forged messages can be used to provide false information about a person or some event. Message forgery can also occur on the Internet if someone wants to smear another person or an organization. Another issue related to this subject is called message **repudiation** or denial that a message was transmitted.

2.7 CERTIFICATES AND CERTIFICATION SYSTEMS

Remember that a digital certificate is a personal digital identifier or digital signature that can be used for authentication, to ensure the sender is validated, and ensure the message has not been altered in transit. This authenticity verification extends to the server, which can prove its authenticity to users who access it. Server users must be assured that malicious people, who may be trying to collect credit card and personal information or distribute bogus and virus-infected copies of software, are not spoofing (impersonating) a site. Certificates can be used in place of credit card numbers for online buying transactions. The Secure Electronic Transaction (SET) scheme, developed by major credit card companies, is designed to hide credit card numbers from merchants by substituting the card number with a digital certificate as discussed in Chapter 4.

A basic certificate contains a public key, an expiration date, serial number, certifying authority, and a name. More importantly, it contains the digital signature of the certificate issuer. The International Telecommunication Union (ITU) X.509 certificate contains the following information:

- User name
- User organization
- Certificate start date
- Certificate end date
- User public key parameter
- Certificate authority name
- Certificate authority signature on certificate

Corporations, government agencies, and universities issue their own certificates, which are signed and issued to clients by a trusted certificate server. A certificate granted to an individual is a signed recognition on the individual's identification and authenticity. Checking the existing certificate against the public key of the certificate server and the public key of the individual can validate it.

2.8 PUBLIC KEYS AND PRIVATE KEYS

A **key** is described as a number of characters within a data record used to identify the data. It may also control the use of that data. Two types of keys are used in the security environment—public and private. A **public key** is used in an asymmetric encryption system. It is used in conjunction with a corresponding private key. A **private key** is the other key that is used in an asymmetric encryption system. For secure communication, only its creator should know this key. The use of a public and private key, where two keys are used, is called an **asymmetrical** type system.

Historically, secure message transmission relied on single key encryption, where the sender and receiver used the same key for encryption and decryption. Different methods had to be developed for computer communications to prevent eavesdroppers from listening in on transmissions and intercepting keys to decrypt private messages. These new **Public Key Encryption (PKE)** methods are stored in databases that are trusted by everyone. Messages are encrypted using the public key of the recipient, who decrypts it with a private key known only to the user.

Public Keys

In computer communications, a basic problem is how to arrange for the sender and receiver to agree on a secret key without someone else learning it. Whitfield Diffie and Martin Hellman discovered public key techniques in 1975. The public key approach, called Diffie-Hellman, uses a secret or private key for decryption and a different, mathematically paired public key for encryption (Figure 2–5). Every user has a pair of keys, one is kept strictly confidential and another that is shared among other users or computers. The two keys are mathematically related, and both are used

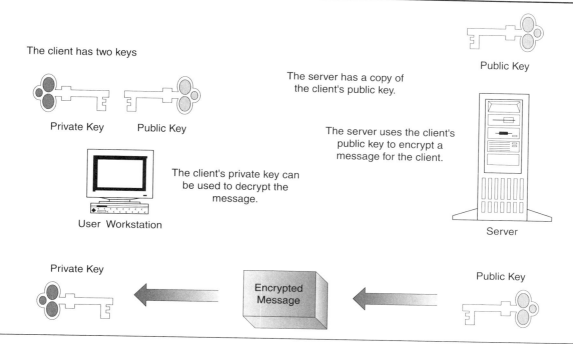

Figure 2-5
Public Key Encryption

in the cryptography process. A message encrypted with the public key can only be decrypted by using the private key. The advantage of this method over secret key technology is that the private key is never shared with other principals. Another public key algorithm, based on Diffie-Hellman is called RSA. RSA and Diffie-Hellman algorithms are described in Chapter 3.

The critical benefit of public encryption is that it is not necessary to trust a communications channel to be secure against eavesdropping. The only requirement is that public keys are associated with their users in a trusted directory. Any user can send a confidential message by using public information, but the message can be decrypted only with a private key, possessed only by the intended recipient.

Public Key Encryption is used in a variety of commercial messaging products and is the basis of several electronic commerce standards, including Secure Electronic Transaction (SET) payment protocol developed by MasterCard and Visa to approve credit card transactions over the Internet.

Private Keys

Private key encryption uses a shared secret key between one or more parties. The possession of the key is used to authenticate one part to the other. A common use of private key technology is for the authentication of user passwords. An encrypted version of a user's password is stored on the computer. The user supplies a user ID and the associated password during login. The password is encrypted, and if the password matches the stored key for that user ID, the user has been authenticated.

Private key technologies are also used to decrypt data. The key is stored in the encrypted file and must be provided to decrypt the data (Figure 2–6). The downside to private key encryption is that the key is shared among two parties, the user and the computer. Session keys, which are generated by a computing process and stored in memory, can be used for secure communications using encryption without the user's participation or knowledge. The keys for the encryption of data that are stored in encrypted format are kept by the user and not by the computer system. The most common private-key encryption standard used today is the Data Encryption Standard (DES), which is described in Chapter 3.

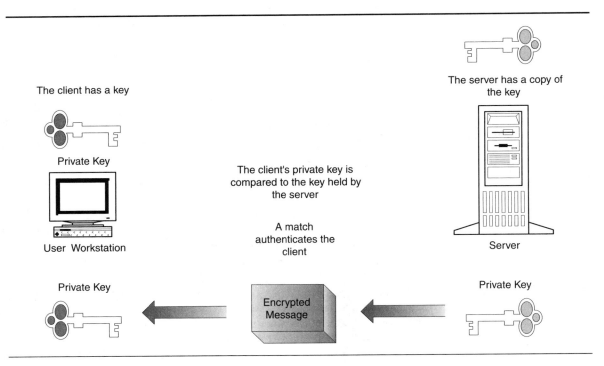

Figure 2–6
Private Key Encryption

Digital Envelopes

Sometimes it is necessary to protect a single message, in which case a secret-key cryptography can be used. The sender and receiver must agree on a session key and each must have a copy. In accomplishing this task there is a risk the key will be intercepted during transmission. Public key cryptography provides an effective solution to this problem within a framework called a **digital envelope** (Muller 2001).

digital envelope

The digital envelope consists of a message that is encrypted using private-key cryptography, but that also contains the encrypted secret key. While digital envelopes usually use public-key cryptography to encrypt the secret key, this is not necessary. If the sender and receiver have an established secret key, they can use it to encrypt the secret key in the digital envelope.

This method can be used to encrypt just one message or an extended communication. A feature of this technique allows the sender and receiver to frequently switch secret keys. Switching keys often is beneficial because it is more difficult for an eavesdropper to find a key only used for a short time. Note that secret-key cryptosystems are faster than public-key cryptosystems.

2.9 DATA SECURITY AND ENCRYPTION

An important issue that must be addressed after a user has successfully gained access to the network and the server is **data security**. As more and more organizations rely on the network for transport of sensitive information, data security is rapidly becoming an area of significant concern. Networks that require a high level of security can use a method called **encryption** (Quinn-Andry 1998).

data security

encryption

Encryption is the conversion of plaintext or data into unintelligible form by means of a reversible translation, based on some algorithm or translation table. It is sometimes called **enciphering**. This process of scrambling cannot be read by anyone except the intended receiver. This means that the data must be decoded or decrypted back to its original form before the receiver can understand it. This encryption process maintains data integrity on the network and assures the confidentiality of the sender and the receiver from other users on the network.

enciphering

Encryption is accomplished by the use of an **algorithm** and a **secret key**. An algorithm is a well-defined set of rules that solve a problem in a finite number of steps. These algorithms take data and scramble it so that it is unintelligible. The secret key is shared by both participants and must remain secret to protect the communication. Figure 2–7 provides a model for the encryption/decryption flow. The key is the value or data string that allows the sender and receiver to actually encrypt and decrypt the transmission. There are several methods used to encrypt and decrypt data.

algorithm
secret key

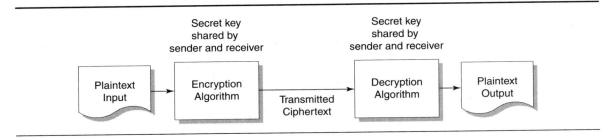

Figure 2–7
Model of Conventional Encryption/Decryption

symmetric encryption
asymmetric encryption
hash functions

The three most commonly used methods are **symmetric encryption**, **asymmetric encryption**, and **hash functions**. Each of these methods will be discussed in detail.

Symmetric Encryption

symmetric algorithm
secret key algorithm

Symmetric encryption is defined as a form of cryptosystem, in which encryption and decryption is performed using the same key. It is also known as conventional encryption. It follows that a **symmetric algorithm**, also called a **secret key algorithm**, uses the same key for encryption and decryption. Data confidentiality is provided when using symmetric encryption. Two end stations agree on which algorithm to use and which secret key to share. Symmetric encryption is most widely used for data integrity, partly because it is the least complex method and requires the least amount of processing on the network devices. Several common symmetric key algorithms used are as follows:

Data Encryption Standard (DES)
International Data Encryption Algorithm (IDEA)
Triple DES (3DES)

- **Data Encryption Standard (DES)**
- **International Data Encryption Algorithm (IDEA)**
- **Triple DES (3DES)**

Asymmetric Encryption

Asymmetric encryption is a form of cryptosystem in which encryption and decryption is performed using two different keys. One key is called the public key and the other is the private key. This system is also known as public-key encryption. This method provides for both data security and sender authentication. Asymmetric encryption is more complex than symmetric encryption because each end station must create a public/private key pair instead of sharing the same secret key. This relationship is summarized in Table 2–2.

Symmetric encryption algorithms are not as computationally intensive as those of asymmetric encryption. The additional processing requirements of asymmetric encryption for data security might reduce network throughput to an unacceptable level. Asymmetric encryption is usually

Table 2–2 Public/Private Key Relationship

Security Type	Sender	Receiver
Data Security	Encrypts with receiver's public key	Decrypts with its own private key
Send authentication	Encrypts with its own private key	Decrypts with sender's public key

reserved for situations where sender authentication is required. Several common asymmetric encryption algorithms include the following:

- **Digital Signature Standard (DSS)**
- **Rivest, Shamir, and Adleman (RSA)**

A combination asymmetric and symmetric encryption algorithm called **Diffie-Hellman** can be used for data security. This algorithm is useful when two end stations need to exchange keys over an insecure transmission channel. This algorithm is implemented in the following steps:

- The asymmetric encryption method is used with public and private keys to establish a channel between two end stations.
- The two end stations then use this channel to exchange the symmetric encryption key, which is the shared secret key.
- The shared key is then used to encrypt and decrypt data.

Digital Signature Standard (DSS)
Rivest, Shamir, and Adleman (RSA)
Diffie-Hellman

Hash Functions

The third method of encryption is called the hash function. A hash function maps a variable-length data block or message into a fixed-length value called a **hash code**. The function is designed in such a way to provide an authenticator to the data or message. This function is also called a **message digest** (Stallings 2000). The hash function is used with symmetric or asymmetric encryption to provide for a higher level of data integrity. The flow of this method is as follows:

- A secure hash function accepts input data.
- These data are passed through a complex mathematical computation (hash function) that is used for encryption.
- The output consists of new encrypted data.

hash code
message digest

It is almost impossible to reverse this process, which would require the output from the algorithm to be used to regenerate the original input data. This would be similar to the scrambled eggs analogy. Crack three eggs in a bowl, scramble them, cook them, serve them, and then reassemble the original three eggs (Quinn-Andry 1998). With hash functions, keys and the same algorithm are not sufficient to decrypt the data. The same hash function must be produced that was originally used to encrypt the data. The secure hash function is usually not used on the data itself because of *hashing*

Message Digest Algorithm 4 (MD4)

Message Digest Algorithm 5 (MD5)

Secure Hash Algorithm (SHA)

its computational complexity. It is usually more useful in proving the sender's identity and ensuring the integrity of the data. Several **hashing** functions include the following:

- **Message Digest Algorithm 4 (MD4)**
- **Message Digest Algorithm 5 (MD5)**
- **Secure Hash Algorithm (SHA)**

Table 2–3 summarizes the various encryption methods, their respective components and functions.

Encryption can take place between the sending and receiving stations, including servers, and at the router. Full encryption security of the network is provided by encryption on the end stations and servers; however, only partial encryption security is provided on the network by router encryption methods. Encryption at the end stations and servers provides the highest level of security, because the data is encrypted the entire time it traverses the network. It is possible to steal data before it enters a router and after it exits a router at the destination. Encryption at the router works well over the wide area network (WAN) or Internet, but it does not provide sufficient benefit for a campus LAN because most traffic does not pass through more than two routers (Quinn-Andry 1998).

Network management personnel must decide what level of encryption is required to preserve the security and integrity of the organization's data. If a secure network is required, then there must be mechanisms in place to control access to the campus (LAN) resources. These mechanisms can take the form of end station, server, and router encryption methods.

Table 2–3
Encryption Method Summary

Encryption Method	Components	Function
Asymmetric encryption	Public/private keys for sender Public/private keys for receiver Encryption algorithm	Sender authentication Data integrity Sender/receiver confidentiality
Symmetric encryption	Shared secret key between sender and receiver Encryption algorithm	Data integrity Sender/receiver confidentiality
Hash function	Mathematical computation providing forward data encryption; reversal extremely difficult.	Verifies data integrity Used with symmetric or asymmetric encryption

2.10 KEY MANAGEMENT

The process of generating, storing, and distributing keys is called **key management**, and all encryption systems must provide a method to accomplish this function. The most difficult issue to address is how to provide secure key management, especially in open systems with a large number of users. This is the task that must be addressed with worldwide electronic commerce applications, discussed in Chapter 13.

    The management of the keys required for the encryption algorithms is one of the complex issues, which must be addressed when selecting an encryption technology. The use of a private key algorithm requires the same key be securely stored and used at all locations that require messages or data to be encrypted or decrypted. Changes to private keys require simultaneous implementation in all locations in a secure manner. The public key approach requires individual key pairs be selected and implemented for all users of the system. This could possibly result in the requirement that a large number of keys be generated, assigned, and used. The basic problem with public key encryption is the proliferation of keys. The problem with private key encryption is that knowledge of the private key is required (Dempsey 1997).

    Key generation and registration is the ability to bind a key to its intended use. The primary problem is how to associate the key to a user. The management of encryption keys requires several specific issues of key management be addressed:

- There must be a secure delivery of the encryption keys to all locations where encryption occurs.
- The activation/deactivation ability must exist to enable and disable keys.
- The ability to change keys in a specific interval- or time-dependent process must exist.
- Assurance must be made that specific keys may only be used for the secure distribution of changed encryption keys.
- Updates or replacements of keys must be possible for an organized change from one encryption key to another.
- The ability for key revocation or termination, which allows the marking of an encryption key as invalid, when a suspected compromise occurs.

The American National Standards Institute (ANSI) has defined a standard to manage the key distribution process in the financial sector. The ANSI X9.17 Financial Institution Key Management standard was defined in 1985. This standard defines a protocol for establishing new keys and replacing existing ones.

key management

key escrow   The requirement for a **key escrow** is another key management issue. It is possible that it will be necessary to hold an encryption key in escrow for extraordinary situations. This might become important if an employee, who has encrypted corporate data, is no longer available. Access to this data will not be possible without access to the encryption key. A provision for holding a key in escrow by a third party may address this situation.

2.11 ACCESS SECURITY OF COMPUTER AND NETWORK RESOURCES

access control   All efforts to secure the organization's resources will be in vain if the facilities that houses them are not protected. **Access control** into computer and network facilities, and buildings, has become a priority issue with many organizations. This is particularly true if the asset is expensive and replacing or recreating it would be difficult and cost prohibitive. This section looks at the issues relating to egress and ingress to buildings that house computer and network resources.

It is essential that only authorized personnel be allowed, unattended, into a facility. Not only is ingress an issue, egress can be just as important, as hardware, software, and documentation tend to disappear if management controls are not present. Unauthorized people can cause several problems, including theft of equipment or data, destruction of equipment or data, and viewing of sensitive data and information. An access control program must be in place that includes internal personnel, contractors, and visitors. Smart cards or security cards are some devices that can be used in the organization's access control program.

Access Control for Internal Personnel

The size of the office and staff can have an impact on the level of security access to resources. In small offices, with limited computer and networking resources, everyone may have access to everything. If this is a stand-alone entity, security implications might be limited; however, if this small office is part of an Enterprise network, the situation changes. A list of questions that should be answered include the following:

- Is the computer system a stand alone system?
- Is the computer system networked? What is the method of communication and where is the communication device located?
- Is the system in a secure room? Who has the keys?
- Who should have access to the processor and console?
- Do employees perform minor maintenance?
- Is other equipment or supplies located with the computer or networking components?

- Where is the office entrance? Is it in a larger building with multiple tenants?
- Is the facility located on the ground floor?
- Are data files stored off-campus? If so, who has access to them?
- Do employees access the facility after hours and on holidays?

Access Control for Contractors

There are a number of support organizations, vendors, and suppliers that might need access to the physical facility. Maintenance personnel will probably require frequent access to the computer and network components. Software engineers and other specialists may also require access during all hours of the day. Communication suppliers and vendors often have a requirement to access the demarcation closet. This is a critical location in the Enterprise network and access must be adequately controlled. Different groups may require a different level of access control. Very few of these people should be allowed unescorted access into sensitive areas. Someone must be responsible for providing authorization for these personnel to access computer equipment and networking facilities.

Activities that might be performed by external organizations include the following:

- Installation and maintenance of network services
- Electrical and communication circuit wiring
- Installation of network and computer devices and components
- Maintenance of computer systems
- Software upgrades
- Office rearrangements
- Furniture additions
- Construction
- Inspections
- Hands-on training
- Cleaning crew

Access Control for Visitors

Visitors can fall into a number of different categories. These include relatives of employees, tour groups, competitors, sales representatives, trainers, and others. None of these visitors should be allowed to roam freely throughout the organization. Trainers and education providers fit in a different category than the others as they may require access to both computer and networking components as part of a training session. The following suggestions may provide some guidance concerning access for the various classes of visitors:

- A clear and concise policy must identify security access requirements.
- An employee must accompany all nonemployees.

- Tour groups must be controlled and monitored by a tour guide.
- A supervisor must accompany sales representatives and competitors.
- Trainers must have a clearance to view and access corporate systems.

Techniques for Theft Prevention

The most successful method of preventing theft of computer and networking resources is to control access. Access control devices can prevent access by unauthorized individuals and record access by those who are authorized. Mainframes, client/servers, and other large devices are difficult to get past the front door, however there are a number of devices and software that are easy to conceal. There are several approaches that can be taken to reduce the incidence of theft of these resources. These include preventing access, restricting portability, and detecting exit.

Guard Solution

A common approach is to employ a human guard, however there are situations where guard dogs are deployed. Guards are traditional, well-understood, and adequate in many situations. There are a number of security services that offer guard solutions. To be effective, human guards must be continually on duty, which implies 7 days by 24 hours/day. A guard must personally recognize a person or some form of identification, such as a badge. People often lose or forget badges, and badges can be forged. Employees and contractors, no longer authorized access, may still have their badges. The guard must make a record of everyone entering and leaving a facility, if any tracking is required. This guard service can be expensive and it is essential to establish the credibility of the guards, who can also be a source of theft.

Lock Solution

The simplest technique to prevent theft is to lock the room or closet that contains the equipment or resource. The lock is easier, cheaper, and simpler to manage than employing a guard. It, however, does not provide a record of entries and exits, and there are difficulties attributed to lost and duplicated keys. There is the possibility that an unauthorized person can walk through a door, which has been unlocked by someone else. There is also the inconvenience of fumbling for a key when one's hands are full. A lock and key solution requires someone be the "keeper of the keys." This person is usually not available when someone needs access to a locked room.

Electronic Solution

More exotic access control devices employ magnetic strip cards, cards with radio transmitters, and cards with electronic circuitry that makes them difficult to duplicate. Because each of these cards interfaces with a computer, lists can be produced of those people entering and exiting a secured location, along with date and time stamps. Some devices operate by proximity, so that people can carry them in their pockets or on a neck rope. It is also easy to update the access list of authorized personnel when someone is added or deleted from the list. New employees, dismissed employees, and lost or stolen cards can be readily noted on the access list.

An electronic solution can also be used as an alert that a resource is removed from a secure location. A theft can be detected when someone tries to leave a protected area with a protected object. This special label is small and unobtrusive, and is used by a number of libraries. It looks like a normal pressure-sensitive label; however, sensors at the entrance and exit can detect the security tag. These tags are available for vehicles, machinery, equipment, electronics, software, and other documents. The detector sounds an alarm, and someone must approach the person trying to leave with a protected object. This may require the addition of a human guard, which can add to the security cost.

2.12 BIOMETRIC ACCESS CONTROL

Biometric identification devices use individual physical attributes such as fingerprints, palm prints, voice attributes, and retina patterns to identify authorized users (Figure 2–8). Biometrics is currently used as a network security technique only at a few U.S. companies. Fingerprint identification is the most popular biometric technique and retina scanning is the most accurate.

biometric

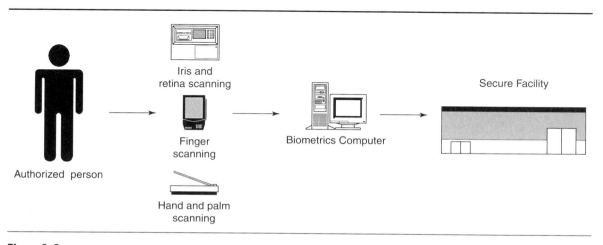

Figure 2–8
Biometric Access Methods

Biometric identification has been expensive to implement; however costs are decreasing. There is considerable end-user resistance to fingerprinting, which is associated with criminal activity. While biometrics promises more secure identification of authorized users than that of passwords, it also can be circumvented. Biometric systems are discussed in Chapter 4.

2.13 SECURITY COST JUSTIFICATION

Since network security measures can require a substantial investment, it is imperative they are made with some confidence that the investments are warranted. One national study has found that large companies and government agencies were able to quantify an average loss of $1 million from security breaches. Security programs and measures, however, can also be very expensive.

One method of determining the degree of security needed is to assess the value of the database element that may be placed at risk. This value would include the cost of collecting the data and re-creating the database. The possibility exists that the original data are no longer available, which means that the database could not be reconstructed. This scenario assumes that the database was not backed up, which is also a possibility. The organization's return on investment in network security would depend on the monetary value assigned to the data and database. This exercise is not a trivial pursuit.

Security Providers

There are a number of suppliers that offer different tools for providing access control into the Enterprise networks. A partial list includes the following:

- Baltimore Technologies
- Cisco Systems
- Entrust
- IPlanet
- RSA Security
- VeriSign

Products offered by these companies address the issues of authentication and authorization and other security functions, such as encryption.

2.14 INTERNET AND WEB RESOURCES

Faulkner	www.faulkner.com
Identix	www.identix.com

Iriscan PR Iris	www.irisscan.com
Key Tronic	www.keytronic.com
XML Digital Signatures	www.ietf.org
VeriSign	www.verisign.com

■ SUMMARY

A hostile environment exists in and around the various Enterprise network and computer configurations. To protect corporate resources from a multitude of security challenges, it is necessary to establish a robust resource-access program. Primary access issues can be addressed through identification, authentication, and authorization methods and techniques. The use of passwords is the first line of defense. The proper implementation of a password program is essential to access control.

In distributed systems, the communications traffic between and among clients and servers is a point of attack for intruders. Vulnerabilities introduced by insecure communications can be counteracted by mechanisms and services, which are part of communications security. These services include encryption algorithms that hide the content of a message, integrity checks that detect message modification, and a means to verify the source and integrity of a message.

Additional access control is implemented through the use of filters, firewalls, and access lists. These controls can be boosted through the use of security cards and other devices.

Digital certificates and signatures can be used to verify the authenticity of the individual users. A certificate authority is responsible for managing this process.

Data security is further enhanced via the use of encryption techniques. These techniques include asymmetric and symmetric encryption and decryption methods.

A key consists of a number of characters that identifies data records. Keys include both public keys and private keys. Key management must be implemented to manage the issuance and maintenance of the keys.

Access to computer and network facilities, buildings, and other assets must be protected from unauthorized personnel and nonemployees. Access control procedures must be available for a number of people and external organizations that might require access to sensitive areas. It is also essential that deterrents be in place to keep hardware, software, documentation, and other resources from disappearing out the door.

Key Terms

- AAA
- Access control
- Access list
- Algorithm
- Anonymous
- Asymmetrical
- Asymmetric encryption
- Authentication
- Authorization
- Biometric
- Certificate
- Certificate authority (CA)
- Checksum
- Closed User Groups (CUGs)
- Confidentiality
- Data security
- Data Encryption Standard (DES)
- Diffie-Hellman
- Digital certificate
- Digital envelope
- Digital signature
- Digital Signature Standard (DSS)
- Enciphering
- Encryption
- Filter
- Firewall
- Hash code
- Hash functions
- Hashing
- Identification
- International Data Encryption Algorithm (IDEA)
- Key
- Key escrow
- Key management
- Message digest
- Message Digest Algorithm 4 (MD4)
- Message Digest Algorithm 5 (MD5)
- One-time password
- Password
- Personal identification number (PIN)
- Private key
- Privilege
- Public key
- Public Key Encryption (PKE)
- Remote Access Server (RAS)
- Remote Authentication Dial-in User Service (RADIUS)
- Repudiation
- Rivest, Shamir, and Adleman (RSA)
- Roles
- Route authentication
- Secret key
- Secret key algorithm
- Secure Hash Algorithm (SHA)
- Security cards
- Smart card
- S/Key
- Symmetric algorithm
- Symmetric encryption
- Terminal Access Controller Access Control System (TACACS)
- Triple-A
- Triple DES (3DES)
- Trust
- Trusted system
- Trust model
- Trust relationship

REVIEW EXERCISES

Questions and Analysis

Provide a short answer or in-depth analysis as required.

1. How is identification accomplished in a security system?

2. Describe the authentication process.

3. What is the relationship between authentication and authorization?

4. Describe the authorization process.

5. Explain the function of a remote access server. Give examples of remote authentication systems.

6. How is a password used in the access to corporate assets?

7. Give examples of passwords and discuss the issues that are related to passwords.

8. What is a one-time password and how is it used?

9. What is the difference between a first-function and second-function authentication?

10. What are the various authorization access types? Where can these be used?

11. What is the difference between a filter and a firewall? Where are they employed?

12. What is a security card? Describe the secure card environment.

13. What is a digital certificate and how is it used?

14. How is a digital signature used? What is it?

15. Describe the function of a certificate authority.

16. What is the difference between a private key and a public key?

17. Describe the differences between asymmetrical and symmetrical systems.

18. What is an algorithm? How is it used in the security systems?

19. What is a key? How is this different from a secret key?

20. What are the methods for encrypting and decrypting messages and data?

21. Describe the data security environment.

22. What are the common symmetric key algorithms?

23. What are the common asymmetric key algorithms?

24. Describe the Diffie-Hellman algorithm.

25. Why is key management necessary?

26. What questions can be asked concerning access control for internal personnel?

27. What activities might external entities need to perform in a secure location?

28. What guidance must be given for allowing access to visitors into a secure location?

29. What techniques can be employed for theft prevention?

30. Discuss security access cost justification.

Definitions and Terms

Provide the most correct answer from the list of key terms.

1. Limiting the flow of information from the resources of a system only to authorized entities of a network.

2. A listing of users and their associated access rights.

3. A prescribed finite set of well-defined rules or processes for a solution to a problem.

4. The process of validating the claimed identity of an end user.

5. The act of granting access rights to a user or system.

6. The process of using hard-to-forge physical characteristics of individuals to authenticate users.

7. A unique collection of information, transformed into unforgeable form, used for the authentication of users.

8. An entity that distributes public and private key pairs.

9. A secret key cryptographic scheme standardized by NIST.

10. A method of scrambling information in such a way that is it not readable by anyone except the intended recipient.

11. A system, based on either hardware or software, used to regulate traffic between two networks.

12. The process of creating a fixed value from arbitrary data, such that if the arbitrary data are altered, the hash value also changes.

13. A secret key, block cipher, used in Pretty Good Privacy.

14. A method of determining the identity of a computer user. It is usually the result of an authentication process.

15. A secret value used to protect data. It might be used to protect files and restrict access.

6. Smart cards include microprocessor chips that perform security functions.

7. A privilege allows a user to enter a computer center facility with a secure card.

8. A certificate is used to verify the user is allowed access to a particular database record.

9. Repudiation means a user has denied the transmission of a message.

10. The use of a public and private key is called an asymmetrical system.

11. A symmetrical system includes both a public and a private key.

12. Encryption takes a document and converts it to unreadable text using an algorithm.

13. The Data Encryption Standard (DES) is an asymmetric encryption algorithm.

14. A hash function and a message digest are different names for the same process.

15. The use of individual physical characteristics for access control is called biometrics.

True/False Statements

Indicate whether the statement is true or false.

1. Triple-A includes the identification of users via authentication, authorization, and accounting.

2. A Remote Access Server (RAS) gives remote users access to all servers on the LAN, which means that authentication is not required.

3. A password is the first line of network security defense, which means that it should consist of a structure that inhibits a hacker.

4. Trust is the composite of security, availability, throughput, and accounting.

5. Filters or access lists can be programmed in bridges and hubs.

Activities and Research

1. Create a process flow that shows how a user or entity would access some enterprise resource. Include identification, authentication, and authorization steps.

2. Develop a list of acceptable and unacceptable passwords. Describe the good and bad attributes for each password developed.

3. Research and identify access control devices that use filters and access lists to perform firewall functions.

4. Research security card and security card system components and develop an access control process flow for a simple computer or network. This design should include all of the physical devices and software that would be necessary to implement such a system.

5. Develop a matrix that shows the attributes and functions of the various encryption methods.

6. Research and develop a comparison report on the various symmetric key algorithms.

7. Research and develop a comparison report of the various asymmetric key algorithms.

8. Schedule a presentation by a security manager concerning assess control issues that are encountered in the presenter's organization.

9. Identify assets and resources in an organization that might require some type of access control protection. Describe how they would impact the operation if destroyed.

10. Prepare a matrix of different theft prevention techniques, along with ease of implementation, and cost.

CHAPTER 3

Cryptosystem Techniques

OBJECTIVES

Material included in this chapter should enable the reader to

- Become familiar with the cryptosystem environment.
- Learn the numerous terms and language associated with cryptology.
- Look at the various cryptographic methods used to convert plaintext to ciphertext.
- Understand the function and operation of the DES encryption algorithm, and other algorithms in the conventional class.
- Become familiar with the Public Key Infrastructure (PKI).
- Look at the activities of a cryptanalyst.
- Identify the various standards organizations that are active in the security arena.
- Become familiar with the differences between public-key and private-key encryption systems.
- Compare and contrast the capabilities of the various security models.

INTRODUCTION

Error detection and correction methods help ensure that users are receiving correct information, however another issue of equal importance is the illegal or unauthorized reception of data. These situations involve a legitimate sender and receiver plus a third party, who intercepts the message. The sender and the receiver may be unaware of the unauthorized reception until the intruder has used the data for some illegal activity. It is obvious that transmission of sensitive information requires some method of ensuring data privacy and integrity. Methods that are used to solve these issues are categorized as cryptosystems.

This chapter presents a high-level overview of those cryptographic techniques that are used in a cryptosystem. The first section sets the stage with cryptology, which is the study of secure communications. Information is presented on the various branches of cryptology, including cryptography and cryptanalysis. Various encryption standards and algorithms are discussed including the Data Encryption Standard (DES). Issues relating to both public-key and private-key technologies, as they relate to cryptology, are presented. The task and efforts of a cryptanalyst are discussed. This chapter should provide a general understanding of these topics, however they will be used throughout the text to further explain the security issues involved. The chapter concludes with an overview of the various security models available to the security specialist.

3.1 CRYPTOSYSTEMS

There are a number of terms associated with a cryptosystem. These include cryptology, cryptography, cryptanalysis, and steganography. **Cryptology** is the study of secure communications, which encompasses cryptography, cryptanalysis, and steganography. **Cryptography** is the branch of cryptology dealing with the design of algorithms for encryption and decryption. These algorithms are intended to ensure the secrecy and/or authenticity of messages and data. **Cryptanalysis** is the branch of cryptology dealing with the breaking of a cipher to recover information, or forging encrypted information that will be accepted as authentic (Stallings 2000). A **cipher** is an algorithm for encryption and decryption. A cipher replaces a piece of information with another object, with the intent of

cryptology

cryptography

cryptanalysis

cipher

steganography

concealing meaning. Typically, a secret key governs a replacement rule. Algorithms and secret key concepts were discussed in Chapter 2. Lastly, **steganography** is a method of cryptology that hides the existence of a message.

3.2 CRYPTOGRAPHY

plaintext
ciphertext

Cryptography is concerned with keeping sensitive data and information private. Data is encrypted to make it private and decrypted to restore it to a readable form. Encryption is performed using an algorithm, which takes some input, called **plaintext**, and converts it to **ciphertext** (Figure 3–1). A key is then applied to the algorithm that affects the ciphertext output. A different key used on the same plaintext will produce different ciphertext. The strength of encryption relies on the key and its length.

Cryptography has its roots in communications security, where the communication of two users over an unsecured channel is subject to an intruder who can read the messages. Cryptology allows users to construct a secure logical channel over an unsecured physical connection. This means that an intruder with access to a physical communications link may not compromise cryptographic protection. Cryptographic mechanisms are the basic building blocks of cryptographic schemes. They are used in cryptographic protocols and rely on a viable key management process to offer effective protection. These schemes include integrity checks, digital signatures, and encryption algorithms (Gollman 2000).

3.3 ENCRYPTION ALGORITHMS

hash algorithm

As stated in Chapter 2, an algorithm is a well-defined set of rules that solves a problem in a finite number of steps. A **hash algorithm** is a check that protects data against modification. Protecting data against undetected modification requires computing a hash of the data and having the receiving party do the same and compare the results. If the result of the hash algorithm is the same, the data are probably secure (Pfleeger 1997).

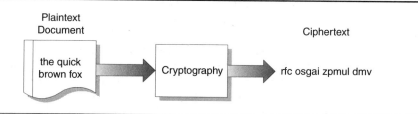

Figure 3–1
Plaintext Document Transformed into Ciphertext

A **hash** is the process of creating a fixed value from arbitrary data, such that if the arbitrary data are altered, the hash value also changes. Hash functions produce a reduced form of a body of data such that most changes to the data will also change the reduced form. The hash process reduces a large amount of data to a smaller result. The result is sometimes called a **digest** or **check value**. Several algorithms that will be introduced include the monoalphabetic cipher, polyalphabetic cipher, transposition cipher, and bit-level cipher.

hash

digest
check value

Mono-alphabetic Cipher

One of the earliest algorithms replaces each plaintext character with another character. The choice of the encrypted replacement is based solely on the plaintext character, which could be as simple as adding the key of "1" to each character. An "A" would become a "B" with a key of "1." This method is called a **monoalphabetic** or **Caesar cipher**, which dates back to the days of Julius Caesar. The decryption process reverses the encryption steps. Both the encryption and keys are identical.

monoalphabetic
Caesar cipher

Although they are simple to describe and seem to yield unintelligible messages, these ciphers are rarely used in serious applications. They are relatively easy to decode without knowledge of the original encryption method because the code does nothing to disguise frequently used letters or combinations.

Polyalphabetic Cipher

One method used to change character frequencies and destroy common sequences is to use a **polyalphabetic cipher**. As with the mono-alphabetic cipher, it replaces each character with another. The difference is that a given plaintext character is not always replaced with the same ciphertext one. The replacement can be chosen depending not only on the actual plaintext character, but also on its position in the message.

polyalphabetic cipher

An example of a poly-alphabetic cipher is a **Vigenere cipher**, which uses a two-dimensional array of characters, where each row contains the letters of the alphabet. The first row contains the letters written from A to Z. Each subsequent row is offset by one character. The last character is wrapped to the first position for each cycle of the process. Table 3–1 provides an example of this array. The Vigenere cipher seems to solve the repetition and pattern problem, but that is only an illusion, since they can still occur.

Vigenere cipher

Row	Characters
0	ABCDEFGHIJKLMNOPQRSTUVWXYZ
1	BCDEFGHIJKLMNOPQRSTUVWXYZA
25	ZABCDEFGHIJKLMNOPQRSTUVWXY

Table 3–1
Vigenere Cipher Key Array

Transposition Cipher

transposition cipher

A **transposition cipher** rearranges the plaintext letters of a message, rather than substituting ciphertext letters. One way to accomplish this is to store plaintext characters in a two-dimensional array with x columns. The first x plaintext characters are stored in the array's first row and so on until the array is complete. A permutation of numbers is determined, either arranged randomly or by some secret method. The columns of the array are manipulated, based on this permutation, in a way that will produce a ciphertext that looks nothing like the original.

The decryption algorithm is a reverse of the encryption algorithm. The problem with the transposition cipher is that it is not very secure, since letter frequencies are preserved. On reception, a cryptanalyst could analyze the ciphertext and perceive the high frequency of common letters.

Bit-level Cipher

Not all transmissions are character sequences, therefore not all encryption methods work by manipulating or substituting characters. One method defines the encryption key as a bit string, where a bit is a "1" or "0." The choice is determined randomly and secretly. The **bit-level** encryption process is as follows:

bit-level

- The bit string is divided into substrings.
- The length of each is the same as the length of the encryption key.
- Each substring is then encrypted by computing the exclusive OR (XOR) between it and the encryption key.
- The encryption step does not reverse the encryption steps—it repeats them.
- Decryption is accomplished by computing the XOR between the encryption key and each of the encrypted substrings. Note the encryption and decryption keys are identical.
- Performing the XOR function twice generates the original bit pattern.

The security of this code depends largely on the length of the key. Using longer encryption keys means longer, but fewer, substrings are created. Like the polyalphabetic cipher, the drawback is the large key that must be communicated to the receiver, which makes the method somewhat unwieldy.

Block Cipher

block cipher

The most commonly used conventional encryption algorithms are block ciphers. A **block cipher** processes the plaintext input in fixed-size blocks and produces a block of ciphertext of equal size for each plaintext block. The two most important conventional algorithms are the data encryption standard and the triple data encryption algorithm. This section also explores the planned advanced encryption standard and other popular conventional encryption algorithms.

Data Encryption Standard

One well-known encryption algorithm is the U.S. government-endorsed **Data Encryption Standard (DES)**, which uses a 56-bit key and a **Data Encryption Algorithm (DEA)** that scrambles and obscures a message by passing it through multiple iterations of an obfuscation algorithm. DES is a block cipher, which takes the plaintext and divides it up into blocks that are processed individually in multiple iterations. Alternatively, there are **stream ciphers** that work on streams of raw bits and are much faster.

IBM developed DES in the early 1970s. It is the industry standard for cryptography systems and the world's most commonly used encryption mechanism. This private-key system is widely deployed in automated machines and point-of-sale devices. It was adopted as a **Federal Information Processing Standard (FIPS)** in 1977 and as an American National Standard (ANSI) in 1981.

The DES process can be executed using either hardware or software, and is available on a number of different platforms. Specialized hardware is also available to execute the encryption algorithm. Some products provide the DES capability on a Smart card or PCMCIA card, which allows the algorithm and encryption keys to remain on the card and be transported from location to location. A hardware solution is desirable when speed is important or the required processing power for software is not available.

The DES encryption algorithm has just recently been successfully broken; however, it is still a secure algorithm. It has been estimated that it would take a single computer 2,000 years at one DES encryption per second to break one key. The ability to break the algorithm is based on the size of the encryption key. Hopefully, the use of **Triple-DES**, using the DES algorithm three times with two keys, will extend the life of the algorithm. Triple-DES is also called the **Triple Data Encryption Algorithm (TDEA)**. Figure 3–2 provides a general depiction of the DES encryption algorithm process. The plaintext contains 64 bits and the key contains 56 bits. The left side of the graphic shows the processing of the plaintext transformation (permutation) in three phases. These phases are as follows:

- The 64-bit plaintext passes through an initial permutation that rearranges the bits to produce the permuted input.
- A phase consisting of 16 iterations follows using the same function. The output of the last iteration consists of 64 bits that are a function of the input plaintext and the key.
- The left and right halves of the output are swapped to produce the pre-output. The pre-output is passed through a permutation that is the inverse of the initial permutation function, to produce the 64-bit ciphertext.

Figure 3–2
General Depiction of DES Algorithm Process

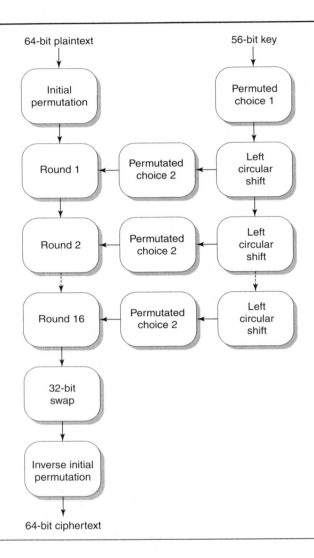

The right side shows the way the 56-bit key is used. These steps are as follows:

- The key is passed through a permutation function.
- For each of the 16 iterations, a subkey (secondary) is produced by the combination of a left circular shift and a permutation.

Concerns about the strength of DES include the algorithm itself and the use of a 56-bit key. There is a possibility of a cryptanalyst exploiting the characteristics of the DES algorithm, because there has already been numerous attempts to crack the algorithm. The days of DES may be numbered because of the speed of computers that can be used to attack the 56-bit key. It is possible to break the key using a brute force attack.

International Data Encryption Algorithm

Another option to DES and Triple-DES is readily available. A replacement developed to replace DES is the **International Data Encryption Algorithm (IDEA)**, a block-oriented secret-key encryption algorithm developed by the Swiss Federal Institute of Technology. It uses a 128-bit key compared with DES's 56-bit key, which makes its resulting ciphertext more secure. It operates on 64-bit message blocks. The algorithm is readily available and does not require licensing. It was designed to be efficient in both hardware and software implementations. The encryption rate is high, and has been implemented on chips that encrypt at 177 Mbps. IDEA can operate in all four modes of ciphertext generation and is considered suitable for electronic commerce. It is implemented in Pretty Good Privacy (PGP); an encryption tool often used in e-mail programs.

International Data Encryption Algorithm (IDEA)

IDEA was one of the earliest of the proposed 128-bit replacements for DES. The object of considerable scrutiny, it appears to be highly resistant to cryptanalysis. IDEA gives no clues to the contents of the plaintext during deciphering, because it spreads out a single plaintext bit over many ciphertext bits, hiding the statistical structure of the plaintext completely.

IDEA does have a minimum requirement since it needs 64 bits of message text in a single coding block to ensure a strong ciphertext. This is great if large amounts of data, such as file transfer protocol (FTP) files, are being encrypted; however, it does not fit in a telnet session that consists of keystrokes (Gonclaves 2000).

Secure Hash Algorithm-1

The **National Institute of Standards and Technology (NIST)** developed the **secure hash algorithm (SHA-1)**. It was published as a FIPS (FIPS PUB 180) in 1993, with a revised version (FIPS PUB 180-1) issued in 1995. The algorithm accepts as input a message with a maximum length of less than 2^{64} bits and produces an output of a 160-bit message digest. The input is processed in 512-bit blocks. The SHA-1 algorithm has the property that every bit of the hash code is a function of every bit of the input. The possibility of two messages having the same message digest is extremely remote (Stallings 2000).

National Institute of Standards and Technology (NIST)

secure hash algorithm (SHA-1)

Message Digest Algorithms

The **message digest 5 (MD5)** algorithm was developed by Ron Rivest and is documented in RFC 1321. MD5 was the most widely used secure hash algorithm until the emergence of brute-force and cryptanalytic concerns. The algorithm accepts as input a message of arbitrary length and produces as output a 128-bit message digest. As processor speeds have increased, the security of this hash code has become questionable (Stallings 2000).

message digest 5 (MD5)

RACE Integrity Primitives Evaluation Algorithm

The **RIPEMD-160** message-digest algorithm was developed under the European **RACE Integrity Primitives Evaluation (RIPE)** project. This resulted from partially successful attacks by a group of researchers on MD4 and MD5 algorithms. RIPEMD-160 is similar in structure to SHA-1. The algorithm accepts an input message of arbitrary length and produces an output of a 160-bit message digest. The input is processed in 512-bit blocks (Stallings 2000).

Blowfish

Blowfish was developed in 1993 by Bruce Schneier, an independent consultant and cryptographer, and quickly became one of the most popular alternatives to the DES algorithm (Stallings 2000). Blowfish is easy to implement and has a high execution speed. It is a very compact algorithm that can execute in less than 5k bytes of memory. Blowfish has a variable key length, which can extend to 448 bits, however 128-bit keys are used.

Blowfish uses *S-boxes* and the *XOR* function, as does DES, however it also uses binary addition. (Note that the **S-box** is a substitution table used to map bit patterns.) Unlike DES, which uses fixed S-boxes, Blowfish's dynamic S-boxes are generated as a function of the key. The subkeys and S-boxes are generated by repeated application of the Blowfish algorithm to the key. A total of 521 executions of the Blowfish encryption algorithm are required to produce the subkeys and S-boxes. This means that blowfish is not suitable for applications where the secret key changes frequently.

Blowfish is one of the most formidable conventional encryption algorithms implemented to-date. This is because a process of repeated applications (recursion) of Blowfish itself, which thoroughly mangles the bits and makes cryptanalysis very difficult, produces the subkeys and S-boxes.

RC5

Ron Rivest developed **RC5** in 1994. He is one of the inventors of the RSA public-key algorithm. RC5 is defined in RFC 2040 and has a number of attributes. It is used in a number of RSA Data Security products (Stallings 2000).

RC5 uses only primitive computational operations commonly found on microprocessors, because it is a simple algorithm and is *word* oriented. The basic operations work on full words of data. The number of bits in a word is a parameter of RC5. This means that different word lengths yield different algorithms.

The number of rounds (cycles) is a second parameter of RC5. It allows for a tradeoff between higher speed and higher security. The key length is a third parameter, which also allows for a tradeoff between speed and security.

RC5's simple structure is easy to implement and eases the task of determining the strength of the algorithm. RC5 is intended to provide high security with suitable parameters. It incorporates circular bit shifts (rotations), which are data dependent, therefore strengthening the algorithm against cryptanalysis. A low memory requirement makes RC5 suitable for smart cards and other devices with restricted memory.

CAST-128
CAST is a design procedure for symmetric encryption algorithms developed in 1997 by Carlisle Adams and Stafford Tavares of Entrust Technologies. One specific algorithm developed is **CAST-128**. It is defined in RFC 2144, and uses a key size that varies from 40 bits to 128 bits in 8-bit increments. CAST is the result of a long process of research and development and has been reviewed extensively by cryptologists. It is used in a number of products, including PGP (Stallings 2000).

CAST uses fixed S-boxes, however they are considerably larger than those used by DES. The S-boxes are designed to be resistant to cryptanalysis. S-boxes, or selection boxes, are a set of highly nonlinear functions, which are implemented as a set of lookup tables. The subkey-generation process used in CAST-128 is different from that employed in other conventional block encryption algorithms previously described. The CAST designers' objective was to make subkeys as resistant to known cryptanalytic attacks as possible. They were convinced that the use of highly nonlinear S-boxes to generate subkeys from the main key provided this strength. The *round* function also differs from round to round, which adds to the cryptanalytic strength.

Skipjack
Skipjack is the NSA-developed encryption algorithm designed for the **Clipper** and **Capstone** chips. The algorithm is an iterative 64-bit block cipher with an 80-bit key, which has 32 rounds of processing per each encrypt or decrypt operation. Each chip also contains an 80-bit key that is split into two parts, with each being stored at separate key escrow agencies. The key fragments can be obtained and combined to recover information encrypted with a particular chip. Skipjack was originally classified, however the NSA has declassified it, making it subject to public cryptanalysis. The developers, however, do not believe that it will be broken in the next 30 to 40 years.

Capstone is a U.S. government program to develop cryptography and security standards for protecting government communications as specified in the Federal Information Processing Standard (FIPS) (Sheldon 2001). There is considerable resistance from concerned citizens and the business sector against the Clipper chip because they perceive it as an invasion of privacy.

Advanced Encryption Standard
The **Advanced Encryption Standard (AES)**, currently under development, will have 128-bit, 192-bit, and 256-bit keys. It should encrypt blocks of 128 bits. This future standard looks to replace DES, however DES is expected

to remain in use for some time since it is imbedded into old software. AES is intended as a new Federal Information Processing Standard (FIPS) publication specifying a cryptographic algorithm for use by United States governmental organizations to protect sensitive information. NIST anticipates that the AES will be widely used on a voluntary basis by organizations, institutions, and individuals outside the United States government, as well as outside the United States. Given the widespread commercial and governmental applications of encryption, AES will come to play a fundamental role ensuring secure data traffic within the realm of daily national and international data communications.

This standard possesses a powerful combination of security, performance, efficiency, ease of implementation, and flexibility. It is consistent in performance both in hardware and software across a wide range of computing environments regardless of its use. AES's very low memory requirements make it suited for restricted-space environments, where it also demonstrates excellent performance.

One-Way Function
Another important encryption scheme is called the **one-way function**. This function is easily computed and takes little processing power, but the calculation of its inverse is infeasible. There is usually no intention of ever deciphering the message. One-way functions can be used to provide for password database security. This technique stores the hash of the passwords, and not the passwords themselves. When a user performs a login, a hash of the password is created and compared to the hash in the password database. If they compare, the user is considered authentic.

Compression Issue
It should be noted that encryption can affect data compression performance. Because **compression** involves the reduction of the number of repetitive bit patterns in a message, additional processing cycles are required in the computer, effectively reducing throughput. Encrypted data is essentially random, because it lacks the redundancy of ordinary text that is normally exploited by compression algorithms. This means that if encryption is followed by data-link compression, the overall throughput will be much lower than it would be with compression and no encryption. In systems where compression is required, data compression should be performed before encryption.

3.4 PUBLIC KEY CRYPTOGRAPHY

An alternative to the private-key DES and IDEA schemes is asymmetrical **public-key** cryptography, discussed in Chapter 2. All users get a set of keys in this scheme. One key, the public key, is freely available, while the other, the private key, is held secretly. The process of sending encrypted messages to another user is simple. A copy of the recipient's public key is obtained, either directly from the intended user or from a public-key server, and used to encrypt the message. This message can only be decrypted with the recipient's private key, which is never made publicly available (Sheldon 1998). The public-key scheme is revolutionizing computer and network security by providing ways to enable electronic commerce, authenticate users, and exchange secure electronic documents.

In many cases, parties in a message exchange do not know or trust each other and have not exchanged any keys in advance. This problem was solved by the **Diffie-Hellman** key exchange, which provides a way for two parties to establish a shared secret key over an insecure channel. Anyone who wants to encrypt and send messages to someone else generates a pair of security keys. One key is kept private and the other is located in a public place. If a user wants to send a private message, the sender obtains the recipient's public key, encrypts the message with this key, and then transmits it. The recipient's private key is used to decrypt the message. Anyone intercepting the message cannot decrypt it with the generally available public key.

Diffie-Hellman

As an example, assume two users want to exchange private encrypted messages over an insecure system like the Internet. An encryption method is chosen that will make the messages unreadable to any other person who happens to capture the transmission. The following flow illustrates this exchange:

- User 1 and user 2 are attached to the same company network.
- A security server for storage of public keys is available to both users.
- Users 1 and 2 generate a set of keys, using commonly available software.
- Their public keys are placed on the company's public-key server. (Note: a certificate authority could provide public-key management.)
- User 1 encrypts the message with user 2's public key from the security server and sends it to user 2.
- User 2 receives the message and decrypts it with the private key.

The public-key scheme solves the problem of passing keys to other parties who need to decrypt incoming messages. This scheme allows any sender to encrypt a message and send it to another user without any prior exchange or agreement. The primary requirement is for both parties to have a way to exchange public keys. Key management, discussed in Chapter 2, provides this function.

RSA Public Key

As presented in Chapter 2, public key encryption, also called asymmetric encryption, uses two separate, but mathematically, connected keys (Figure 3–3). Every user has a pair of keys; one is kept strictly confidential (private), and

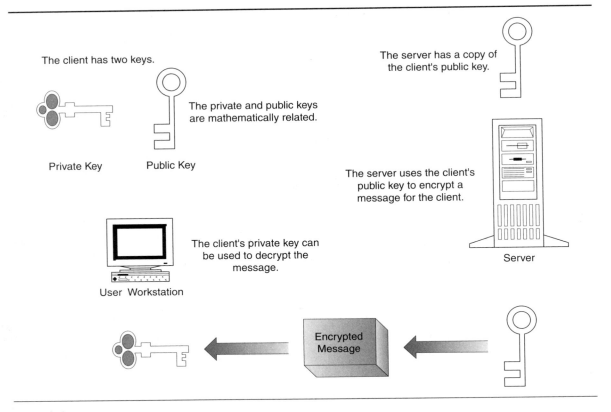

Figure 3–3
RSA Public Key Encryption

the other is shared (public) among other users or computers. A message encrypted with the public key can only be decrypted using the private key. The advantage of this method over secret technology is that the private key is never shared with other principals. Public keys can also be used for the creation of digital signatures, which are used to verify both the sender and contents of an electronic message.

RSA (Rivest, Shamir, and Adleman)

RSA (Rivest, Shamir, and Adleman) is a public key algorithm based on the algorithm developed by Diffie-Hellman. This system uses two related, complementary keys, one of which is kept secret for encryption with the other publicly available and used for decryption. Only the private key is known for the encryption process and must be kept secret. The use of the RSA algorithm is rapidly growing, particularly in the area of electronic mail and electronic messaging. There is a downside to RSA, which is the amount of computer power required for the process, making it slower than the DES process (Dempsey 1997). RSA, however, is stronger than DES because the encryption key can be variable in size, whereas the DES encryption key is fixed at 64 bits.

The RSA technology can also be used to provide a digital signature for document authentication. A message digest is produced using a hashing function performed on the document, which is then encrypted with the private key. This digest is known as the digital signature, which is attached to the message. The decryption of the digital signal using the public key and processing with the same hashing function should produce the same result as the message digest. Hashing and message digest are described in Chapter 2.

The RSA encryption algorithm incorporates results from number theory, combined with the difficulty of determining the prime factors of a target. The RSA operates with arithmetic mod n. The encryption algorithm is based on the underlying problem of factoring large numbers. The RSA method has been scrutinized intensely by professionals in computer security and cryptanalysis, and few flaws have been identified.

3.5 PUBLIC-KEY INFRASTRUCTURE (PKI)

Many security protocols rely on public-key cryptography to provide services such as confidentiality, data integrity, data origin authentication, and nonrepudiation. The **Public-key infrastructure (PKI)** can provide a structure for verifying and authenticating the parties involved in transactions on the Internet, intranets, or extranets.

The purpose of a PKI is to provide a secure infrastructure for managing public keys that allows anyone to exchange private documents with anyone else without knowing these people in advance. A PKI is defined by the Internet X.509 Public Key Infrastructure PKIX Roadmap as the set of hardware, software, people, policies, and procedures needed to create, manage, store, distribute, and revoke certificates based on public-key cryptography (Kaeo 1999). As stated previously, the reason for developing a PKI is essentially to build an infrastructure of trust. If a user wants to send a private message to someone, the user can obtain the recipient's public key, encrypt the message, and send it. The recipient can then decrypt the message with their private key.

A PKI consists of five types of components. A brief description of each component and its function follows:

- Certificate authorities (CA)—issue and revoke certificates.
- Certificate holders—are issued certificates and can sign digital documents.
- Clients—validate digital signatures and their certification paths from a known public key of a trusted certificate authority.
- Repositories—store and make available certificates and certificate revocation lists.
- Organizational registration authorities—vouch for the binding between public keys, certificate holder identities, and other attributes.

Note that **X.509** is an International Telecommunications Union-Telecommunications Sector (ITU-T) standard. Also, the National Institute of Standards and Technology (NIST) is working on certificate authorities and digital certificates based on X.509 standards. The X.509 standard constitutes a widely accepted basis for a PKI infrastructure, defining data formats and procedures related to the distribution of public keys using certificates digitally signed by certificate authorities. RFC 1422 specified the basis of an X.509-based PKI, targeted primarily at satisfying the needs of Internet privacy-enhanced mail. Application requirements for an Internet PKI have broadened tremendously, and the capabilities of X.509 have advanced significantly.

3.6 PRIVATE KEY CRYPTOGRAPHY

Ciphers may include single-key, or **symmetrical**, and two-key, or **asymmetrical** structures. As presented in Chapter 2, symmetric schemes are also called **private-key** or **secret-key** encryption schemes. A single key is used for both encrypt and decrypt messages. If an encrypted message is sent to someone, they must also get a copy of the key. This is a problem in some environments, especially if the recipient is unknown and the key needs to be transmitted over untrusted channels by untrusted personnel. Asymmetric schemes can be used to solve this problem.

3.7 CRYPTANALYSIS

A **cryptanalyst** is a person who analyzes and attacks a cipher. It is through cryptanalysis that one may find a weakness in the cryptosystem that could eventually endanger the secrecy of a message being exchanged.

Cryptanalysis involves analyzing a **cryptosystem**. A system is analyzed to verify its integrity. An attacker may also analyze a system to identify any weaknesses and gain illegal access to systems. There are several methods that can result in a successful attack on a system. One such method is called the **brute force attack**. Every possible key is tried in an attempt to decrypt the ciphertext. A dictionary of common passwords, which can be obtained from the Internet, may be used. A weak password, could include common names, words from the dictionary, or common abbreviations. Brute force attacks are difficult if long keys are used and if the keys consist of a mix of characters and numbers in a nonsense pattern. The more characters contained in the key, the less likely there will be a break, however this too can be overcome by a dedicated attacker.

Factoring is used as a tool to split the modulus into the original integers that were multiplied together to form it. Prime factorization requires splitting an integer into factors that are prime numbers. The strength of the public-key system relies on the fact that multiplying two prime numbers together is easy, however factoring the product is difficult.

It is possible that the cryptanalyst knows something about what is contained in an encrypted message and has the algorithm used to create the ciphertext. The cryptanalyst can analyze the original plaintext, the algorithm, and the resulting ciphertext to identify a pattern or weakness in the system. It is often easy to figure out message content, given some basic knowledge of the information contained in the message. It is possible for a cryptanalyst to find a way to get some text inserted into a sensitive document before it is encrypted, then use the techniques described above to look for this text in the ciphertext.

Some cryptanalytic techniques that could be effective in penetrating or breaking encrypted communications are as follows:

- Guessing text content—Many types of messages use fixed formats. Ordinary documents and letters begin in predictable ways. Learning parts of a message can then be used to guess other parts of the same message.
- Using chosen text—If attackers are able to have some text of their own choice encrypted with an unknown key, the key could possibly be identified.
- CryptoSystem faults—The secret key might be discovered because some algorithm shows faults under some conditions.
- Duplicate block encryption—The same block is encrypted twice with the same key to produce the ciphertext.

An additional technique in which chosen plaintext with a particular XOR (exclusive-OR) difference pattern is encrypted is called **differential cryptanalysis**. The difference patterns of the resulting ciphertext provide information that can be used to determine the encryption key. This technique uses an interactive and iterative process, which allows the attacker to work through many rounds to break ciphertext.

differential cryptanalysis

Computer system weaknesses are the prime reasons that the cryptanalyst is able to crack Enterprise security. The security administrator should be aware of three of the major contributors to this problem. These are as follows:

- Disk drive storage—Portions of encrypted and decrypted messages or temporary files could be left on system hard drives. This residual data might be used break the system.
- Encryption products—Product information is available to anyone who can access the Web, therefore it is not a good idea to publicize which products are being used to protect the Enterprise computer network.
- Cryptographic keys—If an attacker has partial information about a key, it is possible to determine the master key. Some systems are protected by both strong and weak keys, and some use session keys. It is essential that mechanisms are in place to secure access to all keys.

What can be done to counter the efforts of the cryptanalyst? A number of possible solutions are presented throughout this book. The desired security level of the Enterprise computer and network assets depends on the sensitivity of the database resources, the difficulties and costs involved in deploying a security system, and the repercussions resulting from an attack.

Government Encryption Policies

The fundamental goals of data encryption and the government's policies regarding it are similar. They embody information security, national security, economic strength at corporate and national levels, public safety, crime prevention and investigation, privacy, and other intellectual and academic freedoms. As the federal government becomes more dependent on electronic means to conduct its day-to-day business, its policies have changed. The fear of cryptographic code breaking by outsiders and other undesirables has given way to the reality that information must be as secure as technologically possible in the global economy (Traeger 2001).

In January 2000, the U.S. government published new encryption export regulations that represent a fundamental change in U.S. policy and updated them again in October 2000. These regulations implement the policy announcement on encryption made by the White House in September 1999, and make it much easier for companies and individuals in the United States to widely export strong encryption in common products regardless of their strength or the type of technology used.

Our current national encryption policies are based on a technique called "key recovery." **Key recovery** allows the use of sophisticated algorithms with long code keys, while accommodating the ability to break the code under very tightly controlled conditions, specifically, by the owners of the encrypted data and by government officials with a court order of other lawful authorization. It is considered the solution to a fundamental dilemma of encryption: hiding information and secrets for both legal and illegal use.

3.8   STEGANOGRAPHY

Steganography is the practice of hiding information within other information. The term literally means covered writing, and like cryptography, has been used since ancient times. Steganography allows the system to take a piece of information, such as a message or image, and hide it within another image, message, or audio clip. It takes advantage of insignificant space in digital files, in images, or on removable disks. Attackers can also use the extra space in TCP and IP headers to carry information between systems without the knowledge of system administrators.

An attacker can also hide data during transmission. Using **covert channels**, which are hidden communication paths, an attacker can send hidden data across the network. This data could consist of files to be transferred or commands for a backdoor running on a victim computer. Another technique is tunneling, which can carry one protocol on top of another protocol. A number of tools implement tunneling of command shells over various protocols like ICMP and HTTP.

An increasingly popular application of steganography is the **digital watermark**. A digital watermark can be either visible or invisible, and it usually consists of a company logo, copyright notification; or other mark or message that indicates the owner of the document. The owner of a document could show the hidden watermark in court to prove that the watermarked item was stolen.

Digital watermarking capabilities are built into some image-editing software applications such as Adobe PhotoShop. Companies that offer digital watermarking solutions include Digimarc and Cognicity.

3.9 SECURITY MODELS

Formulating a security policy includes a description of the entities governed by the policy and the rules that constitute the policy. **Security models** capture policies for confidentiality and integrity of Enterprise computer networks. Models discussed in this section include Bell-LaPadula, Harrison-Ruzzo-Ullman, Chinese Wall, Biba, Clark-Wilson, and Information-Flow.

Several terms must be introduced before this discussion, including subjects, objects, state, and state transitions. A typical subject is a user or a process. A typical object includes a file or a resource, such as node. A state is a representation of the system under investigation at one moment in time, which should capture exactly those aspects of the system relevant to the problem. The possible state transitions can be specified by a state transition (next step) function, which defines the next state depending on the present state and input. A more complete description of these models can be found in *Computer Security* (Gollman 2000).

Bell-LaPadula Model

The **Bell-LaPadula (BLP)** model was developed at the time of the first concerted efforts to design secure multiuser operating systems. It is a state machine model capturing the confidentiality aspects of access control. Access permissions are defined both through an access control matrix and through security levels. Security policies prevent information flowing downward from a high security level to a low security level (multilevel security). BLP only considers the information flow that occurs when a subject observes or alters an object. BLP has played an important role in the design of secure operating systems and is used as a model for other security systems.

Harrison-Ruzzo-Ullman Model

The BLP model does not state policies for changing access rights or for the creation or deletion of subjects and objects. The **Harrison-Ruzzo-Ullman (HRU)** model defines authorization systems that address these issues.

Chinese Wall Model

The **Chinese Wall** model, proposed by Brewer and Nash, models access rules in a consulting business where analysts have to ensure that no conflicts of interest arise when dealing with different companies. Informally, conflicts arise because companies are direct competitors in the same market or because of the ownership of companies. Analysts must adhere to the security policy that states that there must be no information flow that causes a conflict of interest.

Write access to an object is granted only if no other object can be read, which is in a different company database, and contains unsanitized information. Unlike BLP, where access rights are static, this model requires access rights be re-examined in every state transition.

Biba Model

The **Biba model** addresses integrity in terms of access by subjects to objects using a state machine very similar to that of BLP. There is a mathematical structure, called a lattice, of integrity levels where various integrity levels are assigned to subjects and objects. These levels form the basis for expressing integrity policies that refer to the corruption of "clean" high-level entries by "dirty" low-level entries. In the integrity lattice, information may only flow downward. Unlike BLP, there is no single high-level integrity policy. Instead, there are a variety of approaches, which could yield mutually incompatible policies.

It is possible to state policies where integrity levels never change. The Biba model has two policies that prevent clean subjects and objects from being contaminated by dirty information. Similar to the Chinese Wall model, Biba has two integrity properties that automatically adjust the integrity level of an entity if it comes into contact with low-level information.

Clark-Wilson Model

The **Clark-Wilson model** addresses the security requirements of commercial applications. Included are data integrity requirements that prevent unauthorized modification of data, fraud, and errors. The integrity requirements are divided into internal and external consistency. Internal consistency refers to the properties of the internal state of a system, and can be enforced by the computing system. External consistency refers to the relation of the internal state of a system to the real world and must be enforced by means outside the computing system.

The general mechanisms for enforcing integrity are well-formed transactions and separation of duties. A well-formed transaction means data items can be manipulated only by a specific set of programs and users have access to programs rather than to data items. Separation of duties means users have to collaborate to manipulate data and to collude to penetrate the security system. Separation of duties appears repeatedly in the operation of a secure system. This means that different persons participate in the development, testing, certification, and operation of a system.

Clark-Wilson is basically an application-oriented IT system, whereas BLP is a general-purpose operating system. The Clark-Wilson system exhibits the following features:

- Subjects must be identified and authenticated
- Objects are manipulated only by a restricted set of programs
- Subjects execute only a restricted set of programs
- Audit logs are maintained
- The system is certified to work properly

In should be noted that the Clark-Wilson model is a framework and guideline for formalizing security policies, rather than a model for specific security policy.

Information Flow Model
In the Bell-LaPadula model, information can flow from a high security level to a low security level through a covert channel. **Information-flow models** consider any kind of information flow, not only the direct flow-through access operations that are modeled by BLP. A system is called secure if there is no illegal information flow. The advantage of such a model is that it covers all types of information flow; however, the disadvantage is that it becomes more difficult to design secure systems. Currently, information-flow models are areas of research, rather than the basis of a practical methodology for the design of secure systems.

Information-flow models

3.10 INTERNET AND WEB RESOURCES

The following is a list of Internet and Web resources for cryptosystem techniques:

Faulkner Information Services	www.faulkner.com
Bureau of Industry and Security	www.bxa.doc.gov
Computer Systems Policy Project	www.cspp.org
Critical Information Assurance Office	www.ciao.gov
PKI Forum	www.pkiforum.org
Cryptography FAQ	www.faqs.org/faqs/cryptography-faq
Outguess	www.outguess.org

■ SUMMARY

Cryptology is the study of secure communications, including the areas of cryptography and cryptanalysis. Cryptography looks at the design and implementation of encryption and decryption algorithms. Cryptanalysis is concerned with the breaking of ciphers and forging data and information. Among the terms associated with the cryptosystem environment are cipher, plaintext, and ciphertext, which are algorithms and inputs and outputs to the various algorithms. A number of algorithms, standards, and standards organizations that support them are part of this environment. These include the DES, Triple-DES, SHA-1, MD5, and RIPEMD-160 algorithms.

Public-key and private-key algorithm schemes are used in cryptography. One of these is the Diffie-Hellman key exchange, which addresses the issue of security keys—a major part of a security system.

A cryptanalyst is someone who analyzes and attempts to break ciphers. Various techniques are employed to accomplish this task, including brute force, factoring, and differential cryptanalysis.

Security models set forth the policies and rules that govern security systems. Major security models include Bell-LaPadula, Harrison-Ruzzo-Ullman, Biba, and Clark-Wilson.

Key Terms

Advanced Encryption Standard (AES)
Asymmetrical
Bell-LaPadula Model (BLP)
Biba Model
Bit-level cipher
Block cipher
Blowfish
Brute force attack
Caesar cipher
Capstone
CAST-128
Check value
Chinese Wall Model
Cipher
Ciphertext
Clark-Wilson Model
Clipper
Compression
Covert channels
Cryptology
Cryptanalysis
Cryptanalyst
Cryptography
Cryptosystem
Data Encryption Algorithm (DEA)
Data Encryption Standard (DES)
Differential cryptanalysis
Diffie-Hellman

Digest
Digital watermark
Factoring
Federal Information Processing Standard (FIPS)
Harrison-Ruzzo-Ullman Model (HRU)
Hash
Hash algorithm
Information Flow Model
International Data Encryption Algorithm (IDEA)
Key recovery
Message digest 5 (MD5)
Monoalphabetic cipher
National Institute of Standards and Technology (NIST)
One-way function
Plaintext
Polyalphabetic cipher
Private key
Public key
Public Key Infrastructure (PKI)

RACE Integrity Primitives Evaluation (RIPE)
RC5
RIPEMD-160
RSA (Rivest, Shamir, Adleman)
S-Box
Secret key
Secure hash algorithm (SHA-1)
Security models
Skipjack
Steganography
Stream cipher
Symmetrical
Transposition cipher
Triple Data Encryption Algorithm (TDEA)
Triple-DES
Vigenere cipher
X.509

REVIEW EXERCISES

Questions and Analysis

Provide a short answer or in-depth analysis as required.

1. What are the major components of cryptology?
2. What is the difference between cryptography and cryptanalysis?
3. What is the difference between a monoalphabetic cipher and a polyalphabetic cipher?
4. Give an overview of the DES.
5. What is the difference between a block cipher and a stream cipher?
6. What standards organizations were responsible for cryptographic systems?
7. How does DES relate to Triple-DES?
8. Describe SHA-1.
9. Who developed the SHA-1 algorithm?
10. Describe the MD5 algorithm.
11. Describe RIPEMD-160.
12. Who developed the RIPE algorithm?
13. What are the advantages of the Blowfish algorithm? How does it differ with DES?
14. Describe the features of the RC5 algorithm.
15. What gives the CAST-128 algorithm its strength?
16. How does skipjack relate to clipper and capstone?
17. How is DES related to AES?
18. Why does an organization need to be concerned about data compression?
19. Describe public-key cryptography.

20. Describe the Diffie-Hellman key exchange.
21. What are the security provisions of the RSA algorithm?
22. What is a one-way function?
23. What is the difference between a symmetric and asymmetric structure?
24. What is the difference between a public and a private key?
25. Explain the public key infrastructure.
26. Why should a security manager be concerned with a cryptanalyst?
27. What are major contributors to system weaknesses?
28. What are the goals of government encryption policies?
29. Define steganography and provide an example of an application.
30. What is a security model? Give several examples.

Definitions and Terms

Provide the most correct answer from the list of key terms.

1. A form of cryptosystem in which encryption and decryption are formed using two different keys.
2. An encryption method in which data are encrypted and decrypted in fixed-size blocks.
3. A type of attack under which every possible combination is attempted.
4. A communications channel that enables the transfer of information in a way unintended by the designers of the communications facility.
5. The branch of cryptology dealing with the breaking of a cipher to recover information.
6. The branch of cryptology dealing with the design of algorithms for encryption and decryption.
7. The study of secure communications, which encompasses both cryptography and cryptanalysis.
8. An algorithm that provides a way for two parties to establish a secret key that only they know.
9. An algorithmic representation of a message that is encrypted and appended to a message to be used for integrity checking.
10. An agency of the U.S. government that established national technical standards.
11. A function that is easily computed, but the calculation of its inverse is infeasible.
12. The original message in unencrypted, readable form.
13. A trusted and efficient key and certificate system.
14. The key used in an asymmetric encryption system, where both users share the same key.
15. A method of encryption that hides the existence of a message.

True/False Statements

Indicate whether the statement is true or false.

1. Cryptography is the branch of cryptology dealing with the breaking of ciphers.
2. Cryptanalysis is the branch of cryptology dealing with the breaking of ciphers.
3. Encryption is described as a process to convert a plaintext document to a ciphertext document.
4. A hash is synonymous with a digest.
5. A mono-alphabetic cipher is synonymous with a Vigenere cipher.
6. A Caesar cipher replaces each plaintext character with another character.
7. Two types of commonly used conventional encryption algorithms use block cipher and stream cipher processes.
8. The U.S. government endorses the Data Encryption Standard (DES). Examples of encryption algorithms include Blowfish, RC8, MD5, IDEA, and CAST-128.
9. Compression of data is not an issue with secure data transmissions.
10. Two types of cryptography include public key (symmetrical) and private key (asymmetrical) techniques.
11. A public key infrastructure (PKI) uses certificates to verify and authenticate users.
12. A cryptanalyst often uses a method called brute force to crack an algorithm.

13. Steganography can be used to establish ownership of a document through both visible and invisible digital watermarks.
14. Security models, such as Diffie-Hellman, capture policies for confidentiality and integrity of Enterprise computer networks.

Activities and Research

1. Develop a matrix of the various encryption algorithms. Show the various attributes associated with each method.
2. Do research using various resources, including the Web, for information defining and describing encryption algorithms. Produce a spreadsheet highlighting any differences.
3. Research the topic of "encryption" and develop a five-minute oral presentation concerning state-of-the-art methods.
4. Identify the standards organizations that are involved in the areas of cryptography. Describe their functions.
5. Invite a cryptanalyst to discuss the activities and tools of the trade.
6. Create a dictionary of cryptosystem terms and definitions.
7. Research and prepare a paper concerning the Advanced Encryption Standard.
8. Create a matrix of the various security models. Show the features and benefits of each model.
9. Research the subject of security standards and prepare a paper that identifies any new activity with regards to cryptosystems.
10. Use a cryptography technique to convert a plaintext message to a ciphertext one.

CHAPTER 4

Security Systems

OBJECTIVES

Material included in this chapter should enable the reader to

- Look at the issues associated with security systems design.
- Evaluate the various technical solutions that can be implemented to secure Enterprise systems.
- Identify the collection of secure protocols.
- Understand the functionality of secure sessions and secure transaction systems.
- Become familiar with the client/server security aspects of the Secure Sockets Layer.
- Understand how the Secure Electronic Transaction (SET) system is used to provide security to e-commerce operations.
- Become familiar with the Kerberos protocol.
- Look at the components of Multipurpose Internet Mail Extension (MIME), Secure MIME, and Secure Hypertext Transfer Protocol (S-HTTP).
- Become familiar with the benefits and requirements of intrusion detection.
- Identify the components and utilization of security evaluation and security specifications.
- Understand the various elements of the *Orange Book* and *Red Book* security specifications.

■ INTRODUCTION

Security systems include technical, software, and protocol solutions to ensure the protection of the Enterprise computer and network resources. Included in this chapter will be discussions concerning biometric systems, secure protocols, and secure electronic transactions. Also included will be sections that discuss security specifications and security evaluation.

Security systems design incorporates a number of technologies and techniques to provide for a secure Enterprise computing environment. Biometrics is in the forefront of technology in providing access control techniques for Enterprise computer networks. It is the study and application of biological data and biometric-based authentication for access control.

Secure sessions and transactions can occur after identities have been established and parties are assured there is a secure transmission link. This is possible because a number of secure protocols have been developed to accomplish this task. Protocols that have been developed include Secure Sockets Layer, Transport Layer Security, Private Communication Technology, and Pretty Good Privacy.

With the advent of e-commerce systems, a secure process was required to handle credit card transactions over the network. A major specification that accomplishes this is called Secure Electronic Transaction.

Other security tools available to the Enterprise security include Secure Hypertext Transfer Protocol, Secure Multipurpose Internet Mail Extension, and Kerberos. Also included in this portfolio are password and challenge handshake authentication protocols techniques.

Intrusion detection involves detecting and responding to a myriad of potential attacks and intrusions into the Enterprise computer network. There are network-based and host-based systems that can be deployed to counter these threats. This chapter concludes with an in-depth discussion of security evaluation and security specification documents, such as the *Orange Book* and the *Red Book*.

4.1 SECURITY SYSTEMS DESIGN

This chapter is concerned with the design of the perimeter network and the security policies that must be included in this development. It is essential that the administrator and security manager know how the components function and interact with both internal and external networks. A number of issues must be addressed to successfully protect the organization's computer and networking assets. These include the following (Cisco 2001):

- Identify and understand the enemy
- Measure the cost of security, or lack thereof
- Identify security assumptions
- Look at the human factors
- Identify the organization's weaknesses
- Limit the scope of system access
- Understand the computer and network environment
- Limit trust in software
- Look at physical security
- Make security part of the normal operating procedures

This process can be initiated by conducting a security audit of the organization. (See Appendix B.)

4.2 TRUSTED SYSTEMS

As discussed in Chapters 1 and 2, trust is the composite of availability, performance, and security, which includes the ability to execute processes with integrity, secrecy, and privacy. Technical solutions that provide support to Enterprise trust systems will be presented in the subsequent sections. These include biometric systems, secure protocols, and secure electronic transactions. Each of these systems provides a part of the total solution for protection of the organization's computer and network assets. As depicted in Figure 4–1, the task of the systems administrator is to develop and implement security systems that will protect the Enterprise from the untrusted networks.

Figure 4–1
Trusted and Untrusted Networks

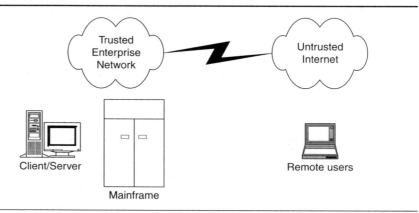

4.3 BIOMETRIC SYSTEMS

Biometrics is the study of biological data and biometric-based authentication, which is an application that uses specific personal traits for access control. Biometrics offers a secure means of limiting access to computer and networking assets. Some of these personal verification methods include iris and retinal scanning, hand and finger geometry, and palm and finger scanning. This "new" technology may increase the accuracy with which security systems can readily identify individuals. The technology relies on measurements of physical characteristics that are unique to an individual. These include behavioral characteristics, handwritten characteristics, and voice recognition. Table 4–1 provides a matrix of techniques along with their comparative accuracy (Falk 2001). There are practical biometric techniques currently available that may be acceptable to users. This technology is not in wide use today, mainly due to the lack of standards and the implementation expense. There is also a lack of trust and recognition from users, who are generally not educated on the technology. These concerns include possible invasion of privacy and the association of finger scanning with criminals. This section will look into the various biometric methods that can be deployed to provide for security access. Biometric methods discussed will include **finger scanning**, finger and hand geometry, iris and palm imaging, face and voice recognition, and finally, signature verification.

biometrics

finger scanning

Finger Scanning

The user's finger is placed on a reader where a picture is taken of the fingerprint. The system then converts this picture into a map of minutiae points, which is then input into an algorithm for creating a binary template. This binary template is stored and compared during the authentication and verification process. Common fingerprint patterns are divided into three main groups, which consist of arches, loops, and whorls. Approximately 5 percent of the patterns are arches, 30 percent are whorls, and 65 percent are loops.

Technique	Comparative Accuracy
Finger scanning	1 : 500
Hand geometry	1 : 500
Iris imaging	1 : 131,000
Retina recognition	1 : 10,000,000
Signature verification	1 : 50
Speaker recognition	1 : 50

Table 4–1
Techniques and Comparative Accuracy

Finger-scanning imaging techniques include optical, thermal, tactile, capacitance, and ultrasound. Optical images can be captured from images made by a finger on a glass platen. Tactile and thermal images can be captured from a pressure or temperature sensor. Capacitance images are generated from capacitance silicon sensors. Sound waves can generate ultrasound images from finger patterns. Many automated fingerprint identification systems are currently being used by a number of organizations. Finger scanning provides high accuracy and fraudulent deception of the system is difficult. The equipment is easy to use and is readily accepted by users.

Finger Geometry

finger geometry

Some biometric vendors use **finger geometry** or finger shape to determine identity. Unique finger characteristics, such as finger length, width, thickness, and knuckle size are measured on one or two fingers. Two techniques are used for capturing the images. The user can insert the index and middle finger into a reader and a camera takes a three-dimensional image, or can insert a finger into a tunnel where sensors take three-dimensional measurements.

Finger geometry systems are simple to use, very accurate, and are impervious to deception. Public acceptance, however, is somewhat lower than for finger scanning, because users must insert their fingers into a reader.

Hand Geometry

Some commercial biometric applications are using the two-dimensional shape of the hand for access control. Users place their hand on a reader, aligning their fingers with specially positioned guides, and a camera captures an image. Measurements center on finger length and the shape of the fingers and knuckles. Commercial systems that perform three-dimensional shape analysis of the hand are under development.

hand geometry

Hand geometry systems are simple to use, accurate, impervious to deception, and are readily accepted by users. Hand geometry, however, is less distinctive than finger scanning techniques, since two people can have similar hand geometry.

Palm Imaging

Measurement of palms is performed by techniques similar to those used for finger scanning. A scanner shaped to accommodate the palm scans the ridges, valleys, and minute details found on the skin of the palm.

Alternatively, latent or ink images of the palm can be scanned, and the minutiae data is extracted, processed, and stored in the system. Palm images are useful in crime detection, and some vendors are developing commercial applications. Like finger scanning, **palm scanning** is simple to use, very accurate, and impervious to deception. User acceptance, however, is not very high.

palm scanning

Iris Imaging
The human iris is a complex structure that is well-suited for unique identification. Each iris contains a complex pattern of specific characteristics such as corona, crypts, filaments, freckles, pits, and striations. A black and white video camera can be used to capture an image of the iris. Unique features of the iris are extracted from the captured image by the recognition system. These features are converted into a unique iris code, which is compared to previously stored iris codes for user recognition.

Artificial duplication of the pattern of an individual iris is virtually impossible and **iris imaging** provides high accuracy. The technique is relatively easy to use, and there appears to be little resistance from users. Fraudulent deception of the system is very unlikely.

iris imaging

Retina Recognition
The retina forms a unique pattern for each individual. The user views a green dot for a few seconds until the eye is sufficiently focused for a scanner to capture the blood vessel pattern. The retinal pattern is captured by the scanner and then compared to previously stored patterns for identification.

Retinal scanning provides very high accuracy, provided the user's eye is properly focused. Reflection from glasses can create interference. There is some resistance among users to this technique because of the infrared light scanner—users are afraid of eye damage. Deception of the system is very unlikely.

retinal scanning

Face Recognition
No direct physical contact is required with this system. The camera captures a face image and a number of points on the face are mapped by the system. From these measurements, a unique representation of the individual's face is created. A complete map of the entire face can be created. There is a downside to this system, since people change over time. Some systems compensate for this by combining recently stored information with previously stored images.

Face recognition offers reasonable accuracy, however poor lighting, glasses, facial hair, and aging affect the accuracy. The system can be retrained and updated to recognize changes in users. The equipment is easy to use and readily accepted by users. Current systems can recognize and identify users from a distance of 2 to 32 meters.

face recognition

Signature Verification
Signature scanning techniques examine the way users sign their names. The system examines the dynamics of the signing process, rather than the signature. Extracted characteristics may include the angle at which the pen is held, the time taken to sign, velocity and acceleration of the signing process, and the number of times the pen is lifted from the document during the signing process.

signature scanning

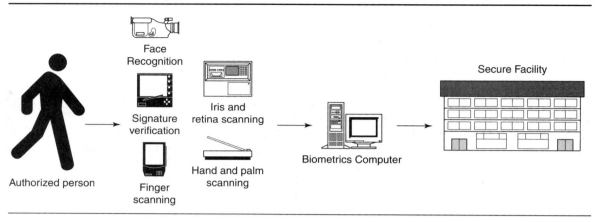

Figure 4–2
Biometric Access Methods

Ordinary forgery techniques do not work, because behavioral characteristics are used rather than a signature. Signature data can be captured with a special pen that contains sensors or with a tablet that senses the motion of a stylus. Acoustic emission measurements of the pen can be captured. A number of signatures can be recorded to build a user's profile.

Signatures are one of the most common methods of establishing identity, therefore are readily acceptable by financial institutions and others that require conventional signatures. Recognition accuracy is high, and the method is considered impervious to deception.

Biometric Security

Biometric technology can be used successfully to identify users, however it is necessary that the application be specifically designed, taking into consideration the ease of use, deployment, accuracy, and cost. Finger scanning appears to be the technology of choice. These systems are moderately accurate and easy to use. Acceptance by users is good and the technique is relatively inexpensive. The biometric methods depicted in Figure 4–2 may be acceptable alternatives for access control.

4.4 SECURE PROTOCOLS

Once keys have been exchanged, clients and servers can engage in secure sessions and transactions. A number of protocols have been developed to handle secure sessions and transactions, which are discussed in this section (Sheldon 1998).

Private Communication Technology

Private Communication Technology (PCT) is a protocol that provides secure encrypted communications between two applications over a TCP/IP

Private Communication Technology (PCT)

network. It works over the Internet, an intranet, or an extranet. Microsoft developed PCT in response to security weaknesses in **Secure Sockets Layer (SSL)**. **SSL** was restricted to 40-bit keys, so Microsoft separated the authentication and encryption functions in PCT to bypass this restriction. PCT allows applications to use 128-bit key encryption for authentication within the United States and 40-bit key for export use. PCT is included with Microsoft Internet Explorer.

Secure Sockets Layer (SSL)

Secure Sockets Layer

As organizations use the Internet for more than information dissemination, they will need to use trusted security mechanisms. An increasingly popular general-purpose solution is to implement security as a protocol that sits between the underlying transport protocol (TCP) and the application. The foremost example of this approach is the Secure Sockets Layer and the follow-on Internet standard called **Transport Layer Security (TLS)**. SSL is built into many Web browsers, including Netscape Communicator, Microsoft Internet Explorer, and numerous other software products.

Transport Layer Security (TLS)

Secure Sockets Layer is a Web protocol that sets up a secure session between a Web client and server. All data transmitted over the communication channel is encrypted. As stated previously, S-HTTP is a similar protocol that encrypts only at the HTTP level, whereas SSL encrypts all data passed between client and server at the IP socket level. SSL was originally developed by Netscape; then it was submitted to the IETF for standardization.

Both SSL and S-HTTP provide security benefits for e-commerce, including protection from eavesdropping and tampering. With SSL, browsers and servers authenticate each other, then encrypt data transmitted during a session. This procedure verifies to the client that the Web server is authentic before it submits confidential information. It also allows Web servers to verify that users are authentic before granting them access to restricted and sensitive information. Digital certificates are required in this scheme.

Both Web browsers and Web servers must be SSL-enabled. When a client contacts a server, the server forwards a certificate signed by a CA. The client then uses the CA's public key to open the certificate and extracts the Web site's public key. SSL consists of the SSL Handshake Protocol, which provides authentication services and negotiates an encryption method, and the SSL record protocol, which performs the task of packaging data, so it can be encrypted.

Although SSL protects information as it is passed over the Internet, it does not protect private information, such as credit-card numbers, once stored on the merchant's server. If the server is not secure and the data is not encrypted, an unauthorized party can access that information. Hardware devices called peripheral component interconnect (PCI) cards can be installed on Web servers to secure data for an entire SSL transaction from the client to the Web server. The PCI card processes the SSL transactions, freeing the Web server to perform other tasks.

Pretty Good Privacy

Pretty Good Privacy (PGP) is an encryption and digital signature utility for adding privacy to electronic mail and documents. PGP is an alternative to RSA's S/MIME. While S/Mime uses RSA public-key algorithms, PGP uses Diffie-Hellman public-key algorithms. They both yield similar results. A similar, but older and less robust privacy protocol is Privacy-Enhanced Mail (PEM), which will be discussed later.

PGP was designed by Phil Zimmermann on the principle that e-mail, like conversations, should be private. Both sender and recipient need assurances that messages are from authentic sources. They are also looking for assurances that messages have not been altered or corrupted and that the sender cannot repudiate or disown the messages. PGP can assure both privacy and nonrepudiation. It also provides a tool to encrypt information on magnetic media.

PGP is designed to integrate into popular e-mail programs and operate on major operating systems such as Windows and Macintosh. It uses a graphical user interface (GUI) to simplify the encryption process. The actual operation of PGP consists of authentication, confidentiality, compression, e-mail compatibility, and segmentation.

The package uses public-key encryption techniques in which the user generates two keys—one public and one private. These keys can then be used to encrypt and digitally sign messages. PGP supports key servers, so users can place their public keys in a central location where other users can access the keys.

There are a number of reasons to use PGP for e-mail and file storage applications. These include the following (Stallings 2000):

- Available worldwide on a number of platforms
- Based on secure algorithms
- Works in a wide range of applicability
- Outside of standards organizations and government control

4.5   SECURE HYPERTEXT TRANSFER PROTOCOL (S-HTTP)

The native protocol that Web clients and servers use to communicate is **Hypertext Transfer Protocol (HTTP)**. This protocol is ideal for open communications, however in its native form does not provide authentication or encryption features. **Secure HTTP (S-HTTP)** works in conjunction with HTTP to enable clients and servers to engage in private and secure transactions. It is especially useful for encrypting forms-based information as it passes between clients and servers. It should be noted that S-HTTP only encrypts HTTP-level messages at the application layer, whereas SSL encrypts all data being passed between client and server at the IP socket level (Sheldon 1998).

S-HTTP provides considerable flexibility in terms of what cryptographic algorithms and modes of operation can be used. Also, as the need for authentication among the Internet and Web grows, users need to be authenticated before sending encrypted files to each other. With S-HTTP, messages may be protected using digital signatures, authentication, and encryption. During the initial contact, the sender and receiver establish preferences for encrypting and processing secure messages.

A number of encryption algorithms and security techniques can be used, including DES and RC2 encryption or RSA public-key signing. Users can also choose to use a particular type of certificate, or no certificate. In a situation where certificates are not available, it is possible for a sender and receiver to use a session key that has been exchanged in advance. A challenge/response mechanism is also available. The IETF Web Transaction Security Working Groups is responsible for developing S-HTTP.

4.6 SECURE ELECTRONIC TRANSACTION (SET)

Secure electronic transaction (SET) is an open specification for handling credit card transactions over a network, with emphasis on the Web and Internet. Secure transactions are critical for **electronic commerce (e-commerce)** on the Internet. Merchants must automatically and safely collect and process payments from Internet clients; therefore, a secure protocol is required to support the activities of the credit card companies. Also impacted by e-commerce requirements are consumers, vendors, and software developers. GTE, IBM, MasterCard, Microsoft, Netscape, and Visa developed SET.

SET is designed to secure credit card transactions by authenticating cardholders, merchants, and banks by preserving the confidentiality of payment data. SET includes the following features:

- Requires digital signatures to verify that the customer, the merchant, and the bank are all legitimate.
- Uses multiparty messages that allow information to be encrypted directly to banks.
- Prevents credit card numbers from getting in the wrong hands.
- Requires integration into the credit card processing system.

SET includes a layer that negotiates the type of payment method, protocols, and transports. This task is the responsibility of the Joint Electronic Payment Initiative (JEPI). Payment methods could include credit cards, debit cards, electronic cash, and checks. Payment protocols, such as SET, define the message format and sequence required for completion of the payment transaction. Transports include such protocols as Secure Sockets Layer (SSL) and Secure Hypertext Transfer Protocol (S-HTTP).

An important part of SET's success will be its overall acceptance by cardholders, credit card issuers, payment processors (acquirers), and

secure electronic transaction (SET)
electronic commerce (e-commerce)

certificate authorities (CAs). Note that CAs provide digital signatures that are critical for verifying the authenticity of cardholders and others involved in the transactions. Microsoft and Netscape have included SET support in their browsers (Sheldon 1998).

Credit Card Transactions

If e-commerce is to succeed, a method must exist for consumers to use credit cards over the Internet. Credit card usage on the Internet is still low, however is likely to grow in the future. SET and SSL is used in the credit card transaction to provide security provisions. Figure 4–3 provides a generic example of a credit card transaction flow.

SSL encrypts a credit card number and other information using a 40-bit key. Because of its size, this key can be hacked; however, it may be adequate for some needs. Even though SSL keeps the credit card number and information private while being transmitted, it does not address the issue of whether the card is valid, stolen, or being used without permission. SET addresses these SSL limitations by using an "electronic wallet" that can identify the user and validate the transaction. An electronic wallet is a type of software application used by the consumer for securely storing purchasing information. Furthermore, SET-based systems have an advantage over other mechanisms, in that SET adds digital certificates that associate the cardholder and merchant with a particular financial institution and the Visa or MasterCard payment system.

SET Overview

Set is an open encryption and security specification designed to protect credit card transactions on the Internet. It is a set of security protocols and formats that enables users to employ the existing credit card payment infrastructure on an open network, such as the Internet, in a secure fashion. SET provides the following three services:

- Ensures privacy
- Provides trust
- Provides a secure communications channel

The business requirements, key features, and the participants in SET transactions are summarized as follows:

Business Requirements
- Provide authentication that a cardholder is a legitimate user of a credit card account.

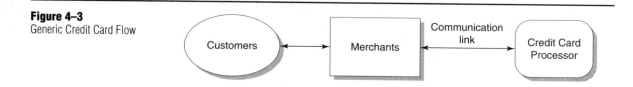

Figure 4–3
Generic Credit Card Flow

- Provide authentication that a merchant can accept credit card transactions through its relationship with a financial institution.
- Provide confidentiality of payment and ordering information.
- Ensure the integrity of all transmitted data.
- Ensure the use of the best security practices and system design techniques to protect all legitimate parties in an electronic commerce transaction.
- Create a protocol that neither depends on transport security mechanisms nor prevents their use.
- Facilitate and encourage interoperability among software and network providers.

Key Features
- Cardholder account authentication
- Merchant authentication
- Integrity of data
- Confidentiality of information

Participants in SET Transactions
The participants in SET transactions include the cardholder and the merchant, in addition to the issuer, acquirer, payment gateway, and a certification authority. Figure 4–4 depicts these SET components and their relationships.

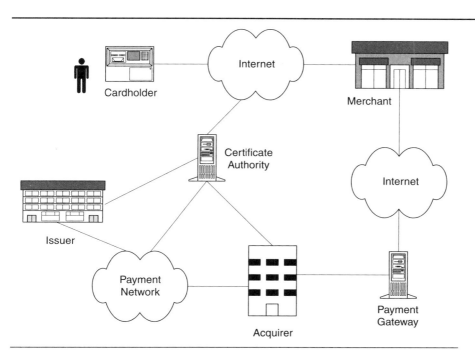

Figure 4–4
SET Components and Relationships

SET Transaction Flow

Figure 4–5 depicts the sequence of events associated with a SET transaction (Stallings 2000).

1. The customer opens an account and obtains a credit card account with a bank that supports electronic payment and SET.
2. The customer receives a certificate after suitable verification of identity. It establishes a relationship between the customer's key pair and the credit card.
3. Merchants have certificates consisting of one key for signing messages and one for key exchange. They also need a copy of the payment gateway's public-key certificate.
4. The customer places an order, which is accepted by the merchant. The order form returned from the merchant includes the items, the cost, and an order number.
5. The customer receives the merchant's certificate.
6. The customer sends the order, payment, and their certificate to the merchant.
7. The merchant requests payment authorization through the payment gateway.

Figure 4–5
SET Transaction Flow

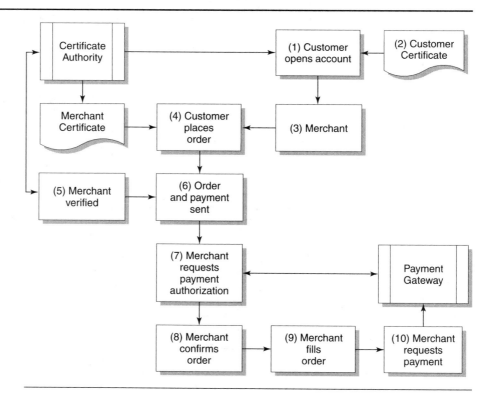

8. The merchant provides the customer with order confirmation.
9. The merchant ships the product or service.
10. The merchant requests payment from the payment gateway, which handles all payment processing.

4.7 SECURE MULTIPURPOSE INTERNET MAIL EXTENSION (S-MIME)

Multipurpose Internet Mail Extension (MIME) is an Internet Engineering Task Force (IETF) standard defined in 1992 for sending a variety of different types of information via Internet electronic mail. E-mail specifications are set forth in RFC 822, RFC 1049, and RFC 1521.

Multipurpose Internet Mail Extension (MIME)

Traditional electronic mail messages handle only text; however, MIME provides ways to encode data types for electronic mail transmission. A data type is a category that identifies the international representation of data. MIME supports binary files, non-ASCII character sets, images, sound, compressed files, and special fonts. MIME was designed to be backward compatible with the previous Internet e-mail standard, which was strictly a text-oriented mail messaging system. This e-mail system worked well at the time, however it has limitations with today's multimedia data types.

A typical e-mail message consists of a header that includes fields for the recipient (TO), the sender (FROM), the subject, and the message. Each part of a MIME message, called body part, can hold a different data type, which includes text, graphics, audio, or video. MIME adds two lines to the header of an e-mail message. The first line indicates the MIME version and the second line indicates how the body parts are formatted in the message. A "boundary marker" separates each part of the message.

S/MIME

If security and privacy are desired, users have a range of e-mail encryption techniques from which to choose. **Secure Multipurpose Internet Mail (S/MIME)** is an RSA data security specification for securing electronic mail. PGP is another encryption product that provides security capabilities for e-mail.

Secure Multipurpose Internet Mail (S/MIME)

S/MIME is an extension of the MIME standard that adds security to protect against interception and forgery. Basically, it is designed to secure messages from unintended viewers. Because S/MIME is an extension of MIME, it easily integrates with existing electronic messaging products. The demand for e-mail security is growing, along with a demand to validate message authenticity. E-mail security allows users to electronically sign messages to prove their origin.

S/MIME uses the RSA public-key algorithms for key exchange, making it easy for users to exchange keys, even if they have never met. The key scheme can also be used to digitally sign e-mail messages. Key management is discussed thoroughly in Chapter 2. Encryption is possible with DES, Triple DES, or the RSA/RC2 algorithm, all of which are discussed in Chapter 2.

S/MIME functionality can be summarized as follows (Stallings 2000):

- Clear-signed data—Forms a digital signature of the content; recipients can view message, however cannot verify signature.
- Enveloped data—Consists of encrypted content of any type and encrypted content encryption keys for one or more recipients.
- Signed data—Forms a digital signature by taking the message digest of the content to be signed and then encrypting with private key of signer; can only be viewed by recipient with S/MIME capability.
- Signed and enveloped data—Encrypted data may be signed and signed data or clear-signed data may be encrypted.

RSA Data Systems promotes S/MIME, and VeriSign has set up a certificate authority (CA) that also supports it. The IETF is considering a version of PGP, called Open PGP, for e-mail security.

4.8 KERBEROS

Kerberos

Kerberos is a secret-key network authentication protocol, developed at MIT, which uses the DES cryptographic algorithm for authentication and encryption. Kerberos Version 5 is an Internet standard, which is specified in RFC 1510. Kerberos was designed to authenticate user requests for network resources. It is based on the concept of a trusted third party that performs secure verification of users and services. This trusted third party is called the **key distribution center (KDC)**. The primary function of Kerberos is to verify that "users are who they say they are," which is the responsibility of an authentication function. A trusted Kerberos server, which issues "tickets" to users, accomplishes this task. These tickets have a limited life span and are stored in the user's credential cache, where they can be used later in place of the standard username and password combination (Kaeo 1999). RFC 1510 describes a ticket as a record that helps a client authenticate itself to a server.

key distribution center (KDC)

Kerberos Authentication

Kerberos' primary function is to provide password authentication services that grant clients access to servers. It runs in a computer, called an **authentication server (AS)**, that is separate from any client or server. Remember from Chapter 1 that a **client** is a device or software that requests information from a **server**, which can be either hardware or software.

authentication server (AS)
client
server

Kerberos provides security for remote logons, and can provide a single logon solution, which relieves the user from the requirement of a new logon every time a new server is accessed. The AS stores passwords for all users in a central database. It issues **credentials** that clients use to access servers within the **realm** of the AS. Credentials are basically a way of establishing, via a trusted third party, that the user is as claimed. The AS is also physically secured and managed by a single administrative staff. Kerberos ensures the AS provides a method of safely distributing encryption keys, called **session keys**, to clients and servers that need to engage in secure communications.

credentials
realm

session keys

A Kerberos realm is a single administrative domain that controls access to a collection of servers (Gollman 2000). A specific security policy is set within the realm. Realms can trust other realms, which means a user authenticated by an AS in one realm, accepts the authentication to access the server in another trusted one. It accepts or trusts that another AS has properly identified and validated a user.

A step-by-step explanation of the Kerberos is as follows (Sheldon 1998):

- The client gets a **ticket-granting ticket (TGT)** from the AS.
- The user is prompted for a password and is validated.
- The client sends a request to a **ticket-granting server (TGS)** to prove authenticity of the user TGT.
- The TGS decrypts the TGT and views its contents and builds a server ticket for the target server and encrypts it with a key shared in common with the target server.
- The server ticket is encrypted again by the TGS with a key derived from the user's password and returned to the client.
- The client decrypts the response to extract the server ticket and the session key.
- The client forwards the ticket to the target server along with the user ID.
- The target server decrypts the ticket and compares the user ID with the ticket to the user ID that was encrypted in the ticket. Access is granted to the server if there is a user ID match.

ticket-granting ticket (TGT)
ticket-granting server (TGS)

Because the ticket contains the session key, which was also sent to the client, both client and server have a session key that can be used to encrypt and decrypt messages transmitted across the network.

Kerberos imposes five assumptions on the environment in which it can properly function. These include the following:

- A Denial of Service (DoS) attack is not solved with Kerberos.
- Principles must keep their secret keys secret.
- Password guessing attacks are not solved by Kerberos.
- Each host on the network must have a clock that is "loosely synchronized" to the time of other hosts.
- Principle identifiers are not recycled on a short-term basis.

There is an enhancement available to the authentication process, which allows users to enter a token ID from a smart card that generates tokens. Users carry this **secure card**, which is similar to a credit card. It displays values that are synchronized with the server, and the user enters these values when logging in.

Even though Kerberos implements symmetric (secret key) encryption techniques, some vendors are supporting public-key authentication. Microsoft is providing this feature in its Kerberos implementation and has submitted a proposal to the IETF that recommends this technique (Sheldon 1998).

4.9 POINT-TO-POINT PROTOCOL (PPP)

Point-to-Point Protocol (PPP) is an 8-bit serial interconnection protocol, which allows a computing device to connect as a TCP/IP host to a network through an asynchronous port. PPP includes error detection and data protection features such as password authentication protocol and challenge handshake authentication protocol (Forouzan 2001).

Password Authentication Protocol

The **Password Authentication Protocol (PAP)** is a simple two-step authentication procedure. The user sends a username and password. The system then checks the validity of both and either accepts or denies the connection. PAP is not sufficient for systems that require a high level of security access control.

PAP authentication can be used at the start of the PPP link as well as during a PPP session to re-authenticate the link. When the PPP link is established, PAP authentication can be carried out over that link. The peer sends a userid and a password in the clear to the authenticator until either the authenticator accepts the pair or the connection is terminated (Kosiur 1998).

PAP is not secure because the authentication information is transmitted in the clear, and nothing protects against playback attacks or excessive repetition by attackers trying to guess a valid userid/password pair.

Challenge Handshake Authentication Protocol

Challenge Handshake Authentication Protocol (CHAP) is a three-way handshaking authentication protocol that provides a higher level of security than PAP. In this method, the password is kept secret since it is never sent online. The process contains the following steps:

- The system sends a challenge packet to the user. This packet contains a challenge value, usually a few bytes.

- The user applies a predefined function, uses the user's password and challenge value, and creates a result.
- The user sends the result in the response packet to the system.
- The system mirrors the previous activity. If the results are the same, the access is allowed.

CHAP is more secure than PAP, especially if the system continuously changes the challenge value. CHAP also may not meet the scalability requirements of large organizations. Even though it does not transmit any secrets across a network, it requires a large number of shared secrets to be passed through the hash function. Organizations with many dial-up users would need to maintain vary large databases to accommodate them.

4.10 INTRUSION DETECTION

Intrusion detection is the art of detection and responding to computer and network misuse. The benefits to the organization include the following security elements (Proctor 2001):

- Deterrence
- Detection
- Response
- Damage assessment
- Attack anticipation
- Prosecution support

intrusion detection

Different intrusion detection techniques provide different types of benefits for a variety of environments, therefore it is essential to select and implement the most viable one. The key to selecting the correct detection system is to define the environment-specific requirements that best satisfy the organization's security needs.

System Configurations

The two categories of intrusion detection systems are network based and host based. Host-based technologies examine events relating to file and database access and applications being executed. Network-based technologies examine events such as packets of information exchanged between networked computers. A hybrid system that includes components from both host- and network-based detection systems offers the best solution.

The intrusion detection industry supplies tools with capabilities and features that go far beyond detecting intruders from outside the organization. Intrusion detection systems may provide the following capabilities:

- Event log analysis for insider threat detection
- Network traffic analysis for perimeter threat detection

- File and database integrity checking
- Security configuration management

Many products provide only one of these capabilities, however hybrid systems are available that are multifunctional. A hybrid system contains the following components:

- Alert notification
- Command console
- Database
- Network tap
- Network sensor
- Response subsystem

Network-based Intrusion Detection

Network intrusion detection is effective at detecting outsiders attempting to penetrate the Enterprise network defenses. The two intrusion detection architectures include traditional sensor and distributed systems. The traditional sensor architecture is easy to deploy and operate, however is limited by high-speed, switched, or encrypted networks. The distributed architecture addresses these issues, but is significantly more difficult to deploy and manage.

Host-based Intrusion Detection

Host-based intrusion detection systems are distributed systems that gather and process event logs and other data from computers in an Enterprise. The data may be processed locally at the target location or transported to a centralized site for processing. Operationally, host-based systems can be effectively used for surveillance, damage assessment, intelligence gathering, and compliance.

Managing a host-based system is considerably more difficult than managing a network-based system. It is critical to have developed a good policy management and an efficient audit policy before implementing either system, which can significantly reduce the operational performance overhead.

Network management and monitoring issues are presented in Chapter 5. The basics concerning an Enterprise Network Management System (NMS) are presented in Chapter 7. An in-depth understanding of the issues regarding intrusion detection can be gained by reading *The Practical Intrusion Detection Handbook* (Proctor 2001).

Intrusion Detection Systems

There are a number of providers of intrusion detection systems. There are different levels of support for acquisition, deployment, and maintenance that can be provided by these companies. Examples of these system providers are as follows:

- Axent Technologies, Inc.
- Cisco Systems, Inc.
- Computer Associates
- CyberSafe Corporation
- Internet Security Systems
- Network Associates
- Network Ice
- Network Security Wizards
- ODS Networks
- Pentasafe

4.11 SECURITY EVALUATION

How does an administrator know that the security system deployed will perform as promised? Reliance can be placed on the word of the service provider or on an independent, impartial evaluator. The process of **security evaluation** involves an assessment of the security properties of a particular computer system or communications network. A standard set of requirements is used, as a metric, for determining and comparing how effective different systems will respond in mitigating the effects of malicious attacks. This determination and comparison is usually performed by an independent organization, which has no vested interest in the outcome of the assessment.

security evaluation

The U.S. Department of Defense (DoD) does not reserve security evaluations for assessments of products oriented toward specific operating environments. If specific operating environments are considered, the term **security certification** is reserved. The notion of a particular government organization or agency making a decision that a given product or system possesses sufficient security protection is called a **security accreditation** (Amoroso 1994). The specific goals that can be associated with security evaluation include a uniform measure of security, an independent assessment, criteria validity, and evaluation assurance.

security certification

security accreditation

Comparing the security effectiveness of different systems is difficult if varied approaches are taken to mitigating attacks. Security evaluation that includes a standard, well-defined set of security requirements provides a means of comparing the effectiveness of different security approaches. To achieve this goal of uniformity, there must be minimal subjective judgment to determine compliance.

A security evaluation offers the opportunity for an independent organization to provide an assessment of the security properties of a system. This will remove the inherent biases that can occur if a development organization performs an evaluation on its own product. The security evaluation process can be viewed as part of quality assessment processes that are common in typical software and system engineering efforts.

When a standard metric is established for determining security effectiveness, computer systems and network vendors should meet these requirements in the development efforts. These requirements must be balanced by resources and cost considerations. If the security specifications are too stringent, the vendors will ignore them.

System tests, formal methods, and other life-cycle activities can be used as evidence that a particular system is secure. Security evaluation can be used as additional evidence that a system is secure. Security evaluation may provide the most convincing assurance, because it serves as a means for reviewing, assessing, and summarizing the results of all other assurance activities.

4.12 SECURITY SPECIFICATIONS

Several security protocols, initiatives, and specifications are being developed. These are addressed in this section (Sheldon 1998).

Trusted Computer System Evaluation Criteria/Orange Book

The U.S. National Security Agency has outlined the requirements for secure products in a document entitled *Trusted Computer System Evaluation Criteria* **(TCSEC)**. TCSEC is commonly called the *Orange Book*. This standard defines access control methods for computer systems that computer vendors can follow to comply with Department of Defense (DoD) security standards. TCSEC is a collection of criteria used to grade or rate the security offered by computer systems. The documents are printed with different colored folders, thus the name *Rainbow Series*.

The *Orange Book* was the first evaluation criteria to gain wide acceptance. Manufacturers of secure systems still quote the *Orange Book* ratings of their products. A number of other criteria have since been developed to improve on it, based on new and different situations. These are published in the following documents (Gollman 2000):

- Canadian Trusted Computer Product Evaluation Criteria
- Common Criteria for Information Technology Security Evaluation
- Federal Criteria for Information Technology Security
- Information Technology Security Evaluation Criteria

Security evaluation can best be explained by looking at the different components or elements of evaluation. These include

- Targets—products and systems
- Purpose—evaluation, certification, and accreditation
- Methods—product and process oriented
- Structure—functionality, effectiveness, and assurance
- Framework—repeatability and reproducibility

- Benefits—improvement and perception
- Costs—time and resources

The *Orange Book*

The *Orange Book* was the first guideline for evaluating security products for operating systems. Security evaluation examines the security-relevant part of a system. The initial efforts were concentrated in the national security sector. The purpose of the *Orange Book* was to provide (Gollman 2000)

- a basis for specifying security requirements when acquiring a computer system
- a yardstick for users to assess the degree of trust that can be placed in a computer security system
- guidance for manufacturers of computer security systems

The evaluation classes of the *Orange Book* are designed to address typical patterns of security requirements. These security features and assurance requirement categories include the following classes:

- Security policy—mandatory and discretionary access control policies expressed in terms of subjects and objects
- Accountability—audit logs of security relevant events maintained
- Assurance—operational; security architecture; and life-cycle: design methodology, testing, and configuration management
- Marking of objects—labels specify the sensitivity of objects
- Identification of subjects—individual subjects identified and authenticated
- Documentation—guidance to install and use security features; test and design instructions
- Continuous protection—tamperproof security mechanisms

The security classes of the *Orange Book* are defined incrementally, which means that all requirements of one class are automatically included in those of all higher classes.

The four security divisions of the *Orange Book* include minimal, discretionary, mandatory, and verified protection. Minimal protection (Class D) includes products that are submitted for evaluation, but do not meet any *Orange Book* requirement.

Discretionary protection, which is basically a need to know, carries a classification of C1 (**TCSEC Class C1**). These systems are intended for an environment where cooperating users process data at the same level of integrity. Discretionary access control based on individual users or groups enables users to share access to objects in a controlled fashion. Users must identify themselves and their identities must be authenticated. Controlled access protection carries a classification of C2 (**TCSEC Class C2**). This system makes users individually accountable for their actions. Audit trails of security-relevant events, as specified in the definition of the C2 class, must be maintained.

Mandatory protection, which is based on labels, is identified as a class B1 (**TCSEC Class B1**) security division. It is intended for products that handle classified data and enforce mandatory **Bell-LaPadula** policies. There are labels for each subject and object, constructed from hierarchical levels and non-hierarchical categories. The integrity of these labels must be protected. Identification and authentication contribute to determining the security label of a subject.

The Bell-LaPadula model is a formal description of the allowable paths of information flow in a secure system. The goal of the model is to identify allowable communication where it is important to maintain secrecy. The model has been used to define the security requirements for systems concurrently handling data at different sensitivity levels. The model is a formalization of the military security policy (Pfleeger 1997).

Class B2 (**TCSEC Class B2**) structured protection increases assurance by adding requirements to the design of the system. Mandatory access control also governs access to physical devices. There must be a trusted path for login and initial authentication.

Verified design, or class A1 (**TCSEC Class A1**), is functionally equivalent to class B3 and achieves the highest assurance level through the use of formal methods. Formal specification of policy and system, consistency proofs, show with a high degree of assurance that the **trusted computing base (TCB)** is correctly implemented. Very few products have been evaluated to class A1.

Table 4–2 shows the trusted computer system evaluation criteria. The criteria has been grouped into four major categories—Security policy, accountability, assurance, and documentation (Pfleeger 1997).

Trusted Network Interpretation/Red Book

Secure networking is defined in the *Red Book* or *Trusted Network Interpretation (TNI)*. The Red Book is part of the Rainbow Series of documents published by the National Computer Security Center (NCSC) that describe the requirements in the TCSEC. It describes TCSEC in terms of computer networks. It also attempts to address network security with the concepts and terminology introduced in the *Orange Book*. The *Red Book* may be viewed as a link between the *Orange Book* and new criteria, which has been proposed in later years. The *Red Book* differentiates between two types of networks—independent and centralized structures.

A network with independent components consists of different jurisdictions, management, and policies. Enforcing security in this distributed environment is a very difficult problem. Centralized networks consist of a single accreditation authority, policy, and network trusted computing base. The *Red Book* only considers the centralized network. In a computer network, security mechanisms may be distributed among different net-

Table 4–2
Trusted Computer System Evaluation Criteria

Criteria Categories	Criteria
Security policy	Device labels
	Discretionary access control
	Exportation of labeled information
	Labeling human-readable output
	Mandatory access control
	Object reuse
	Subject sensitivity labels
Accountability	Audit
	Identification and authentication
	Trusted path
Assurance	Configuration management
	Covert channel analysis
	Design specification and verification
	System architecture
	System integrity
	Security testing
	Trusted distribution
	Trusted recovery
Documentation	Design documentation
	Security features user's guide
	Test documentation
	Trusted facility manual

work components. New security problems and issues arise because of the vulnerability of the communications paths and the method of concurrent and asynchronous operation of the network components.

In networks it becomes obvious that different entities, such as users, service providers, network operators, and system administrators, are responsible for setting the security policy. In the *Red Book*, the responsible entity who states the security requirements, defining the security policy, and submitting a system for evaluation, is called the sponsor. Security policies are concerned with security and integrity, and control the establishment of authorized connections. Mandatory access control in networks includes mandatory integrity policies, such as integrity labels. An integrity label could indicate whether an object had ever been transmitted between network nodes.

Other security service designs supported by the *Red Book* include encryption and protocols. Each service requires a specification for functionality, strength, and assurance. Strength indicates how well a mechanism is expected to meet its goal. Assurance addresses good software engineering practices, validation, and verification. Table 4–3 provides a summary of the network security services available in the *Red Book* (Gollman 2000).

Table 4–3
Red Book Network Security Services

Network Security Service	Components
Communication integrity	Authentication
	Communication field integrity
	Nonrepudiation
Denial of service	Continuity of operation
	Protocol-based protection
	Network management
Compromise protection	Data confidentiality
	Traffic confidentiality

Information Technology Security Evaluation Criteria (ITSEC)

Information Technology Security Evaluation Criteria

The TCSEC was exclusively U.S. development. European countries recognized the need for a criterion and methodology for evaluations of security-enforcing products, which resulted in the **Information Technology Security Evaluation Criteria (ITSEC)**. ITSEC is a logical progression from the lessons learned in various *Orange Book* interpretations. A vendor or sponsor must define a target of evaluation (TOE), which is the focus of the evaluation. The TOE is considered in the context of an operational environment, which contains a set of threats, and security enforcement requirements. An evaluation can be made of a product or a system. Rating categories include the following:

- System security policy or rationale—Do the chosen functions implement the desired security features?
- Definition of product mechanisms—How is security enforced for the product or system?
- Strength of the mechanisms—What are the results of the ratings?
- Target evaluation level—What are the levels of functionality and effectiveness?

The evaluation procedure determines the suitability and binding of functionality, ease of use, and strength of mechanism. The results of these subjective evaluations determine whether the evaluators agree that the product or system deserves its proposed functionality and effectiveness rating.

Common Criteria

European security evaluation criteria responded to the problems exposed by the *Red Book* by separating function and assurance requirements and considering the evaluation of entire security systems. The flexibility offered by the ITSEC may sometimes be an advantage, however there are

drawbacks. The next link in the evolutionary chain of evaluation criteria is the United States' Federal Criteria for Information Technology Security. This "Combined Federal Criteria" was issued only once, in initial draft form. The U.S. team joined forces with the Canadian team and the ITEC to produce the **Common Criteria** for the entire world. The Common Criteria merged ideas from its various predecessors and abandoned the total flexibility of ITSEC. It followed the Federal Criteria by using protection profiles and predefined security classes. The definition of these classes contains information about security objectives, rationale, threats, and threat environment. The Common Criteria defined topics of interest to security that are shown in Table 4–4 (Pfleeger 1997).

Common Criteria

Common Data Security Architecture

Common Data Security Architecture (CDSA) is a security reference standard that provides a way to develop applications that take advantage of software security mechanisms. The **Open Group** has accepted CDSA for evaluation, and IBM, Intel, and Netscape are refining it. It addresses security issues relating to Internet and intranet applications. It also provides an interoperable standard and an expansion platform for future security elements.

Common Data Security Architecture (CDSA)
Open Group

The Open Group is an international consortium of vendors, government agencies, and educational institutions that develops standards for open systems. It was formed in 1996 as the holding company for **Open Software Foundation (OSF)** and X/Open Company, Ltd. These two organizations work together to deliver technology innovations and widespread of open systems specifications. An **open system** is a computer and communications system capable of communicating with other like systems, transparently. The objective is that each system implements common international standards and data communication protocols.

Open Software Foundation (OSF)

open system

Functionality	Assurance
Communication	Configuration management
Identification and authentication	Delivery and operation
Invocation of security functions	Development
Privacy	Life-cycle support
Protection of trusted security functions	Guidance documents
Resource utilization	Testing
Security audit	Vulnerability assessment
Trusted path	
User data protection	

Table 4–4
Common Criteria Classes

4.13 ISO 17799 STANDARD

ISO 17799

The **ISO 17799** standard provides information concerning compliance and implementation of audit data and procedures, maintenance of security awareness, and it conducts a postmortem analysis for this standard. It is divided into 10 major sections:

- Business continuity planning
- System access control
- System development and maintenance
- Physical and environmental security
- Compliance
- Personnel security
- Security organization
- Computer and network management
- Asset classification and control
- Security policy

A general description of the ISO 17799 sections are presented in Appendix E. Other references on creating security policies can be found in *Generally Accepted System Security Principles (GASSP)*.

4.14 ENCRYPTION EXPORTING

Many corporations and organizations depend on the ability to transport private and confidential communications on an international basis. In the United States, export of encryption software is tightly controlled, and exportable encryption technology was limited to a 40-bit key. However, 40-bit keys can be cracked in a matter of minutes and do not offer an acceptable level of security. Businesses want to use stronger encryption technology, whereas the government wants to exercise greater control over international encrypted communication.

The current regulation holds that U.S. companies with international subsidiaries may export 56-bit key-based encryption technology provided they establish, within two years, a key recovery mechanism that will offer a backdoor into encrypted data for the government. After the key recovery mechanism is in place, the companies will be allowed to export keys of any length. The new regulation transferred responsibility for encryption product export from the U.S. State Department to the U.S. Commerce Department (Goldman 1999).

4.15 INTERNET AND WEB RESOURCES

Phil Zimmermann'
  PGP page   www.gpg.com
  PGP V6.5.8 freeware site   web.mit.edu/network/pgp.html

PGP information and link sites	thegate.gamers.org/~tony/pgp.html
Internet Mail Consortium	www.imc.org
Electronic Messaging Association	www.ema.org
IETF drafts (search for "mime")	www.internic.net
Douglas W. Sauder's MIME	www.fwb.gulf.net
MIME FAQ	www.cis.ohio-state.edu/text.faq
MIME media types	www.isi.edu/in-notes/iana/assignments/media-types/media-types
RSA's S/MIME Central Page	www.rsa.com/smime
VeriSign, Inc.	www.verisign.com
Visa's SET information	www.visa.com
RSA's SET Central page	www.rsa.com
IETF Web Transaction Security Working Group	www.ietf.org/html.charters/wts-charter.html
Entrust Technologies' Web Security Primer (SSL and S-HTTP)	www.entrust.com/primer.htm
Netscapes' SSL document	www.netscape.com/newsref/std/SSL.html
Biometrics Consortium	www.biometrics.org
International Biometric Group	www.biometricgroup.com
Kerberos V.5	www.faqs.org/rfcs/rfc1510.html
Secure Electronic Trans LLC	www.setco.org
Open SSL Project	www.openssl.org
Cisco Systems	www.cisco.com
Cybersafe	www.cybersafe.com
GASSP	web.mit.edu/security
Network Associates	www.nai.com
ODS Networks	www.ods.com
SANS	www.sans.org/newlook/resources/policies/policies.htm
Security Dynamics	www.rsasecurity.com
Tripwire Security	www.tripwiresecurity.com
ISO 17799	www.iso17799software.com

■ SUMMARY

Security systems are designed to protect the Enterprise organization against both internal and external attacks, including both trusted and untrusted systems. Biometrics is one method that can be employed to accomplish this task. Biometrics includes finger scanning, finger and hand geometry, iris and palm imaging, face and voice recognition, and signature verification.

Major secure protocols include SSL, TLS, and PGP. An open specification for secure credit-card transactions is called SET. It is a primary security mechanism for e-commerce transactions. HTTP, S-HTTP, MIME, and S/MIME protocols and standards are used in the unsecure Internet to provide a level of protection against network attackers. Kerberos is a password authentication service that grants client access to servers, in addition to logon protection. The Point-to-Point Protocol provides security through the CHAP and PAP authentication procedures.

Intrusion detection is responsible for detecting and responding to computer and network misuse. Either host- or network-based detection systems can be deployed. There are a number of commercial systems available that provide intrusion detection.

Security evaluation involves the assessment of a computer network to identify and certify security effectiveness and completeness. A number of tests and formal methods are used to accomplish this task.

Security specifications include the *Orange Book*, *Red Book*, and CDSA. These documents outline the requirements for secure products. The ISO 17799 standard is used to create security policies.

Key Terms

Authentication server (AS)

Bell-LaPadula

Biometrics

Challenge Handshake Authentication Protocol (CHAP)

Client

Common Criteria

Common Data Security Architecture (CDSA)

Credentials

Electronic commerce (e-commerce)

Face recognition

Finger geometry

Finger scanning

Hacker

Hand geometry

Hypertext Transfer Protocol (HTTP)

Information Technology Security Evaluation Criteria (ITSEC)

Intrusion detection

ISO 17799

Iris imaging
Kerberos
Key distribution center (KDC)
Multipurpose Internet Mail Extension (MIME)
Open Group
Open Software Foundation (OSF)
Open System
Orange Book
Palm scanning
Password Authentication Protocol (PAP)
Point-to-Point Protocol (PPP)
Pretty Good Privacy (PGP)
Private Communication Technology (PCT)
Rainbow Series
Realm
Red Book
Retinal scanning
Secure card
Secure Electronic Transaction (SET)
Secure Hypertext Transfer Protocol (S-HTTP)
Secure Multipurpose Internet Mail Extension (S/MIME)
Secure Sockets Layer (SSL)
Security accreditation
Security certification
Security evaluation
Server
Session key
Signature scanning
TCSEC Class A1
TCSEC Class B1
TCSEC Class B2
TCSEC Class C1
TCSEC Class C2
Ticket-granting server (TGS)
Ticket-granting ticket (TGT)
Transport Layer Security (TLS)
Trusted Computer System Evaluation Criteria (TCSEC)
Trusted computing base (TCB)
Trusted Network Interpretation (TNI)

REVIEW EXERCISES

Questions and Analysis

Provide a short answer or in-depth analysis as required.

1. List issues that must be addressed when designing a security system.
2. Identify the various biometric methods available to protect against illegal entry.
3. Discuss finger scanning and finger geometry authentication methods.
4. Discuss hand geometry and palm imaging for entry control.
5. Describe iris imaging, retinal recognition, and retinal scanning techniques.
6. What is the function of the PCT protocol?
7. How are SSL and TLS protocols used?
8. What is PGP and how is it used?
9. Describe the features and functions of SET.
10. How is SET used with e-commerce applications?
11. Describe a SET transaction flow.
12. Describe Kerberos authentication.
13. Describe the PAP authentication procedure.
14. What are the steps of a CHAP authentication session?
15. Why would an organization implement intrusion detection?

16. What are some capabilities of intrusion detection systems?
17. What is the difference between host-based and network-based intrusion detection systems?
18. Who are some suppliers of intrusion detection systems?
19. What is security evaluation?
20. What are the benefits of a security evaluation?
21. Describe the *Orange Book*.
22. What is the purpose of the *Orange Book*?
23. What are the classifications set forth in the *Orange Book*?
24. Describe the *Red Book*.
25. What are the services available through the *Red Book*?
26. What is Common Criteria?
27. Describe the CDSA architecture.
28. What is an open system?
29. What are the ten major sections of the ISO 17799 standard?
30. Why should an organization be concerned with encryption exporting?

Definitions and Terms

Provide the most correct answer from the list of key terms.

1. A security framework for developing security and authentication application programs.
2. A person who "hacks" away at a computer until access is successful.
3. A secret-key network authentication protocol that uses the DES cryptographic algorithm for encryption and a centralized key database for authentication.
4. The Department of Defense Trusted Computer System Evaluation Criteria.
5. A simple authentication method used with PPP.
6. An encryption scheme, developed by Phillip Zimmerman, based on the RSA encryption algorithm.
7. The Department of Defense Trusted Network Interpretation Environments Guide.
8. A message-oriented communications protocol that extends the HTTP protocol.
9. A cryptographic protocol, designed by Netscape, which provides data security at the socket level.
10. A temporary encryption key used between two principals.

True/False Statements

Indicate whether the statement is true or false.

1. Biometrics is the study of biological data and biometric-based authorization.
2. Retinal recognition has a comparative accuracy greater than finger scanning.
3. Finger geometry, or finger shape, uses finger length, width, thickness, and knuckle size in its measurements.
4. Protocols that have been developed to handle secure sessions and transactions include PCT, SSL, and TLS.
5. A major difference between SSL and PCT is the size of the key.
6. PGP is an encryption and digital signature utility for adding security to TCP/IP packets.
7. The only difference between HTTP and S-HTTP is the number of ports and sockets supported.
8. SET is a proprietary specification for handling credit card transactions over a network.
9. MIME provides ways to encode data types for electronic mail transmission.
10. Both S/MIME and PGP provide an encryption product for securing e-mail.
11. Kerberos is based on the concept of a trusted third party that performs secure verification of users and services.
12. PPP is an 16-bit serial interconnection protocol, which allows a computer to connect as a TCP/IP host to a network through a synchronous port.
13. PAP is a simple, three-step authentication procedure that uses userid, username, and password entries.
14. CHAP is a three-way handshaking authentication similar to PAP, however it provides a higher level of security.
15. Intrusion detection can provide for deterrence, detection, response, and other security elements.

Activities and Research

1. Develop a matrix comparing biometric methods.
2. Research and develop an analysis of biometric products.
3. Develop a comparison analysis of the various secure protocols.
4. Compare and contrast the MIME and S/MIME standards.
5. Research e-commerce and develop a transaction flow for some credit card transaction.
6. Create a hypothetical organization and show how intrusion detection can thwart computer network misuses.
7. Research intrusion detection providers and develop a list of suppliers and their product offerings.
8. Research and identify organizations that provide security evaluation services. Compare their offerings.
9. Develop a comparison between the *Orange Book* and the *Red Book*.
10. Discuss the provisions found in the GASSP policies.

CHAPTER 5

Enterprise Security Management

OBJECTIVES

Material included in this chapter should enable the reader to

- Become familiar with the various components and requirements for Enterprise security policies, planning, and assessment.
- Identify the security threats and faults that can invade the Enterprise network.
- Identify countermeasures for defeating network threats.
- Become familiar with the various tools available to the administrator for combating security intrusions.
- Identify the components of a network management system.
- See how network management systems are used to ensure the integrity and security of the Enterprise network.
- Become familiar with the various terms and definitions that are part of Enterprise security.
- Look at the various access precautions and techniques that can be implemented in the Enterprise network.
- Look at techniques for providing security to data and database management systems.

INTRODUCTION

Internetworks may involve large-scale, multiprotocol networks that span multiple time zones or may involve simple single-protocol, point-to-point connections in a local environment. The trend is toward an increasingly complex environment, which involves multiple media types, multiple protocols, and often some connectivity to a private network service or an ISP. It is, therefore, possible that control of the components, throughput, and management of these networks is in someone else's hands. More complex network environments mean that the potential for performance and availability issues in internetworks is high, and the source of problems is often elusive. The keys to maintaining a secure, problem-free network environment, as well as maintaining the ability to isolate and fix a network fault quickly, are documentation, planning, and communication. This requires a framework of procedures and personnel in place before the requirement for problem solving and recovery occurs.

Network problems can typically be resolved in one of two ways, either proactive prevention or after-the-fact reaction. It is possible to prevent problems from occurring by a program of computer and network resource planning and management. The alternative is repair and control of damage that can be accomplished by reactive responses. Network management and planning should combine to form an overall security plan. This network plan combination should include

- Cable diagrams
- Cable layouts
- Documentation on computer and network device configurations
- Important files and their layouts
- Listing of protocols and network standards in use
- Network capacity information
- Software

The Network management section concludes with a presentation on network management support systems and network management products and services, including NetView, OpenView, SunNet Manager, and RMON.

Security troubleshooting is often characterized as an art that can only be gained from experience in the trenches. This experience, along with some documented methodology, allows the administrator to successfully solve difficult network issues. The methodology keeps the troubleshooter from skipping or overlooking the obvious and the experience helps to draw appropriate conclusions and to check for obscure problems.

Before someone starts this troubleshooting process, it is essential for the network professional to establish a baseline or reference point for system comparisons. Information should be gathered and documented on the network devices, software elements, facility components, network traffic, performance levels, and security systems. This information will prove invaluable later when a problem develops in the network.

Establishment of policies and procedures that apply to the enterprise network must be developed during its planning stages and continue throughout the network's life. Such policies should include security, hardware and software standards, upgrade guidelines, backup methods, and documentation requirements. Through careful planning, it is possible to minimize the damage that results from most predictable events and control and manage their impact on the organization.

5.1 ADMINISTRATION

It is essential that some person in management be responsible for computer and network security administration. This responsibility is often assigned to the network administrator, since many of the duties required fall under the auspices of network management. This position should not be taken lightly as the success or survival of the organization may depend on this person (Thompson 2000).

Common duties for a network administrator's office include the following:

- Software installation and upgrades
- Database access approval and maintenance
- Login script and menu creation and maintenance
- Login Ids and password assignment
- Backup and restoral processes
- Training and documentation support
- Facilities monitoring and other audits
- Future expenditure requests and justifications

The security administrator should participate in the organization's planning council activities.

5.2 CORPORATE PLANNING

Enterprise security goals must be set and supported by the organization's upper management. The "real" leadership must define the vision and allocate sufficient resources to completely and successfully implement all of the security elements necessary to protect the organization's assets. A clear message must be sent to all personnel and users that the computer and network resources are valuable and must be properly protected (Goldman 1998).

Contingency planning scenarios should be a part of the development of a security program. As an example, a plan must be in place if a laptop capable of accessing sensitive corporate resources is stolen. It is necessary to look at different security-oriented events and develop a solution to counter particular threats and attacks.

The entire organization must be aware of the need for a security plan. By making employees aware of the ramifications for leaving a workstation unattended or writing a password on a post-it note attached to the monitor, awareness and compliance is more likely to occur. Security must be part of the hiring process. If security is emphasized from day one, employees are more likely to develop habits conducive to a secure operation. It should not be a surprise that the Information Technology (IT) department is the most sensitive area of the organization. IT personnel have access to all levels of information, so it is incumbent on management to monitor activities in this area.

In any security system there will be a weak link that is human and not technological. A common problem is personnel turnover. Former employee's network access must be denied promptly to avoid possible security violations. There must be a process where the human resource department notifies IT of an employee's departure, so that network access can be removed. Delays can result in data theft or corruption. It is common practice in many organizations to escort former employees to the door and collect all corporate items such as smart cards and credit cards. Having security measures in place means that someone must administer those measures. This task is usually the responsibility of the IT department, which means that these employees must be trustworthy, competent, and conscientious.

With today's heterogeneous networking environments, it is necessary to know whether security protocols and software work together. It is useless to have only part of the network secure—the entire network must be secure from both internal and external threats. A successful computer network security system requires a marriage of technology and process, which is part of a security requirement assessment.

5.3 SECURITY REQUIREMENTS ASSESSMENT

As with any assessment program, a formalized process must be followed to accomplish the stated goals. This usually takes the form of an iterative process where a number of steps are executed many times until the process has been refined. Each of these steps must be clearly defined, including

Figure 5–1
Security Requirements Development Life Cycle

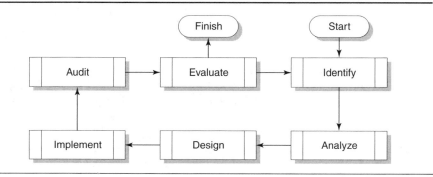

roles and responsibilities for computer operations processes and network-related activities. This means that process definition and setting of the organization's security goals and standards must precede technology evaluation, selection, and implementation.

There is a simple and straightforward life-cycle approach that can be used to accomplish this task. The application of this structured process ensures that all potential user group and information combinations have been considered. When successfully completed, this implies that appropriate security processes and technology has been determined that allows legitimate access into any of the organization's computer and network resources. Figure 5–1 depicts a security requirement development life cycle process. This process includes the following steps that are executed repetitively until an evaluation is successful:

- Identify the organization's security issues.
- Analyze security risks, threats, and vulnerabilities.
- Design the security architecture and the associated processes.
- Audit the impact of the security technology and processes.
- Evaluate the effectiveness of current architectures and policies.

Note that evaluation processes validate the effectiveness of the original analysis steps and feedback from these evaluation steps causes a renewed analysis with possible ripple effects of changes in architecture or implemented technology.

5.4 MAINTAINING NETWORK INTEGRITY

Protecting network assets and operations is a continuing task with results that can never be certain. When intelligently applied, protective efforts can reduce, but never completely eliminate, the chances of losses caused by security breaches. Although most network security violations take place within corporate networks and are initiated by authorized users, most funding for security programs is allocated to measures that guarantee that only authorized users are allowed to access the network. These funds are also allocated to prevention and detection of external invasions.

A primary task of maintaining network integrity involves troubleshooting the network environment. Past results indicate that a systematic approach is the most effective. The administrator should be able to identify problems from symptoms and initiate corrective action based on this systematic approach. If change is not managed, a great deal of time will be spent fighting fires instead of preventing them. A major part of this structured approach is network management and planning, which can be accomplished by addressing the following issues:

- Data backup
- Documentation procedures and methodology
- Hardware and software standards
- Network baseline
- Preemptive troubleshooting
- Security policies
- Upgrade guidelines

Network Baseline

A baseline for network performance must be established if network monitoring is going to be used as a preemptive troubleshooting tool. A **baseline** defines a point of reference against which to measure network performance and behavior when problems occur. This baseline is established during a period when no problems are evident on the network. A baseline is useful when identifying daily network utilization patterns, possible network bottlenecks, protocol traffic patterns, and heavy-user usage patterns and time frames. A baseline can also indicate whether a network needs to be partitioned or segmented, or whether the network access speed should be increased. The three components that must be created to establish a baseline are

- Current topology diagrams
- Response-time measurements of regular events
- Statistical characterization of the critical segments

These three components will require some effort in developing; however, the payoff will come when a problem occurs in the network. A small amount of time spent each week by a number of personnel who are assigned portions of the network can accomplish this task in a relatively short time.

Security Policies

All security policies set forth in a network plan should be detailed and followed closely. The security policies will depend on the network size, the organization's security standards, and the value and sensitivity of the data. The security plan must include physical, network, and computer security. The five most significant issues involving security of a computer network, that should be addressed, include the following:

- Identification/authentication—Users are accurately identified.
- Access control/authorization—Only legitimate users can access a resource.
- Privacy—Eavesdropping is not an issue and transmissions are private.
- Data integrity—Activities on a database are controlled and protected.
- Nonrepudiation—Users cannot deny any legitimate transactions.

Network security can be enhanced by a number of user-name and password requirements and resource access requirements. Standards for user name and passwords include the following suggestions:

- Establish minimum and maximum password lengths for user accounts.
- Provide the users with the details in reference to character restrictions.
- Determine the frequency for changing passwords.
- Decide if and when passwords can be reused.
- Decide if there will be exceptions to the policy.

Resource access is generally granted only to those who specifically require it. It is always easier to grant new access to users than to take it away. For dial-in users, special security arrangements will probably be necessary. Many organizations require a security card, which provides a code that must be entered for dial-up access. It is essential that the number of users who perform network administration tasks be limited to the absolute minimum. The more users with access to administrative functions, the more likely security problems will occur.

Another user issue that can cause a security breach is leaving the work area without logging out of the system. This not only allows anyone to use that persons access rights, but also could put that employee's job in jeopardy. The system can assist operators by logging off any user who has not entered a transaction within a certain time period, such as five minutes. Users who leave their workstations for more than five minutes would have to go through the Triple A process again. This can become a performance issue if transaction-processing time is sporadic and the workstation is frequently logged off. A transaction log can be used to identify the optimum automatic logoff value (Stamper 1999).

Access to many Enterprise systems is via dial-up using modems. Because this technique is available to everyone in the world that has a dial-up modem, special attention must be devoted to limiting the access to legitimate users. A security policy that is both usable and enforceable is a requirement in all Enterprise computer-networking systems. This policy should include some, if not all, of the following elements (McClure 2001):

- Provide protection for analog telephone numbers. Don't advertise them to the world. Use numbers that are different from regular Enterprise telephone numbers.
- Maintain an inventory of all dial-up telephone lines. If the line is not being used, make sure that it is not connected to a modem.
- Don't have modems scattered around the premises. Home all modems into a central rack where they can be controlled. Use a modem pool to consolidate them into a controlled environment.
- Dial-up lines and modems must be located in a secure closet. These lines and modems must be tagged with circuit Ids.
- Monitor all dial-up activity. Look for failed login attempts. CallerID may be useful for identifying calling telephone numbers.
- Don't display any banner information that is presented on connection. This is a place where a warning message can be displayed to callers.
- Require a dial-back authentication procedure if feasible. This may be impractical with roving users.
- Provide help desk and PBX employees with specific instructions about social engineering tactics. Remote access credentials cannot be reset without the approval of a supervisor.
- Connectivity for all analog facility activity, including fax and voice-mail, must be controlled by a central authority. This includes the standard POTS line.
- Provide for an audit activity that monitors all of the above elements.

Hardware and Software Standards

To make hardware and software easier to manage, all network components should follow established standards. Several different levels of configurations can be established, depending on the requirements of the desktop users. These standards should cover both hardware and software configurations.

Standards should also be established for networking devices, including manufacturers, and operating systems, including versions. This also includes standards for server configurations and server types.

When establishing hardware and software standards, keep in mind the pace of industry change and obsolescence. Regular evaluations of network standards will be required to ensure that the network remains current.

Upgrade Guidelines

Upgrades for hardware and software and new networking products are a fact of life. It is necessary to establish guidelines for handling these upgrades. They can be handled easier if the user community has advance notice of such an upgrade, and they should not be performed during normal working hours.

It is a good idea to test upgrades through stand-alone platforms or through a group of technically astute network users. When performing upgrades, always have a plan for backing out of the upgrade if it fails to perform as expected. Through careful planning and testing, the upgrade process can be relatively painless.

Intrusion Detection

Intrusion detection techniques can be used to monitor network traffic, looking for patterns that suggest known types of computer attacks. An alternative approach, designed to discover previously unknown types of attacks, requires monitoring network statistics for unusual changes that might indicate attack activity. Intrusion detection monitoring may be accomplished using a network host computer, or detection facilities may be distributed at various points throughout the network. These systems often employ both pattern and statistical methods, and use both host-based and distributed monitoring. Chapter 14 presents information on distributed networks and the security issues involved.

The most effective intrusion detection monitoring occurs dynamically as information packets travel through the network. If this activity is in-band, network traffic could experience slow throughput. To meet intrusion detection needs, vendors are now offering security chips capable of examining as many as 20 million packets per second.

Attackers often turn to scanning systems to gather information on the target network. Scanning tools have the ability to automatically check a target network for possible computer and network vulnerabilities. These tools employ a database of known configuration errors, system bugs, and other problems that can be used to infiltrate a system. Defending against vulnerability scanners require the administrator apply system patches on a regular basis and periodically conduct internal vulnerability scans (Skoudis 2002).

Testing to assess the network intrusion vulnerability may help administrators discover security weaknesses. Such testing can be performed in-house or through a security consultant. When holes in network security are discovered through vulnerability testing, proactive action can be taken to close them.

A security scanner consists of software that will audit remotely a given network and determine whether bad guys may break into it, or misuse it in some way. Nessus is an example of a vulnerability scanner that might be used to improve the security of the Enterprise.

Security Threats

Security is concerned with protecting data and data systems, and include security techniques for both physical facilities and software. It is also concerned with ensuring the integrity of the **network operating**

network operating system (NOS)

system (NOS). This NOS contains the software that runs on a server and controls access to files and other resources from multiple users. Physical security is concerned primarily with providing secure access, while software security includes authentication, authorization, access controls, and user logon. This section addresses security controls for operating and networking systems.

Threats are the reason it is necessary to be concerned about security. These include the following destructive activities (Sheldon 1998):

- Internal users may try to access unauthorized data systems.
- Internet users may try to attack systems available to the network.
- Unauthorized users may try to access personal user accounts.
- Attackers may gain access to an account and lockout or restrict the legitimate users.
- Attackers may modify data values or destroy data.
- Proprietary information may be copied for resale to competitors.

A number of these security attacks can be attributed to **hackers**. Historically, a hacker was defined as a person who "hacks" away at a computer until access is gained or until a software function was successful. Students initially performed these activities, primarily for fun and sport. Today the term has a more positive meaning and has been replaced with the term **cracker**, which is reserved for the individual who willfully breaks into computer systems with the purpose of wreaking havoc.

The most common security breach is access to unauthorized user accounts. This activity occurs when the attacker impersonates or masquerades as the legitimate user by obtaining a username and password. These can often be obtained from posted notes, shoulder surfing, or from a network monitor. Because some users may use some personal object or a relative's name as a password, the possibility exists for a **password cracker** to guess usernames and passwords. Another technique that is successful is the **dictionary attack**, where the cracker has access to a very large dictionary of words commonly used for passwords. This threat can be countered by locking a user account after a number of failed logon attempts, usually three.

The most serious attack, however, is when someone gains access to an administrator or **superuser** account. A superuser is a UNIX system administrator who has high-level access privileges. With this access, attackers may lock out the legitimate account owner and perform destructive activities on the system or databases.

Another technique that can have significant repercussions is **eavesdropping**. Network traffic can be monitored with sniffers or wiretaps. A **sniffer** is a LAN protocol analyzer that supports a variety of hardware to include Ethernet and Token Ring topologies. The attacker may monitor the network for long periods of time and record valuable or sensitive information that can be used for future attacks. Information captured can be present in a **replay attack**, which results when a service

hackers

cracker

password cracker
dictionary attack

superuser

eavesdropping
sniffer

replay attack

already authorized and completed is forged by another duplicate request in an attempt to repeat authorized commands. It is possible for an attacker to replay an authentication routine to gain illegal access to a system or network. Sequencing and time-stamping packets can avoid these situations. An eavesdropper might capture packets, modify them, and reinsert them into the data stream, which can be unknown to the legitimate sender and receiver.

The last threat discussed in this section concerns the **denial of service** attack. This involves an attack that attempts to deny corporate computing resources to legitimate users. The attacker can cause a server or processor to slow or stop operations, using techniques that overwhelm it with some useless task. It is also possible to corrupt the operating system of the processor. This type of attack is becoming commonplace across the Internet.

Securing Data Systems

Network security features such as access controls can be used to protect resources from unauthorized access. Many directories and network resources contain **access control lists** that contain entries, which specifically identify access parameters. These lists identify which users and groups have object access and the respective permissions for that access. An object is an entity or component, identifiable by the user, that may be distinguished by its properties, operations, or relationships. These objects can have specific access levels, which include read, write, and execute functions. These functions will be discussed later in more detail.

Most network operating systems have an auditing system designed to record network activities. These systems can be available for real time or almost real time network traffic monitoring. The auditing records can be reviewed on a regular basis to determine if the system is being compromised. This auditing system can record activities such as logons, file access, network traffic, and account access. It is essential that auditing records be protected to prohibit an attacker from modifying them, which could effectively cover up an illegal activity.

Authentication Techniques

As presented in Chapter 2, authentication methods identify users. Once users are identified and authenticated, it becomes possible to access resources based on an authorization. An authentication system may be the most important part of a NOS. It requires a user supply some identification, such as a username and password. Simple password systems may be sufficient for small systems, however if a higher level is required, then more advanced security may be necessary. These systems may provide dial-back capabilities, or use biometric or token authentication devices. Also included are global positioning and Kerberos systems. Each of these systems will be discussed in detail.

Dial-back Systems

An increasingly mobile workforce often requires remote access to the computer resource. Remote network access can assist employees in gaining access to databases, facilitate a telecommuting system, or help traveling executives maintain contact with corporate headquarters. This type of dial-up access should be granted cautiously and judiciously, and only to those that require it, because network resources are easily compromised by this access method. This type of doorway into the network invites other non-employees to try the locks on the door. There are many technical solutions that can be implemented to provide the required level of security and ensure that only authorized individuals gain access to the corporate network using such a door.

A **dial-back** system is a solution that can provide some level of protection for these remote dial-in users. Using a dial-back modem, the remote user calls the computer resource and the dial-back modem calls the user back at a predefined telephone number to make the connection. The normal identification and authentication process occurs over this connection (Blacharski 1998).

dial-back

Biometrics

Biometrics is the process of using hard-to-forge physical characteristics of individuals, such as fingerprints, voiceprints, and retinal patterns to authenticate users (Kaeo 1999). This is called a **biometric-based authentication**, or third-factor authentication, which is "something you are." Minute measurements such as the timing of a pattern of keystrokes or a written signature can be used to differentiate users.

biometrics

biometric-based authentication

Since a scanner or sensor is used to capture biometric data, there is a possibility of obtaining incomplete or incorrect data for authentication. The performance of biometric methods is based on the percentage of false rejections and false acceptance. A false acceptance occurs when the authentication process accepts an invalid user. A false rejection occurs when the process rejects a valid user. Biometric authentication over a network can still be subject to capture and replay of additional encryption is not used for the biometric data.

Token Authentication

Token-based authentication is a security technique that authenticates users who are attempting to login to some secure computer or network resource. This method helps eliminate insecure logons—logons that send users' passwords across the network where they can be observed in the clear. Someone capturing the password could use it to repeatedly masquerade as the legitimate user and access a secure system resource. This situation can be countered by not sending passwords in any form across an insecure channel. An alternative process is to use a **token**, which resembles a credit card.

token-based authentication

token

These devices are microprocessor-controlled **smart cards**, which are used to implement **two-factor authentication**. The user supplies the logon password and the one-time value generated by the smart card. The smart card generates one-time passwords that are good for only one logon attempt. Two-factor authentication identifies a user and then authenticates the user. Organizations often assign tokens to remote and mobile users who need to access internal network resources from outside locations.

Security Dynamics offers a token called SecurID that uses a time-based technique for displaying a number that changes every minute. This card (token) is synchronized with a security server at the corporate location. When users login, they are prompted for a value from the SecurID card. Because the value changes every minute, someone who manages to capture it on-line cannot reuse it (Sheldon 1998).

Software-based token devices, which run on portable computers, are also available, and they provide much the same functionality as hardware tokens. It has been suggested, though, that software tokens are less secure than hardware ones, and can be more easily compromised.

The use of an authentication device, which is protected by a password, is superior to a single password since the loss of one of the authentication factors will not allow system access. The likelihood of losing both factors is remote.

Global Positioning

The **global positioning system (GPS)** can verify the physical location of a user anywhere on the planet. This can provide the network with the assurance that the call is not originating from an unauthorized location. The government is spending billions to build and maintain the system.

GPS is based on satellite ranging. This means that the position on earth of the entity being tracked is determined by measuring the distance from a group of satellites in space. The satellite transmits its own position, time, and a long pseudo-random noise code (PRN). The receiver uses this noise code to calculate the range. There are currently two PRN strings being transmitted, a course acquisition code (CAC) and a precision code. At the present time, civilian users are only authorized to use the CAC code, called the GPS Standard Positioning Service, which maintains an accuracy of about 5 meters.

5.5 NETWORK MANAGEMENT SYSTEM

Network management is a collection of activities that are required to plan, design, organize, maintain, and expand the network. A network management and control system consists of a collection of techniques, policies, procedures, and systems that are integrated to ensure that the network delivers its intended functions. At the heart of the system is a database of information, either on paper or mechanized. The database consists of several related files that allow the network managers to have the information they need to exercise control over the database's functions. A network control system has five major functions:

- Managing network information
- Managing network performance
- Monitoring circuits and equipment on the network
- Isolating trouble when it occurs
- Restoring service to end users

Network management is generally concerned with monitoring the operation of components in the network, reporting on the events that occur during the network operation, and controlling the operational characteristics of the network and its components.

Monitoring

Monitoring involves determining the status and processing characteristics currently associated with the different physical and logical components of the network. Depending on the type of component in question, monitoring can be done either by continuously checking the operation of the component or by detecting the occurrence of extraordinary events that occur during the network operations.

Reporting

The results of monitoring activities must be reported, or made available, to either a network administrator or to network management software operating on some machine in the network. It is essential that some positive action be taken on the results of these reports.

Controlling

Based on the results of monitoring and reporting functions, the network administrator or network management software should be able to modify the operational characteristics of the network and its components. These modifications should make it possible to resolve problems, improve network performance, and continue normal operation of the network.

5.6 NETWORK MANAGEMENT ARCHITECTURE

Most network management architectures use the same basic structure and set of relationships. Managed devices such as routers and other network devices run software that enables them to send alerts when they recognize problems or predefined situations that might warrant attention. On receiving these alerts, management entities are programmed to react by executing some action including operator notification, event logging, system shutdown, or automatic attempts at system repair.

Management entities can also poll end-network devices to check the values of certain variables. Polling can be automatic or user initiated, but agents in the managed devices respond to all polls. Agents are software modules that first compile information about the managed devices where they reside, then store this information in a management database, and then provide it to management entities within the NMS. These subjects are discussed in detail later in this chapter.

Network Management Requirements

An important requirement for Enterprise networking is the ability to manage large networks that consist of hardware and software from a number of providers and manufacturers. This section describes the protocols and standards that are available to the Enterprise network manager.

Well-known network management protocols include the Simple Network Management Protocol (SNMP) and Common Management Information Protocol (CMIP). Management proxies are entities that provide management information on behalf of other entities. Figure 5–2 depicts a typical network management architecture. SNMP, CMIP, and the other network management elements are discussed in Chapters 5 and 7. The following are brief descriptions of each of the SNMP architectural components:

- Network Management Station—an end system in the network that executes a network management application. A system administrator typically uses a NMS to monitor and control the network.
- Network Management Application—the program running in a network management station that monitors and controls one or more network elements.

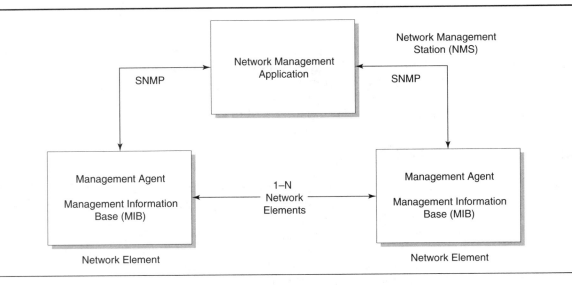

Figure 5–2
Network Management Architecture

- **Network Element**—a component in the network that maintains a management agent and a portion of the **Management Information Base (MIB)** that contains manageable objects. Examples of network elements are end systems and routers.

    Management Information Base (MIB)

- **Management Agent**—a program running in a network element that is responsible for performing the network management functions requested by a network management application. A management agent operates on the objects stored in the portion of the MIB that is maintained in that network element.

As presented in the network management architecture section, an SNMP-managed network consists of four key components. The relationship between these components is shown in Figure 5–3.

These components interact with each other to provide the data and information necessary to manage the Enterprise network. To understand how this works, a discussion follows that shows the interaction between the managed device, the agent, the network management system, and the network management station.

A **managed device** is a network node that contains an SNMP agent and resides on a managed network. Managed devices collect and store management information and make this information available to the network management systems using SNMP. Managed devices, called network elements, can be routers, access servers, switches, bridges, wiring hubs, and computers.

managed device

A **Managed object** is a variable of a managed node. This variable contains one piece of information about the node. Each node can have several objects. The management station monitors and controls the node by reading and changing the values of these objects.

Managed object

An agent is a network management software module that resides in a managed device. An agent has local knowledge of management information and translates that information into a form compatible with SNMP. Each agent keeps a database of information, including the number of packets sent, the number of packets received, packet errors, the number of

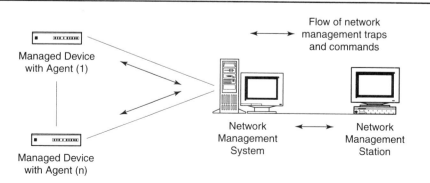

Figure 5–3
SNMP-managed Network

connections, and others. An agent's database is called an MIB. The agent responds to requests for information from a management station for actions from the management station. The agent may also provide the management station with important, but unsolicited, information. The agent can either be resident on the node or be a proxy residing elsewhere that acts on behalf of the node.

The network management system consists of hardware and software implemented in existing network components. The software used in accomplishing the network management task resides in the host computers or servers and communications processors. A network management system is designed to view the entire network as a unified structure. It contains addresses and labels assigned to each network interface and the specific attributes of each element and link known to the system. HP Openview is an example of such a software system.

The network management station is typically a stand-alone device that serves as the interface for the human network manager into the network management system. The management station contains the following components:

- An interface by which the network manager can monitor and control the network.
- The capability of translating the network manager's requirements into the actual monitoring and control of remote network elements.
- A set of management applications for data analysis and fault recovery.
- A database of management information extracted from the databases of all the managed entities in the network.

In a traditional centralized network management scheme, the network management station is located at the central site. Remote locations are accessible to the network management system via a number of routers. MIB agents are resident in the routers and other managed devices that are components on the various topologies. There may be, however, one or two other management stations in a backup role. As networks grow in size and traffic load, the centralized approach to network management becomes unworkable. This decentralized approach provides for two network management stations. This type of architecture spreads the processing burden and reduces total network traffic.

5.7 SNMP MANAGEMENT FUNCTIONS

Simple Network Management Protocol (SNMP)

The **Simple Network Management Protocol (SNMP)** includes management, security, and operations elements. SNMP and network management software are discussed in Chapters 5 and 7.

SNMP Management
SNMP is a distributed-management protocol. A system can operate exclusively as with a network management station or an agent, or it can perform the functions of both.

SNMP Security
SNMP lacks any authentication capabilities, which results in vulnerability to a variety of security threats. Because SNMP does not implement authentication, many vendors do not implement "Set" operations, thereby reducing SNMP to a monitoring facility.

SNMP Operations
SNMP is a simple request/response protocol. Responses to messages from agents do not have a set send/receive structure. The sender can send multiple requests without receiving a response.

Monitoring with SNMP
The Simple Network Management Protocol (SNMP), as presented in Chapter 7, is part of the TCP/IP suite that is used for network management. SNMP is an industry standard that is supported by most networking equipment manufacturers. A common NMS environment, presented in Figure 5–4, illustrates managed network devices that contain software agents that are managed by SNMP. Each agent monitors network traffic and device status and stores the information in a MIB.

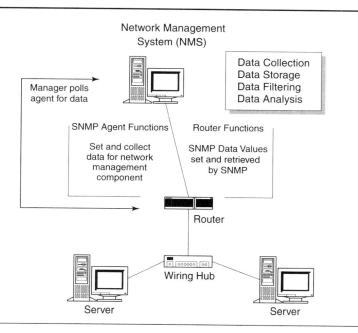

Figure 5–4
Common NMS Environment

The information collected by the agents is provided to a network management computer system that is present on the network. A management station on the network communicates with the software agents and collects data that has been stored in the MIBs on the managed network devices. The NMS can combine information from all networking devices and generate statistics or graphics detailing current network conditions. SNMP provides for the setting of thresholds and the generation of messages that provide information on the status of the network and the devices.

5.8 COMMON MANAGEMENT INFORMATION PROTOCOL (CMIP)

As mentioned, the International Standards Organization's approach to network management is the **Common Management Information Protocol (CMIP)**. This protocol defines the notion of objects, which are elements to be managed. The key areas of network management as proposed by the ISO are divided into five Specific Management Functional Areas (SMFAs), which include Fault Management, Configuration Management, Performance Management, Security Management, and Accounting Management.

Fault Management
Fault management concerns the ability to detect, isolate, and correct abnormal conditions that occur in the network environment. Central to the definition of fault management is the fundamental concept of a fault. Faults are distinguished from errors. A fault is an abnormal condition that requires management attention to fix. A fault is usually indicated by a failure to operate or by excessive errors. It is usually possible to compensate for errors using the error-control mechanisms of the various protocols.

Because faults cause downtime or unacceptable network degradation, fault management is the most widely implemented of the ISO network management elements. Fault management involves (1) determining symptoms and isolating the problem, (2) fixing the problem and testing the solution on all-important subsystems, and (3) recording the detection and resolution of the problem.

Configuration Management
Modern data communications networks consist of individual components and logical subsystems that can be configured to perform many different applications. Configuration management is concerned with the ability to identify the various components that comprise the network configuration; to process additions, changes, or deletions to the configuration; to report on the status of components; and to start up or shut down all or any part of the network.

Each network device has a variety of versions associated with it. A workstation could have version information concerning the operating system, communications hardware and software, and network protocols. Configuration management subsystems store this information in a database for easy access. When a problem surfaces, this database can be searched for clues that may help solve the problem.

Performance Management

Data communications networks are composed of many and varied components that must intercommunicate and share data and resources. Performance management concerns the ability to evaluate activities of the network and to make adjustments to improve the network's performance. Performance variables that might be provided include network throughput, user response times, and line utilization.

Performance management involves three main steps:

- Performance data are gathered on variables of interest to network administrators.
- The data are analyzed to determine normal (baseline) levels.
- Appropriate performance thresholds are determined for each important variable so that exceeding these thresholds indicates a situation requiring attention.

Performance management of a computer network comprises two broad functional categories: monitoring and controlling. Monitoring is the function that tracks activities on the network. The controlling function enables performance management to make adjustments to improve network performance.

Management entities continually monitor performance variables. When a performance threshold is exceeded, an alert is generated and sent to the network management system. Performance management also permits proactive methods using network simulation software. Such simulation can alert administrators to impending problems so that counteractive measures can be taken.

Two measurements that are necessary in performance management are system effectiveness and system availability. A system is effective if it provides good performance, is available when needed, and is reliable when it is used.

Security Management

Security management concerns the ability to monitor and control access to network resources, including generating, distributing, and storing encryption keys. Passwords and other authorization or access-control information must be maintained and distributed. Security management is involved with the collection, storage, and examination of audit records and security logs (Stallings 2000B).

Security management subsystems work by partitioning network resources into authorized and unauthorized areas. For some users, access to

security management

any network resources is inappropriate, because such users are usually company outsiders. For internal users, access to information originating from a particular department, such as payroll, is inappropriate.

Security management subsystems perform several functions. They identify sensitive network resources and determine mappings between sensitive network resources and user sets. They also monitor access points and log inappropriate access to sensitive network resources.

Accounting Management

Accounting management concerns the ability to identify costs and establish charges related to the use of network resources. In many Enterprise networks, individual divisions or cost centers are charged for the use of network services. These are internal accounting issues rather than actual cash transfers, but are important to the participating users. The network manager must be able to track the use of network resources to ensure that the network is functioning efficiently and effectively.

Steps toward appropriate accounting management include

- Measuring utilization of all-important network resources
- Analysis of the results for insights into current usage patterns
- Setting of usage quotas
- Corrections made to reach optimal access practices
- Ongoing measurement of resource use to yield billing information

Thirteen CMIP supporting functions are mapped onto the five SMFAs. These supporting functions are (Martin 1996)

Alarm reporting

Access control

Accounting meter

Event-report management

Log control

Object management

State management

Security-alarm reporting

Security-audit trail

Test management

Summarization

Relationship management

Workload monitoring

Trap Protocol Data Unit

A protocol data unit (PDU) is a set of data specified in a protocol of a given layer and consists of protocol control information, and possibly user data of that layer. Basically, a PDU is the OSI model terminology for packet.

The PDU is also known as a header or a trailer that contains specific information, which is sent from one layer on the source computer to the same layer on the destination computer. This type of communication is called peer-to-peer communication. The SNMP message contains the message header and the PDU.

Traps

A **trap** is a message sent without an explicit request by an SNMP agent to a network management station, console, or terminal to indicate the occurrence of a significant event, such as a specifically defined condition or a threshold that has been reached.

Traps are defined in RFC1157. Each SNMP trap contains a generic trap number and a specific trap number. The generic trap numbers range from 0 to 6, where numbers 0 to 5 are defined by SNMP. The user can define an enterprise trap by combining the generic enterprise-specific trap 6 with a specific user-generated trap number.

The managed device responds to commands from the network management station with a Get response. When the managed device needs to report an event in response to some "trigger," it initiates a trap message to the network management station. A trigger is some attribute that has been set in the managed device that is being monitored. An example would include such attributes as interface status (up/down) and operational status (errors/traffic).

Programs can be executed that use the data contained in traps to provide a number of benefits to a network surveillance operations center. Software products such as Seagate Nervecenter and Cisco CiscoWorks (Figure 5–5) perform such functions.

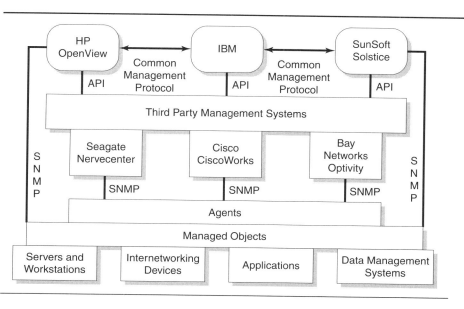

Figure 5–5
Network Management Interaction

5.9 NETWORK MANAGEMENT STANDARDS

Many organizations today deal with network management standardization. The roles played by these organizations range from setting the network management standards to promoting acceptance of the standards. The organizations that play a role in network management include

- American National Standards Institute (ANSI)
- Corporate for Open Systems (COS)
- International Organization for Standards (ISO)
- Institute of Electrical and Electronic Engineers 802 Committee (IEEE 802)
- Internet Activities Board (IAB)
- International Telecommunications Union Telecommunications Sector (ITU-T)
- National Institute of Standards and Technology (NIST)
- **Open Systems Foundation (OSF)**

The OSF has developed several network management standards, including DMI, DME, and DCE, as discussed in the following sections.

Desktop Management Interface

The Desktop Management Interface (DMI) is a proposed network management standard being developed by the Desktop Management Task Force, a group of vendors in the desktop computing marketplace. DMI is designed to provide a standardized way to manage the various hardware and software components that are part of a network consisting of primarily personal computers and desktop workstations. DMI is designed to be complementary to standard network management protocols such as CMIP and SNMP.

Distributed Management Environment

The OSF has defined a set of standards for management services called the Distributed Management Environment (DME). DME Services are complementary to those defined by the OSF DCE. DME Services provide common services that are useful to management applications of all kinds. In a distributed computing environment, varying types of hardware and software, located in widely disparate locations, must all interoperate. To achieve efficient and effective operation, management services must be available for monitoring and controlling all the sources connected to the network. DME Services are designed to provide the tools necessary to manage network resources, and include software distribution services, event services, license management services, subsystem management services, and personal computer services.

Distributed Computing Environment

The OSF has developed an important architecture for client-server, network-based computing called the Distributed Computing Environment (DCE). The OSF DCE defines services that fall into many of the OSI application layer service categories such as print, file, electronic-mail, directory, and network management services.

DCE is intended to provide a set of standardized services that can be made available across a variety of different system environments, so a distributed application developed using DCE services can support different operating systems, different network software subsystems, and different transport protocols.

The six services included in the DCE are

- Directory Service
- Distributed File Service
- Distributed Time Service
- Remote Procedure Call Service
- Security Service
- Threads Service

5.10 NETWORK MANAGEMENT SUPPORT SYSTEMS

A Network Management System executes applications that monitor and control managed devices. It provides the processing and memory resources required for network management.

Enterprise network management systems must be able to gather information from a variety of sources throughout the Enterprise network and display that information in a clear and meaningful format. One of the difficulties with implementing Enterprise network management systems is a lack of interoperability between different software systems. In addition to the network management systems are third party, or vendor-specific (proprietary), applications that provide NMS support.

Performance management information can be communicated to Enterprise management systems such as HP OpenView, SunSoft Solstice, or IBM SystemView in the proper SNMP format. Figure 5-5 depicts the interaction between these three systems and third-party network management software (Goldman 1997). SNMP interfaces with and provides a flow of network management system information between these software systems and the agents in the managed devices. Note the common management protocol between the various NMS and an **Application Program Interface (API)** between the NMS and third party software. API is software that an application program uses to request and execute basic services performed by a computer.

Application Program Interface (API)

A Management Information Base (MIB) is a database of network performance information that is stored on a network agent for access by a network management station. Information from the standard MIB is of limited use with these APIs. The number and diversity of devices necessitate extensions to this MIB. Private MIBs are often provided by hardware manufacturers and can be seen in the SNMP **MIB tree** structure under Private Enterprises. Each manufacturer is issued a unique number to define MIB information.

MIB tree

Management software analyzes the data returned by the agent. First a device must be discovered or created, and then it must be determined if it contains an SNMP agent. MIB definitions are "compiled" into the Enterprise management software allowing it to query the agent for information specific to that device.

Historically, the SNMP management protocol has been used as a monitoring tool. SNMP management stations feature an icon that represents a computing component, which turns red to indicate it has gone offline. From there, control is turned over to the application associated with administering the device and making the necessary fix. The monitoring role of SNMP reflects the simplicity of the SNMP message-based protocol.

The administrator uses unique data requests to confirm a device's state. Data logging allows operators to gather values for selected attributes over time. Once captured, the data may be displayed or logged for future analysis.

Event requests allow operators to monitor changes in attribute values based on predefined thresholds. If a threshold is exceeded, the management software alerts the operators with audio, visual, e-mail, or programmatic responses. Traps are events that might represent the crossing of a threshold boundary (e.g., "interface s01 receive errors > 1000") or an event that impacts some MIB variable.

Enterprise management software offers a graphical representation of the network topology. This is a device-based view with little or no connectivity information or monitoring, because the SNMP protocol is device based, not connection oriented. The Remote Network MONitoring (RMON) standard allows connection monitoring.

RMON MIB-II

RMON not only employs SNMP, but also incorporates a special database for remote monitoring, called **RMON MIB-II**. This database enables remote network nodes to gather network analysis data at virtually any point on a LAN or WAN. The remote nodes are agents or probes. Information gathered by the probes can be sent to a management station that compiles it into a database. RMON—MIB-II standards are currently in place for FDDI, Ethernet, and Token-Ring networks.

Network Management and Monitoring

As stated previously, network management is a collection of activities that are required to plan, design, organize, control, maintain, and expand the communications network. There are many software solutions available to assist with network management. This software can help identify conditions that may lead to problems, prevent network failures, and troubleshoot problems when they occur.

After a baseline for the network has been developed, it will be possible to monitor the network for changes that could indicate potential problems. It is essential to establish what is "normal" in the network so "abnormal" situations can be readily identified. Network monitoring software can gather information on events, system usage statistics, and system performance statistics. Information gathered from these monitors can assist the network administrator in the following ways:

- Monitoring trends in network traffic and utilization
- Developing plans to improve network performance
- Providing forecasting information for growth
- Identifying those network devices that create bottlenecks
- Monitoring events that result from upgrades

Network Management Tools

Network management tools allow for the performance of management functions such as monitoring network traffic levels, monitoring software usage, finding efficiencies, and finding bottlenecks. Table 5–1 shows the top nine network management platforms currently used to support network management.

Table 5–1 Network Management Platforms

HP OpenView	21.3
IBM NetView (mainframe)	15.7
Sun Soltice/SunNet Manager	11.2
Tivoli Enterprise	7.9
IBM NetView (AIX)	6.7
CA Unicenter TNG	5.6
Compuware Ecotools	3.4
BMC Patrol	3.4
Boole & Baggage Command/Post	1.1

Datapro 1999 Network Management Survey.

Additional tools used to conduct network management operations are

- CiscoWorks
- Network General Sniffer
- RMON Probes
- Bay Optivity
- Cabletron Spectrum
- Locally developed software

Network managers need a comprehensive set of tools to help them perform the various network tasks. Most tools can be classified as primarily hardware or software, and most hardware test instruments are supported by software. Network management software is categorized into three different types: device management, system management, and application management.

The most common categories of hardware tools for network management are cable testers, network monitors, and network analyzers. Other sophisticated network management tools can be used for daily network management and controls. These management tools typically have three components:

- Agent—the client software part of the management tool. The agent resides on each managed network device.
- Manager—a centralized software component that manages the network. The management software stores the information collected from the managed devices in a standardized database.
- Administration system—the centralized management component that collects and analyzes the information from the managers. Most administration systems provide information, alerts, traps, and the ability to make programmable modifications to the network components.

The two main management protocols used with network management systems are Simple Network Management Protocol (SNMP) and Common Management Information Protocol (CMIP). SNMP allows management agents that reside on the managed devices to provide information to SNMP management software. SNMP uses a management information base (MIB) for maintaining the statistics and information that SNMP reports and uses. Typically stored measurement items include network errors, system-utilization statistics, packets transmitted/received, and numerous other items of information that may be useful for a particular network component.

Network administrators use SNMP to manage devices such as wiring hubs and routers. Management tasks might include

- Network traffic monitoring
- Remote management capabilities

- Port isolation for testing purposes
- Automatic disconnection of nodes
- Automatic reconfiguration based on time-of-day

Network managers can provide network topology maps, historical management information, traps and alerts, and traffic monitoring throughout the network.

Web-Based Management

The **Web-Based Enterprise Management (WBEM)** consortium is currently developing a series of standards to enable active management and to monitor network-based elements. Over sixty vendors of hardware and systems management software have endorsed this proposal. The key purpose of the WBEM initiative is to consolidate and unify data provided by existing management technologies. At the heart are these two initiatives:

Web-Based Enterprise Management (WBEM)

- The definition of an implementation-independent, extensible, common data description/schema called HyperMedia Management Schema (HMMS), allowing data from a variety of sources to be described and accessed in real-time regardless of the data source. HMMS development is under the auspices of the Desktop Management Task Force (DMTF).
- The definition of a standard protocol called HyperMedia Management Protocol (HMMP), over which these data may be published and accessed, allowing Enterprise management to be platform independent and physically distributed across the Enterprise. HMMP development is under the auspices of the Internet Engineering Task Force (IETF).

Preemptive Precautions

Taking preemptive precautions may be costly in the short term, however they save time and resources when problems arise, prevent equipment problems, and ensure data security. A preemptive approach can prevent additional expense and frustration when trying to identify the causes of failures. The ISO has identified five network management categories that can be used to identify preemptive measures:

- Accounting management—records and reports usage of network resources.
- Configuration management—defines and controls network component configurations and parameters.
- Fault management—detects and isolates network problems.
- Performance management—monitors, analyzes, and controls network data production.
- Security management—monitors and controls access to network resources.

Data Management

Data management is concerned with the distribution of data to users and the protection of that data from theft, unauthorized access, or destruction. Elements covered under the data management umbrella are data backup, archiving, data migration and warehousing, database management systems, and security. The major components of resource security—integrity and safety—apply to all of these elements.

Data Backup and Archiving

A comprehensive backup program can prevent significant data losses. A backup plan is an important part of the network plan and should be revised as the needs of the network grow. Six major action parts of this plan include the following:

- Determine what data should be backed up and how often.
- Develop a schedule for backing up the data.
- Develop a plan for storing data in a secure location.
- Identify the personnel responsible for performing backups.
- Maintain a backup log listing of backed-up data.
- Test the backup system regularly.

Types of backups include the full backup, copies of files, incremental backups, daily copies, and differential (changed) backups. Each of these methods has advantages and disadvantages, so the administrator may elect to use different methods for different data files.

Figure 5–6 shows a data backup and archival system. Mirrored servers provide real-time backup for the current state of all files. All files are transported to the backup server, which can be used to restore the mirrored

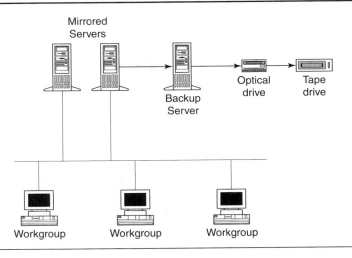

Figure 5–6
Data Backup and Archival System

servers. Files on the backup server are moved to the archival drive on a scheduled basis. These files are then stored on some media for off-site storage. An important part of this system is an inventory of the off-site storage media.

Data Migration and Warehousing

Data migration is an archiving process that moves stagnant files to a secondary storage system, such as optical disks, Zip disks, or streaming tapes. The files may be imaged documents or historical information that will be required at some future date. Migration works in conjunction with strategies, and regular backups are still required. These files remain available off-line, however they are accessible to users over the computer network.

Data Resource Security

The data resource is a very important asset to most organizations. In many cases, data cannot be replaced, once lost. This is a primary reason for backing up the data asset on a regularly scheduled timeframe. If the proper backups are not available and a disaster occurs, the organization could lose a considerable amount of money and/or prestige. In addition to protecting current data, many organizations must store it on a permanent basis for legal reasons. There are a number of measures that can be taken to lessen the chance of data loss. These losses can be caused by intruders, malicious employees, or contractors, or can be caused by nature or accidents.

Protect Against Intruders

Intruders will be described as unauthorized visitors who are not members of the normal user community. These people can cause three problems—theft and destruction of equipment, software, or data; and the viewing of sensitive data and information.

Intruders can use various methods to gain access to a computer or network facility. Access can be physical or electronic, and there are methods to counter both. Intruders can be prevented from accessing a LAN by ensuring that users protect passwords, logoff when not at the workstation, and take security seriously. There are operating system parameters that can be set to limit time-of-day access and device specific logons. Users must be aware of the consequences of leaving a workstation logged on. Most workstations allow for a screensaver login. User accounts can be set to require a scheduled changing of passwords.

Theft Protection

Theft can be the result of espionage by competitors, other governments, internal personal, partners, and contractors. The bottom line is the asset must be protected from theft from any source, because the alternative can be devastating to the organization's operations. This protection must cover the data and the hardware that is used to store that data. Not only is the equipment valuable, but it can also take a long time to acquire new hardware and reload the data. Reloading databases and verifying the integrity of the data can be very time consuming. There are a number of obvious solutions to the theft problem, however the solution should not cost more than the resource is worth, and the value may not be in real dollars.

It is necessary for the administrator to keep track of users. The administrator must know when users leave the company or change roles within the company. Accounts need to be modified to reflect current access levels. Audit trails must be maintained that can identify system disruptions. Theft of storage devices can be detected if they are removed from the network. There are a number of software surveillance tools that can accomplish this task. Although it may not be popular, surveillance cameras are an option at entrance points.

Natural Disaster Protection
There are some natural disasters of such magnitude that a good plan is worthless. The best plan is to allocate sufficient protection based on normal circumstances. It is possible to determine where flood and earthquake zones exist. Although hurricanes and tornadoes cannot be predicted, there are locales that are prone to them. The physical construction of the facility should be a consideration. Computer rooms and network demarcations should be located on second floors. Fire detection and suppression devices should be installed. Electrical failures should be considered and backup systems installed. The bottom line is spend no more than either the asset is worth or the cost to replace it. Last but not least, off-site backup of data is a must.

Database and DBMS Protection
Access to a database or the database management system can be controlled by the use of access rights. Logon restrictions and directory or file access rights are important techniques administrators and supervisors use to protect data against malicious or accidental loss or corruption. Users should never be given more access rights than needed to do their work. Many users only need the right to access and read in a program directory. Giving users excessive or blanket rights to access can lead to such problems as overwrites, corruption, and virus attacks on the database. Although it may seem like extra work, an off-site backup program is required for database restoration in the event of security breaches.

Documentation Procedures
A well-documented network includes everything necessary to review history, understand the current status, plan for growth, and provide comparisons when problems occur. The following list outlines a set of documents that should be included in a network plan:
- Address list
- Cable map

- Contact list
- Equipment list
- Network history
- Network map
- Server configuration
- Software configuration
- Software licenses
- User administration

It is essential to take the time during installation to ensure that all hardware and software components of the network are correctly installed and accurately recorded in the master network record. This documentation also applies to the hardware and software configurations for each of the network components. Being able to quickly find a faulty network component and have documentation on the configuration will save a lot of personnel effort and time.

Documentation should be kept in both hard-copy and electronic form so that it is readily available to anyone who needs it. Complete, accurate, and up-to-date documentation will aid in troubleshooting the network, planning for growth, and training new employees.

Methodology
The first step in solving a problem is to collect and document as much information as possible to have an accurate description of the situation. It is necessary to collect peripheral information that might not appear to have any impact on the situation. This includes environmental conditions that could contribute to the problem, such as heat, humidity, and possible sources of electromagnetic interference (EMI) and radio frequency interference (RFI).

If the problem involves user interaction, it will be necessary to check the sequence of steps the user took. Before troubleshooting the physical devices and system software, attempt to recreate or confirm the problem to ensure that the problem is not user (cockpit) error.

From the information gathered for the trouble report, a structured approach can begin to list, prioritize, and examine possible causes of the reported incident. It is essential that the problem be logically evaluated, starting with the most basic causes. The first step is to check if the network cable and power cable are connected. Now is the time to call a technician.

After checking the obvious, it would be very useful to have a troubleshooting device, such as a Flukemeter, that can be used to check the network cable, the network interface card (NIC), the port (interface) of the device, and the wiring in the installation. If such a device is not available, another option is to troubleshoot by using a replacement. Using a replacement can take longer because of the time necessary to remove and replace the components to see if the problem can be resolved. Since it is necessary to make only one change at a time, this process could become time consuming, especially if a replacement component is not readily available. Both replacement and network management tools can solve problems, but the network management tool will probably allow for a quicker resolution.

Troubleshooting should progress logically. If a single device is malfunctioning, start troubleshooting with that device. If multiple devices are malfunctioning, look for a common cause, such as a centralized wiring hub. If the entire network is malfunctioning, check the network statistics against the baseline. In every case, start with the most basic and work toward the most complex. Once the problem is solved, ensure that documentation is completed and includes a list of the conditions observed, the steps taken to solve the problem, and a summary of the entire service call.

Figure 5–7 illustrates the process flow for troubleshooting and problem solving of some network issue. This model assumes that the organi-

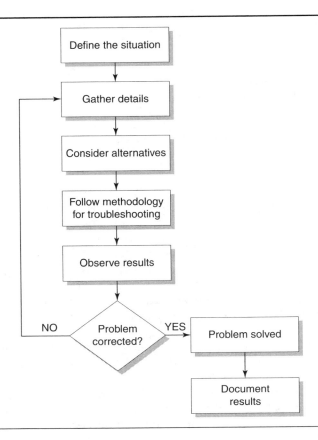

Figure 5–7
Standard Process Flow for Troubleshooting

zation has formally established and documented all elements of the plan as presented in this chapter. This process flow is recursive because the flow repeats until the problem is solved. The following steps detail the activities that occur at each stage of the process flow:

- Definition of the problem—The problem should be defined in terms of a set of symptoms and potential causes. By using the baseline, it is possible to identify the location of the problem or at least know where to start the troubleshooting effort.
- Details of the problem—Information can be collected from sources such as network management systems, protocol analyzer traces, network monitors, and network surveillance personnel.
- Alternatives assessment—Now is the time to consider the possible problems based on the facts that have been gathered. The obvious non-issues can be eliminated at this time.
- Problem-solving methodology—An action plan can be developed and the most likely cause of the problem can be identified. The plan will allow for one variable to be changed at a time until all options are exhausted. Changing one variable at a time allows for the reproduction of a given solution for a specific problem. Also, if more than one change is made before a test is made, it will be impossible to determine which change solved the problem.
- Results observation—Tests must be made at this module to ensure that the modifications are correcting the problem and not creating other problems. This process reiterates the process for each test and may require additional details for each cycle.
- Problem resolution—When the problem has been solved, the next and last step of the process flow is to document the situation. If the problem has not been solved, the process is repeated.
- Documentation—The final step is important because the problem may occur again. It is essential that troubleshooting efforts not be duplicated unnecessarily because of inadequate documentation.

Troubleshooting Documentation

Some type of electronic tracking or journal can be maintained to accumulate troubleshooting and problem-solving information. This ensures that time will not be wasted by repeating work that has already been completed and an audit trail will be developed for each problem. The information developed can also be used in requests for additional equipment, personnel, and training, and will also be useful for training future network support personnel.

A typical method for administering such a database is to assign some unique identifier to each problem or trouble report. A trouble report would be generated for each incident and cross-referenced for recurring incidents to the same network element. The trouble report documents issues and requests for service from network users and can be used to ensure that a consistent troubleshooting methodology is being followed. A typical trouble report might include the following elements:

- A trouble report identifier
- A preliminary description of the situation
- Investigation and analysis of the situation
- The service actions taken to resolve the issue
- A summarization of the incident

Another source of information that can be useful in troubleshooting the network are the data, usually in the form of traps or alerts, which are collected from the managed network devices. It is essential that data collected from the managed devices be maintained in a database so that problems can be tracked and trends can be analyzed.

5.11 NETWORK TROUBLESHOOTING

Many network problems can be solved by first verifying the status of the affected computers or networking components. Taking several initial steps can help the administrator in resolving network problems. These are

- Identify possible cockpit problems and user errors
- Ensure that all physical connections are in place
- Verify that the NIC is working
- Warm-start the device (reload the software)
- Cold-start the device (power cycle off and on)

It is essential that a structured approach be taken when troubleshooting. These simple steps include the following:

- Prioritize the problem in relation to all other problems in the network
- Develop information about the problem
- Identify possible causes
- Eliminate the possibilities, one at a time
- Ensure that the fix does not cause other problems
- Document the solution

The Security Audit

The tasks involved in maintaining a secure network and computer system can be both time-consuming and difficult (Hudson 2000a). A **security audit** should be part of the ongoing operating processes in the Enterprise. There are a number of questions that must be formulated and answered to determine what resources need protection. The following list contains sample questions that can be used in the development process for such an exercise:

- What types of security breaches and attacks have occurred?
- What elements of the organization have been compromised?

- What security tools are operational?
- Who have been the perpetrators?
- Where did the attacks originate?
- Has the monetary or prestige loss been calculated?
- Are all users aware of security policies and procedures?
- Is there a security classification system for files and documents?
- Is security important to the organization?

Appendix B provides a guide for conducting such a security audit. It is essential that all computer and network assets be afforded some level of protection. These assets include the building, equipment rooms, computer and network devices, software, and documentation. Disaster plans for these components are discussed in Chapter 9.

Security Probes

Instead of passively gathering network statistics like auditing tools, security probes actively test various aspects of Enterprise network security, report results, and suggest improvements (Goldman 1998). Such a security probe is known as **Security Analyzer Tool for Analyzing Networks (SATAN)**.

SATAN is able to probe networks for security and weak spots. It was written to analyze UNIX and TCP/IP based systems. Once it has found a way to get inside an Enterprise network, it continues to probe all TCP/IP machines within that network. After all vulnerabilities have been identified, it generates a report that details the results. It also suggests methods for eliminating the vulnerabilities.

Although SATAN was developed as a tool for network managers to detect weaknesses in their own networks, it is widely available on the Internet and can easily be employed by hackers seeking to attack weaknesses in target networks.

Security Analyzer Tool for Analyzing Networks (SATAN)

Point of Presence

The **point of presence (POP)** is the legal boundary for the responsibility of maintaining communications equipment and transmission lines. From a troubleshooting point of view, if a problem has been isolated and it is not on the user's side of the POP, then the carrier or LEC is obligated to provide troubleshooting and repair according to the terms of the maintenance contract. It is, therefore, very important that the user determine that the problem is not part of CPE before calling for help from the communications company. It is the policy of many carriers to charge for repair service when it has been determined that the problem was on the user's side of the POP.

The network administrator can use special troubleshooting tools in addition to the experience that must be gained in the trenches. Since many networking problems occur at the lower layers of the OSI model, it is essential that physical layer tools be employed for problem isolation and identification. These tools include digital voltmeters, time-domain reflectometers (TDRs), cable testers, oscilloscopes, network monitors, and protocol analyzers.

point of presence (POP)

5.12 TEST EQUIPMENT AND RESOURCES

Test equipment is very important in the development of telecommunications equipment and, in particular, the implementation of communications protocols. First, it is necessary to ensure the protocols work correctly, and the medium is capable of carrying the bits without errors. Second, it is important to test the interoperability of different vendor's equipment, because not all implementations of a "standard" will interoperate. Additionally, during the design of one piece of equipment, it is often necessary to simulate the other network devices with which this device will communicate.

There is no substitute for experience and the proper test equipment when troubleshooting or tuning networks. Techniques that work well in one situation may not provide the same results in others. Changes of one parameter may well have an adverse effect on some other parameter. In many cases it is necessary for empirical data to be collected over a long time period.

Analysis and monitoring of the physical media can take many forms. Breakout boxes can be used to ensure that the interface and physical line are operating properly. A BERT or BLERT can be used to provide information on the number of bit errors and the patterns in which they occur. Physical media testers determine whether an appropriate signal quality is being maintained over a given transmission facility. Media simulators, on the other hand, allow real or prototype devices to be tested over simulated transmission lines.

A digital voltmeter or a volt-ohm meter (VOM) can be used to determine a cable break by measuring resistance in the cable. The VOM can also be used to determine if a ground exists between the central core and the shielding in thinnet or thicknet cabling.

A **time domain reflectometer (TDR)** can also be used to determine the existence of a short or break in a cable. A TDR is more sophisticated and can be used to pinpoint the distance from the device where the break is located, and to determine the length of installed cables. TDRs are available for both electrical and fiber-optic cables.

Oscilloscopes contain a video screen that shows information on signal voltages over time. When used with a TDR, an oscilloscope can help to identify cable breaks, shorts, crimps or sharp bends in a cable, and attenuation problems.

Protocol analyzers monitor the traffic being carried on a facility or test implementations of a protocol. Protocol analysis and monitoring is essential to track the performance of a network and to ensure that different vendors' equipment will interoperate. Protocol analyzers can also be used for conformance testing, which is typically done during the product development cycle, prior to connecting a new device onto a live network, to assess the implementation of a protocol.

Because routers and gateways operate at both LAN and WAN layers of the OSI model, network analyzers must have the capability of capturing all of the protocols used, and present an analysis, which can be used to determine the health of the network. Problems that occur with routers and gateways, which can be addressed with network analyzers, include the following:

- Filtering problems
- Protocol mismatches
- Routing path problems
- Routing protocol configurations
- Security problems

These analyzers can look at the contents of a data signal and allow an experienced operator to diagnose and troubleshoot problems. There are basically two levels of analyzers: Low-end devices consist of a circuit board that plugs into a personal computer, and expensive stand-alone devices that have their own chassis and monitor. The major functions of these devices are

- Data trapping
- Device emulation
- Interface lead monitoring
- Performance measurements

Network test equipment has the capability to perform analog tests on both voice and data circuits. These devices can contain multiple function sets and can measure a wide range of network circuit variables and parameters. The following is a list of the principal analog circuit tests that can be performed by these devices:

- Circuit continuity
- Circuit transmission loss
- Capability of the circuit to handle high-speed data signals
- Capability of the circuit to handle data
- Delay of different frequencies propagated over an analog circuit
- Steady-state noise of the circuit
- Frequency response
- Noise burst measurements
- Variation in the phase of a signal on the circuit

Network Monitors, Network Analyzers, and Cable Testers

Network monitors, network analyzers, and cable testers each provide the network manager and technician with a different suite of tools. The capabilities of these devices are listed next.

Network monitors provide the following features:

- Collect network statistics
- Provide diagnostic tools
- Store and opens network packets

Network analyzers provide the following features:

- Collect network statistics
- Use SNMP to query managed devices
- Generate traffic for an analysis
- Analyze network statistics
- Analyze types of devices on the network
- Analyze types of collisions
- Analyze types of errors
- Identify top broadcasting hosts
- Provide diagnostic tools
- Troubleshoot managed network devices

cable testers

Cable testers provide for the following measurements:

- Wire map—provides the basis for checking improperly connected wires, including crossed wires, reversed pairs, and crossed-wire pairs.
- **Attenuation**—the loss of signal power over the distance of the cable. To measure attenuation on a cable, the cable tester must be assisted by a second test unit called a signal injector, or remote.
- Noise—unwanted electrical signals that alter the shape of the transmitted signal on a network. Noise can be produced by fluorescent lights, radios, electronic devices, heaters, and air-conditioning units.
- Near-end crosstalk (NEXT)—a measure of interference from other wire pairs. The signal bleeds from one set of wires to the other, causing crosstalk.
- Distance measure—EIA/TIA–568A specifies maximum cable lengths for network media. Cables that are too long can cause delays in transmission and network errors.

attenuation

EIA/TIA and IEEE specify standards for these measurements, which gives a reference point for acceptable measurements. There is a requirement to perform tests on various types of media including twisted pair, coaxial cable, and optical fiber. The tests previously mentioned can be performed with an instrument such as a Fluke DSP-100 Cable Meter. This handheld tester can perform crosstalk measures and use a TDR function to identify faults on the cable. An optical TDR (OTDR), such as a Tektronix Ranger TFS 3031 OTDR, can be used for troubleshooting optical fiber cable.

5.13 SECURITY REFORM ACT

The Government Information Security Reform Act, also called the **Security Act**, was passed in 2000 as part of the FY2001 Defense Authorization Act. The Security Act focuses on program management, implementation, and evaluation aspects of the security of unclassified and national security systems. It codifies existing OMB security policies and reiterates security responsibilities outlined in the Computer Security Act of 1987 (Ledford 2001).

Security Act

The purpose of the Security Act is to improve program management and evaluation of agency security efforts. The major elements are as follows:

- Provides a comprehensive framework for establishing and ensuring effective controls over Federal information technology (IT) resources.
- Requires interoperability between Federal systems in a cost-effective manner, and provides for the development and maintenance of controls required to protect Federal information systems.
- Provides a mechanism for improved oversight of Federal agency information security programs.

The major requirements of the act are as follows:

- Annual agency program reviews required.
- Annual inspector general of independent party evaluations required.
- Annual OMB report to congress required.
- Agencies must incorporate security into their information systems.
- Security assessments of systems must be performed on systems used by outside contractors.
- Reports by agencies must include implementation plans, budgets, staffing, and training resources for security programs.

In addition to those requirements, the Security Act

- directs agencies to identify, use, and share best practices,
- develops agency-wide information security plans,
- incorporates information security principles and practices through the life cycles of the agency's information systems,
- and ensures that the agency's information security plan is practiced throughout all life cycles of the agency's information systems.

The Security Act pertains to existing systems as well as contractor systems. Once these security principles are established for all information systems, taking steps to ensure the security plan and procedures are practiced by all users will further safeguard security. Analysts advised that personnel are one key to ensuring security practices become effective. Additional reasons information systems are breached include unsafe configurations, failure to fix vulnerabilities, and errors in system settings.

5.14 INTERNET AND WEB RESOURCES

Faulkner	www.faulkner.com
CIO Council	www.cio.gov
GovExec	www.govexec.com
SANS Institute	www.sans.org

■ SUMMARY

Enterprise security management requires many different activities to ensure the organization's computer network is protected. Functions must be performed relating to network administration, planning, assessment, and integrity. It is essential that a baseline be developed as the foundation for security policies and procedures implementations.

Efforts must be directed at identifying network intrusions and other threats and attacks from both internal and external sources. Attacks can originate from hackers and crackers. Types of attacks can include eavesdropping, denial of service, or replays.

Data systems can be protected using techniques such as access control lists and dial-back systems. These can include biometric systems or token-authentication systems.

A network management system can be used to monitor network activities, report breaches, and control the operation and access of the resources. A network management system includes various protocols and software products such as SNMP and CMIP. The major SMFAs include fault, configuration, performance, security, and accounting management. Network management components include an MIB database, an agent in the managed devices, a network management workstation, network management computer and operating system, and the SNMP software.

Data management includes functions necessary to provide integrity and security to the Enterprise's data and database resources. Activities include data backup, archiving, data migration, and warehousing. Security issues include theft protection, natural disaster protection, and protection from intruders.

Formal processes are required to ensure the integrity, privacy, and security of the organization's resources. These include documentation procedures for media, equipment, software, licenses, and network configurations. Network performance and security must be monitored and audited on a continuing basis. There are a number of network management tools that can be employed to assist in problem identification and resolution. These include both hardware and software products.

The Security Reform Act of 2000 was enacted to improve security policies and responsibilities for Federal information systems. There are a considerable number of requirements for adhering to this act.

Key Terms

Access Control Lists (ACL)
Application Program Interface (API)
Attenuation
Baseline
Biometrics
Biometric-based authentication
Cable tester
Common Management Information Protocol (CMIP)
Cracker
Denial of Service
Dial-back
Dictionary attack
Eavesdropping
Global positioning system (GPS)
Hacker
Managed device
Managed object
Management Information Base (MIB)
MIB tree
Network management
Network Management System (NMS)
Network operating system (NOS)
Open Systems Foundation (OSF)
Oscilloscope
Password cracker
Point of presence (POP)
Protocol analyzer
Replay attack
RMON MIB-II
Security Act
Security Analyzer Tool for Analyzing Networks (SATAN)
Security audit
Security management
Simple Network Management Protocol (SNMP)
Smart card
Sniffer
Superuser
Time domain reflectometer (TDR)
Token
Token-based authentication
Trap
Two-factor authentication
Web-Based Enterprise Management (WBEM)

REVIEW EXERCISES

Questions and Analysis

Provide a short answer or in-depth analysis as required.

1. Identify the common duties that could be performed by the network administrator.
2. Discuss the issue of security goals and planning in the Enterprise network.
3. What are the steps in a security requirements' development cycle?
4. What are the issues when addressing network integrity?
5. What are the five most significant issues involving security of a computer network?
6. What are some guidelines for establishing passwords?
7. Why are hardware and software standards required in a network?
8. Hardware and software upgrades should be established for which components?
9. What are examples of security threats that might compromise an organization's resources?

10. What is the difference between a hacker and a cracker?
11. How does a sniffer accomplish eavesdropping?
12. How do dial-back systems work?
13. What are the five major functions of a network control system?
14. Discuss monitoring, reporting, and controlling activities.
15. Describe the SNMP architectural components.
16. What are the five SMFAs of CMIP.
17. Identify organizations that play roles in network management issues.
18. What is the desktop interface?
19. What is the distributed management environment?
20. What are the six services included in the distributed computing environment?
21. Identify APIs that support network management functions.
22. What is Web-based Enterprise management?
23. Why would an organization wish to employ preemptive precautions?
24. What are the five network management categories for preemptive measures?
25. What are the six major action parts of data backup and archiving?
26. How would the Enterprise protect against intruders?
27. Discuss theft protection in the Enterprise.
28. Why is natural disaster protection an issue to the systems administrator?
29. Why do the data and database resources need to be included in a security plan?
30. What are documents that need to be included in a network security plan?
31. Discuss the methodology that can be used to troubleshoot a computer network problem.
32. What is the standard process flow for troubleshooting a computer network issue?
33. Information gathered from monitoring activities can assist the administrator in what ways?
34. Identify network management tools currently available.
35. How can SNMP be used in network management?
36. What elements might be contained in a trouble report?
37. Describe the troubleshooting environment.
38. Describe a security probe. What is SATAN?
39. What is a POP? Why is it important?
40. Identify test equipment that could be deployed in an Enterprise network.
41. What is the function of a protocol analyzer?
42. What are some features of network analyzers?
43. What is a cable tester and why is it useful?
44. Discuss the Security Reform Act of 2000.

Definitions and Terms

Provide the most correct answer from the list of key terms.

1. Software that an application program uses to request and carry out lower-level services performed by the computer's OS.
2. Authentication that is based on a specific personal trait such as a fingerprint.
3. An authentication technique for modem users that disconnects the line and dials the user back at a predetermined telephone number.
4. An entity that contains or receives information, such as a record.
5. A method that seeks to discover passwords by passing a dictionary of terms through a password algorithm.
6. An attack in which a service that is already authorized and completed is forged by another "duplicate request" in an attempt to repeat authorized commands.
7. Security administrators tool for auditing networks.
8. A standard protocol used to manage network devices and computing systems.

9. A credit-card-sized device with an embedded computer chip that can store digital certificates.

10. Authentication that is provided by the possession of a unique physical card or device that supplies one component of the authentication process.

True/False Statements

Indicate whether the statement is true or false.

1. The administration of the Enterprise network does not have to be part of the overall organization's planning process.

2. The first step of security requirements' assessment is to identify the organization's security issues.

3. Data backup is a major part of the organization's efforts in maintaining network integrity.

4. A baseline is used to determine when database backups must occur.

5. Baselines can be created from topology diagrams, response time measurements, and statistical characterizations.

6. Intrusion detection is only effective in monitoring and identifying remote computer activities.

7. A hacker and a cracker have the same motive.

8. A dictionary attack is usually associated with a security breach of user accounts.

9. A sniffer or a wiretap can be used to accomplish eavesdropping.

10. Access control lists are used to determine the time and frequency that users are allowed into a database.

11. A dial-back system does not require a predefined telephone number to originate a call.

12. A biometric-based authentication is called a second-factor authentication.

13. Network management is responsible for monitoring, reporting, and controlling network functions.

14. CMIP is responsible for five SMFAs, including security management.

15. Traps are important components of network management as they are used to track down network problems.

Activities and Research

1. Research and prepare an overview of the techniques that can be employed by the security or network administrator when managing the Enterprise's assets and resources.

2. Invite a representative of a provider of network management systems software to present the company's offerings.

3. Research network management systems and prepare a matrix comparing benefits, functions, and cost of each.

4. Invite a security or network administrator to speak on the subject of Enterprise security management.

5. Use the standard process flow presented in Figure 5-5 to document some computer network problem.

6. Research and prepare an overview and matrix of troubleshooting and testing tools. Include company name, tool name, function, benefit, and cost of each.

7. Research the topic of security regulations that might impact the management and operation of the Enterprise computer network. Develop a paper outlining the findings.

8. Identify current providers of security certification and training programs. Use Appendix F as a starting point. Develop a cost matrix for the various alternatives.

CHAPTER 6

Computer and Network Systems

OBJECTIVES

Material presented in this chapter should enable the reader to

- Become familiar with the hardware components associated with an Enterprise security system.
- Identify the major components of a computer system and its peripherals.
- Look at the functions of firewalls, routers, gateways, and proxy servers.
- Identify the security issues relating to client/server resources.
- Become familiar with the devices that can be used to analyze and monitor networking traffic.
- Recognize the importance of protecting storage-technology devices.
- Learn how to implement a security strategy for hardware-based computer and networking systems.
- Identify the security and integrity risks associated with computer and networking systems.
- Look at the security issues relating to voice communication systems.

■ INTRODUCTION

Computer system and networking components comprise a major portion of an organization's physical assets. These devices, along with their respective operating systems, present an enormous opportunity for security breaches. This chapter describes the major hardware components that are part of an Enterprise's computer and networking operations and offers suggestions to protect these assets from both internal and external threats.

The aim of computer networking security measures and countermeasures is to limit damage to an organization's assets. Enhancement to security efforts includes the deployment of hardware devices such as firewalls, gateways, and routers. Sophisticated software products control these hardware devices.

A number of system vendors such as IBM, Cisco, Sun Microsystems, Symantec, Network Associates, and Hewlett-Packard provide hardware and software solutions oriented toward Enterprise security. A representative sample of these include the following:

- Antivirus protection
- Authentication methods
- Backup and recovery mechanisms
- Firewalls
- Intrusion detection
- Logging, auditing, and reporting
- Vulnerability assessment

The advent of the client/server environment increases the need for security measures to protect these Enterprise resources. These devices provide the processor and storage capabilities for a number of network systems such as e-commerce and distributed computing. Securing mechanized transactions has become a number-one priority.

Voice communications systems play a major role in the Enterprise network. These systems provide access to a number of sensitive resources that must be protected from both internal and external attacks and intrusions.

6.1 COMPUTER SYSTEMS

computer system

A **computer system** consists of a processor, main memory, input/output devices, and the interconnecting devices among these major components. In addition to the hardware components, there is the software component that includes an operating system, utility programs, and numerous application programs. The material presented in this chapter is oriented toward two categories of computers—mainframes and client/servers. The issues relating to security and integrity are relevant for both small client/server and very large mainframe systems.

Mainframe Computer System

mainframe

A **mainframe** is a large, powerful computer, almost always linked to a configuration of peripheral devices such as disk controllers and high-speed printers (Figure 6–1). It is used in a multipurpose environment at the corporate or headquarters level. It contains megabytes of memory and gigabytes of disk storage, and processes data in the millions of instructions per second (MIPS).

Client/Server System

client/server

A **client/server** is a computer located on a local area network that splits the workload between itself and desktop computers. In most cases the "client" is a desktop computing device or program "served" by another networked computing device. Computers are integrated over the network by an application, which provides a single system image. The server can be a minicomputer, workstation, or a larger computer system. Multiple servers can serve a single client (Figure 6–2).

Organizations may institute security policies that constrain how clients and servers communicate, such as partitioning computers into secure and nonsecure subnets. To prevent client/server programs from com-

Figure 6–1
Mainframe Computer System

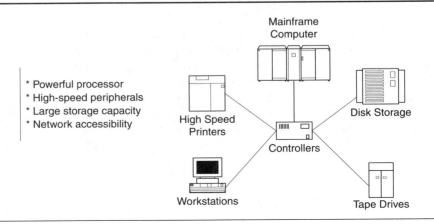

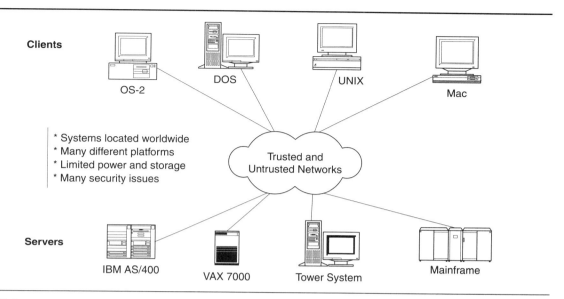

Figure 6–2
Client/Server Systems

promising security, the administrator places policy constraints on connectivity, which allows computer-to-computer communication, while restricting access to the nonsecure portion of the network. Although such policies might increase security, they can make it more difficult to design and program in the client/server environment (Comer 2001).

6.2 MAINFRAME COMPONENTS

Data processing equipment includes communication controllers, disk drives, tape drives, printers, and front-end processors. These devices are normally part of a mainframe computer system. Operating systems include the software that controls the operation of the computer hardware components.

Operating Systems
Operating systems (OS) are software programs that manage the basic operations of a computer system. Examples are OS/2, MS-DOS, and UNIX. An operating system can be thought of as a resource manager or resource allocator. The following resources are usually managed and allocated among competing applications:

- CPU processing time
- Memory access disk storage

operating systems (OS)

- Input/output devices
- Retrieval and file system
- Security

Controllers

controller

A **controller** is a device that controls the operation of another piece of equipment. Specifically, in data communications, a controller resides between a host and terminals and relays information between them. Controllers can be housed in the host, be stand-alone, or reside on a file server. These devices possess an extremely fast transfer rate. Interfaces and interface cards that provide OSI Layer 1, physical connectivity reside in these devices.

Disk Technology

disk storage

Servers require enormous amounts of **disk storage**. Web servers store thousands of Web pages and associated image, sound, and video files. Storage devices can include small computer system interface (SCSI) disk drives, Fibre Channel, and redundant array of inexpensive disk (RAID) drives. High-end PC servers use SCSI drives, which are somewhat faster but more expensive than the standard enhanced integrated drive electronic (EIDE) disk drive. For extremely high-speed disk access, high-end servers are moving toward Fibre Channel disk access. Whereas SCSI hosts must be within about 12 meters of their disk drives, Fibre Channel allows disk drives to be hundreds of meters, and even kilometers, away from their hosts. RAID uses multiple disk drives controlled by a single controller board.

Redundant Array of Inexpensive Disks

redundant array of inexpensive disk (RAID)

Redundant array of inexpensive disk (RAID) devices are used to improve performance and automatically recover from a system failure. They can maintain backup files and protect data by continuously writing two copies of the data to a different hard drive. There are several levels of RAID available, including RAID 0, RAID 1, and RAID 5 that provide for **disk striping**, **disk mirroring**, and **duplexing** (Andrews 2001).

disk striping
disk mirroring
duplexing

RAID 0 is one method of disk striping in which more than one hard drive is treated as a single volume. Some data are written to multiple drives, however they are treated as one logical drive. Windows NT and Windows 2000 support RAID 0.

RAID 1 is designed to protect data from a hard drive failure by writing data twice—once to each of two drives. This type of disk mirroring allows two hard drives to use the same adapter card. Also supported with RAID 1 is duplexing, in which each hard drive has its own adapter card.

RAID 5, which is disk striping with parity, is a combination of RAID 0 and RAID 1. It improves fault tolerance and drive capacity.

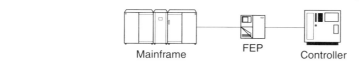

Figure 6-3
Front-End Processor Connections

Front-End Processor

The **Front-End Processor (FEP)** acts as a traffic cop of the mainframe data communications world. It typically resides in front of a mainframe computer and is designed to handle the telecommunications burden so the mainframe computer can efficiently process its programmed functions. Figure 6-3 shows the connectivity of an FEP between a mainframe processor and a communications controller. Cable distance is a major concern in this connectivity. Smaller systems use input/output processor (IOP) modules to accomplish this same task.

Front-End Processor (FEP)

Fibre Channel Management Systems

Many companies are returning to the concepts of centralized management of data storage, even within distributed IT architectures. Storage management requires a global perspective. An IT manager must look at the whole enterprise as well as the parts that make up the whole. **Fibre Channel** supports both needs. Fibre Channel networks provide an integrated storage system with the performance and efficiency of a distributed architecture and the reliability and manageability of centralized data storage. Fibre Channel devices provide information to manage the storage systems individually.

Fibre Channel

Storage must be viewed as a system, delivering services and protecting data assets. Proper management of this system provides highly available data access, improved performance, complete data security, and storage growth at a reasonable cost. A storage system consists of online storage, near-line storage, archival storage, and backup storage. Storage management software moves data among these elements as required to meet the Enterprise's storage management strategy. Fibre Channel removes barriers associated with implementation of this strategy.

Fibre Channel devices use **Simple Network Management Protocol (SNMP)** and **SCSI Enclosure Standard (SES)** for management. The Fibre Channel standard supports SNMP over IP or directly using the Fibre Channel native protocol. Fibre Channel manufacturers normally provide a point solution for SNMP that can be integrated into an Enterprise management system. Fibre Channel is designed to be self-managing, and most of this management activity is status monitoring.

Simple Network Management Protocol (SNMP)

SCSI Enclosure Standard (SES)

SES is similar to SNMP but is designed to obtain information from storage devices that only use SCSI. Typical information from a storage device includes the following:

- Manufacturer's name
- Status on what slots are occupied

- Cooling information
- Power supply status

Centralized storage simplifies management. Fibre Channel provides the ability to view storage as if it were centralized, even if it is physically distributed. Fibre Channel supports critical business applications with performance, reliability, fast data access and transfer, and managed storage and storage networks.

Storage Management

Storage is a service that users want available at all times as an unlimited resource. It is more than a collection of disks and tapes. It is an interconnected system of storage devices, software, and servers receiving and delivering data. The role of storage management is to maximize performance, availability, and security at a reasonable cost.

Fibre Channel brings new levels of performance for reliability, connectivity, security, and availability that enhance an IT organization's ability to manage storage. This involves the following techniques:

- Automated backup and archiving is made faster and safer with gigabit Fibre Channel networks that span long distances.
- On-line diagnostics can be run to determine the health of a storage system using Fibre Channel.
- Enterprise data management provides backup and restoration across a heterogeneous environment using a Fibre Channel network.
- Fibre Channel provides improved reliability over SCSI-connected storage.
- Fault-tolerant Fibre Channel architectures enable highly available storage systems.
- Hierarchical storage management can be implemented across the Enterprise with Fibre Channel networks.
- Media and library management software can report results of media monitoring to a central location over the Fibre Channel network.
- System management is enabled using SNMP or SES.
- Storage networks link existing SCSI-based storage and new Fibre Channel storage with servers and workstations for data storage and retrieval.

Fibre Channel storage and networks bring new capabilities to network storage. Multiple servers and workstations directly share storage devices without an intermediate device such as an NFS server. Thus, direct storage access benefits expand beyond the mainframe to the entire enterprise.

The benefits of central management and distributed computing are made possible with Fibre Channel. Storage management and network management work together using gigabit Fibre Channel solutions for increased levels of performance, security, and availability. Following are examples of Fibre Channel benefits:

- Management of Fibre Channel systems melds storage and network management, bringing new capabilities to IT systems.
- Fibre Channel combines the speed and reliability of a channel with the flexibility and distance of a network.
- Fibre Channel storage devices connect with servers and workstations more like network devices.
- The protocol used to link with these storage devices is SCSI. Fibre Channel is protocol independent and runs SCSI as easily as IP or digital audio/video. This brings new levels of management capability to IT managers. They can effectively use SCSI or IP to implement functions like automated backup on physically distributed devices as if they were centrally located.

6.3 CLIENT/SERVER COMPONENTS

A **server** has a hardware and software component. It is a shared computer on the LAN and a program that provides some service to the other client programs. The server is a component of the client/server environment and connects to a hubbing device. Servers are gatekeepers to systems and information files. The following server systems make the client/server environment possible:

server

- File server
- Print server
- Database server
- Communications server
- Application server

Server Issues

To effectively secure servers, planning and coordination are required. It is necessary to determine possible vulnerabilities in terms of operating systems, communication software, and applications before implementing a security program. It is also necessary to determine different security levels that assign specific access levels on a "need to know" basis. This means that one level of security might be "read only," with other levels ascribed to mission critical and proprietary data. Management must determine who can read what information, who has authorization to modify data, and how the levels relate.

There are several areas of concern when developing a security scheme, most of which revolve around software designed to make more efficient use of the computing environment that can compromise the security at the same time. These include automatic directory listings, symbolic link following, server side includes, and user maintained directories.

The more a remote hacker can learn about a system, the easier it is to compromise. Automatic directory listings provided on a number of vendor's servers are convenient, however they allow potential access to sensitive information. Information could include CGI scripts, symbolic links, and source-code control logs. As a precaution, users should remove unwanted files from the document root directory.

Some servers allow the extension of the document tree with symbolic links. This is a beneficial feature, however it can lead to security breaches when a user accidentally creates a link to a sensitive area of the system. An explicit entry in the server's configuration file would be a more appropriate method to accomplish the same thing. Some servers allow for turning off the symbolic link following feature. It is also possible to enable symbolic link following only if the owner of the link matches the link's target.

The "exec" forms of server side includes (additions) are a major security hole. Their use should be restricted to trusted users or they should be turned off completely. Allowing any user on the host system to add documents to a Web site is considered dangerous. Publishing files that contain sensitive system information, CGI scripts, server side includes, or symbolic links that can compromise security must not be allowed. When creating a home page, it is best to restrict usage to only a portion of the document root. Whether home pages are located in user's home directories or in a piece of the document root, it is prudent to disallow server side includes and CGI scripts in this area.

Processing the public key operations required by the Secure Sockets Layer (SSL) can slow down the performance of a server under heavy traffic conditions. Hardware to speed security processing allows servers to perform such processing very rapidly so commercial transactions are not delayed. There are a number of devices on the market that improve the overall throughput of the enterprise servers and also enhances their security features.

Server Security
Maintaining secure servers is an absolute necessity, since they are the gatekeepers to systems and databases. Computer security involves more than setting up a program of firewalls and passwords—it is an on-going, organization-wide awareness that someone could break into the network and steal information and other resources. Failing to implement protocols and procedures mitigates the effectiveness of the best-designed security program.

Remember that severs consist of a hardware and a software component. The hardware selected will determine the level of performance provided, whereas the software can determine the level of security of a site.

Two approaches to server security include purchasing secure server software packages or developing a variety of packages in-house. Whichever solution is selected, certain elements should be included in this security system. These elements include password protection, encryption, firewalls, and SSL technology. All of these elements offer some degree of security, however, and when combined, they can produce a formidable security scheme.

Server Security Methods

There are a number of ways to secure a server, given the objectives and requirements (Faulkner 2001). The most common security method is access control, which includes user passwords, card or key systems, biometrics, and data encryption. Complex security mechanisms usually increase the inconvenience level for authorized users, which could result in employee resistance. The objective is to develop a protection strategy that does not cause employee complacency or circumvention.

A wide range of access control mechanisms has evolved over time, and many are related to an organization's internal policies. Many of these techniques use authentication mechanisms to establish the identity of a user. An alternative mechanism employs security labels to provide input to rules-based access control policy decisions. Rules-based policies could rely on other inputs, such as time and date, when making access control decisions. In addition to controlling access to databases through a secure server, administrators must evaluate all shared resources that are accessible through the network so that their respective access restrictions can be implemented.

Password protection, discussed in Chapter 2, is the most common means of securing server access. Passwords provide a minimal level of security without requiring special devices such as keys or cards. The downside in implementing only password protection is user ambivalence. If the password method is implemented, several precautions should be mandated:

- Passwords should not contain easily obtainable information.
- They should be dynamic, with mandatory periodic changes.
- A multilevel password protection scheme should be implemented if hierarchical security is an issue.
- A master password file should be maintained and surveyed periodically.

Magnetic card systems allow users to access data from any available workstation with an attached card reader. The card key system allows selective access for each user as opposed to each workstation. Smart cards offer a wide range of security considerations, including eliminating manual passwords. They are viable for both local and remote access control.

Biometrics is identification based on the uniqueness of an individual's physical attributes and traits, which include fingerprints, handprints, voice quality, and retinal patterns. The biometrics security system is presented in Chapter 4. The biometric method requires a considerable amount of computer processing, which makes it slower than some other mechanisms. Some governments are currently using it to protect national security servers.

As discussed in Chapter 2, data encryption is a mathematical method of scrambling data to disguise its original content. This is the most popular security mechanism for protecting sensitive information during transmission between computers. Data confidentiality is achieved by encrypting the message at the sender and decrypting at the receiver. Since attackers, including those using network monitors (sniffers), cannot read encrypted data, it has a high probability of remaining secure.

An additional security technique called inactivity timeout can be used for servers that perform remote access. Servers can be configured to automatically terminate calls when there has been no activity for a specified time interval. All of these different approaches in combination provide a high level of security at the server level.

6.4 SECURING MECHANIZED TRANSACTIONS

A number of technology companies have developed protocols for transacting business in a secure environment. This is in addition to securing the operating system (OS) and preventing unauthorized access to the server. Some types of communications require protection between the Web browser and the secure server. The most common mechanisms, which are discussed in Chapter 4, include the following:

- Secure Socket Layer (SSL) provides for server authentication, data encryption, and data integrity.
- Secure HyperText Transfer Protocol (S-HTTP) allows servers to encrypt responses to browsers, digitally sign replies to browsers, and authenticate the identity of browsers.
- Private Communication Technology (PCT) provides privacy between a client and server and authenticates the client and server.
- Generic Security Service Application Program Interface (GSSAPI) provides for mutual authentication and data encryption capabilities in Web browsers and servers.

There is an option that allows for the combining of HTTP and S-HTTP, which offers the freedom, flexibility, and efficiency of HTTP, while using S-HTTP to protect sensitive parts of a transaction. This is possible because S-HTTP and HTTP are different protocols that use different ports. This allows merchants to offer catalog information to anyone, while providing protection to the server and client during order entry.

Another issue concerns the use of a firewall component between the Internet and the server. Firewalls, discussed in the next section, are used for creating an internal site or private network that is isolated from the general population of Internet users. Using a server for commercial purposes exposes it to the entire untrusted Internet outside the firewall's perimeter of protection. Therefore, placing a commercial server within a protected network is self-defeating unless potential clients are granted accessibility on an individual basis.

6.5 NETWORK COMPONENTS

A LAN in the organization is used to connect a number of devices including PCs and printers. The operating systems on the PCs are user friendly, however they are not designed for security functions. Enterprise network components, which are security candidates, consist of firewalls, routers, servers, gateways, and workstations. These devices are connected via some media to form the physical network.

Firewalls

A common method of ensuring a network infrastructure is with a firewall, which is intimately linked with Internet security. The objective of a firewall is to protect trusted networks from untrusted entities. A **firewall** is a combination of hardware and software, which limits the exposure of a computer or group of computers to an attack from an outside source. In its most simplistic sense, it controls the flow of incoming and outgoing network traffic. It also ensures that information about the trusted network, such as an address list, is not disclosed to an untrusted network. A firewall can manifest itself as a single device or a number of different components, such as routers and computers. Typical tasks of a firewall are as follows (Gollman 2000):

firewall

- Authentication based on the source of network traffic
- Access control based on the service requested
- Access control based on sender or receiver addresses
- Logging of Internet activities
- Virus checking on incoming files
- Hiding of internal network topology and addresses

A firewall scans each packet of data to determine if it is to be rejected or forwarded to the corporate resource. It is basically a hardware or software barrier that provides protection against unauthorized entry. There are basically three types of firewalls—screen host, screen subnet, and dual-homed gateway (Miller 2000).

A screened host uses a screening router that directs all Internet traffic to a bastion host, which is located between the local network and the Internet. The bastion host's function is to decide which access to block and which to allow. A screened subnet firewall uses a network of screened hosts. On the secured side of each bastion host is a router connecting one or more networks to hosts. The screened subnet firewall operates like the screened host for larger networks. An application gateway inspects data before allowing connection in the first place.

The dual-homed gateway is an applications layer firewall that checks all traffic using a bastion host between the network and the Internet. The firewall blocks applications from use rather than individual accesses within an established application. A circuit-level gateway authenticates TCP and user datagram sessions when setting up sessions between two hosts. This method maintains tables of authenticated users and terminates sessions when finished.

The firewall architecture usually involves the use of a pair of routers and two computer systems. Router one, called the external router, connects to the Internet using a WAN interface. It is called a dual-homed bastion host, and is used to filter and process incoming network traffic. Router two controls the connection to the internal network. It is used to perform Internet services such as electronic mail, network time protocol (NTP), and domain name service (DNS). The external router provides the initial connection to the Internet, and is used to filter all unauthorized traffic between the bastion hosts and the Internet. The internal router limits both the nature and origin of traffic from inside the internal network. Both routers limit traffic, based on the source and destination address of the traffic, and the type of traffic, based on the TCP/IP source and destination ports.

A scheme called the three-part firewall system is often used to provide security for the LAN resources. This arrangement consists of three parts or layers, that include an outside filter, the isolation LAN, and the inside filter. The outside filter provides basic packet filtering, which allows particular traffic types to pass into the isolation LAN. The isolation LAN is sometimes called the **demilitarized zone** or **DMZ**. This buffer zone is the location of systems that are accessible by external or untrusted users. The systems in the DMZ, called bastion hosts, are secure hosts accessible by external users. Lastly, the inside filter provides the strongest security, controlling access from the isolation LAN to the internal network (CCDA Cisco Certified Design Associate 2000).

One technique to enable security on a firewall/router is to create rules, called **access lists**, which permit or deny various types of traffic across the network. These permission or denials of traffic can include specific network services. Three functions provided by access lists include the following (Teare 1999):

- Control the amount of traffic on the networks to improve performance
- Control whether network traffic is forwarded or blocked at a router's interface
- Provide a basic level of security

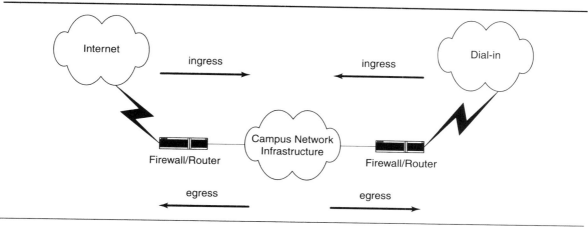

Figure 6–4
Firewall/router Deployment

Typically, firewalls are deployed at critical ingress and egress points of the network infrastructure. Figure 6–4 shows a typical firewall/router deployment.

A primary function of a firewall is to filter traffic, which is the process of selecting the traffic that will be allowed into a certain portion of a network, such as a LAN, MAN, or WAN. **Filters** can be set to prevent internetworking of several types of messages. They may be set to block all packets originating from a specific location (address), or can be set to exclude traffic from specific protocols.

Three classifications of firewalls encompass different filtering characteristics. These include **circuit filtering**, **application gateways**, and **packet filtering**.

filters

circuit filtering
application gateways
packet filtering

Circuit Filtering

These firewalls control access by keeping state information and reconstructing the flow of data associated with the traffic. A circuit filter will not pass a packet from one side of the device to another unless it is part of an established connection.

This filter is a type of proxy service, usually residing on a bastion host, which acts on requests from clients, and if valid, forwards the requests to appropriate servers on the Internet. Special software, like SOCKS (discussed in Chapter 7), must be installed and configured on the client. Circuit filtering may apply controls on both the sources and destination of a request, but do not examine the nature or content of a request. A user's request to use FTP can be controlled, however the files transmitted using FTP are not examined. A user could be transferring unauthorized material, but a circuit filter has no method of controlling this activity. To apply intelligent controls, based on the content of the communication, requires the use of an application gateway proxy.

Application Gateways

These firewalls process messages specific to particular Internet Protocol (IP) applications. They are tailored to specific protocols and cannot easily protect traffic using newer protocols. An application gateway proxy has the ability to control connections and the nature of the connection. It can pass a user's request to an application service, where decision criteria can be applied by the application to the request.

One use of this type of proxy involves control of electronic mail. E-mail messages can contain Trojan horse programs, viruses, and other unwanted material. An application gateway proxy can intercept all mail messages and examine the contents for illicit material. They also permit a finer granularity in the use of audit trails. FTP connections, if implemented through an application gateway, could log every keystroke of a user's connection.

Packet Filtering

These firewalls rely solely on the Transmission Control Protocol (TCP), User Datagram Protocol (UDP), Internet Control Message Protocol (ICMP), and IP headers of individual packets to permit or deny traffic. The packet filter looks at a combination of traffic direction (inbound or outbound), IP source and destination address, and TCP or UDP source and destination port numbers.

Before a network administrator can decide on the best method to implement, it is necessary to determine which traffic flow control fits the environment. Traffic control is based on a combination of the following characteristics (Kaeo 1999):

- Authentication
- Application content
- Direction of traffic
- IP address
- Port numbers
- Traffic origin

At some ingress points to trusted networks it might be prudent to authenticate users before they can access particular services, such as File Transfer Protocol (FTP), HyperText Transfer Protocol (HTTP), or Telnet. Available authentication mechanisms vary, but all aid in controlling network use and auditing who is accessing which services.

It can be useful to look at applications and determine certain controls. It might be necessary to filter certain Uniform Resource Locators (URLs) or specific content types, such as Java applets. Network traffic can be filtered in either the inbound or outbound direction. Inbound traffic usually originates from an outside, untrusted source, whereas outbound traffic originates from inside the trusted network.

The source or destination address can be used to filter certain network traffic. This approach is useful for implementing precursory controls to help avoid spoofing attacks. **Spoofing** is an attempt to gain access to an automated information system by posing as an authorized user. In electronic mail applications, spoofing occurs when one person impersonates another in order to gain access to that person's e-mail.

spoofing

TCP and UDP source and destination port numbers are used to recognize and filter different types of services. It is necessary to determine which services are supported in the network. A computer or network typically runs a number of different processes. In the TCP/IP network environment, these services are available at ports. The most common services are available at predefined port numbers that serve as unique identifiers for those services. As an example, an HTTP Web server uses port 80.

Whether network traffic originated from an inside, trusted network or an outside, untrusted one, can be a factor in managing traffic flow. It may be necessary to allow certain UDP packets to originate from inside the trusted network, but not allow Domain Name Server (DNS) requests to arrive from the outside, untrusted network. The administrator may want to restrict TCP traffic to outside untrusted networks, if the TCP session was initiated from the inside trusted network. Each organization will have a different set of needs.

There are a number of limitations to using packet filtering as a firewall technique. These include the following issues (Dempsey 1997):

- A router can limit traffic based on header information, however it cannot apply any intelligence to the content of the packet information in the routing decision.
- Router access control lists can be cryptic in their configuration, leading to mistakes that could allow unauthorized traffic or no traffic at all.
- Routers can be configured to contain information about the internal network, which can be a point of compromise.
- Many TCP/IP-based services open communication through a known port, however designate the use of higher numbered ports for later communication. These port numbers are not always predictable.

There are additional issues that must be considered when implementing firewalls in the Enterprise. It is important to recognize that firewalls and all of their security provisions cannot protect against every type of security breach. Those breaches, such as viruses brought into the organization on floppy disks, cannot be blocked by the firewall. Other internal sources of attack by employees, maintenance personnel, partners, or contractors would bypass the firewall capabilities. It is, therefore, essential that a firewall be only one part of the total security umbrella.

Bastion Host

screened subnet

bastion hosts

A perimeter defense system, called a **screened subnet**, can be created between the internal network and the Internet. This special sub-network allows a special class of computers called **bastion hosts**. Traffic between these hosts, and the internal and external networks, can be controlled by routers. They can limit the nature of network traffic and force both incoming and outgoing traffic through special computer systems of the screen sub-network of the bastion hosts. Figure 6–5 shows a typical bastion host environment. Bastion hosts are used to collect valid incoming traffic and forward it to appropriate locations on the internal network. They also act as single connection points for internal traffic destined for the Internet and may also be able to accommodate the need for some TCP/IP services to communicate on unknown port numbers by selectively opening the required port for a period of time. Internal users can access applications or TCP/IP network services on a proxy basis, which allows for additional controls (Dempsey 1997).

Separating the packet filtering from other tasks performed by a firewall results in a less complex router and provides better performance. This allows hardware optimization for the routing task and a higher degree of assurance for its security. A proxy server may offer more relevant features.

Proxy Servers

proxy

proxy server

A **proxy** is the authority to act for another. Proxy service providers are transparent to the users, who are usually unaware that an intermediary is acting on their behalf, when accessing network services. A **proxy server**, like a firewall, is designed to protect internal network resources that are

Figure 6–5
Bastion Host Environment

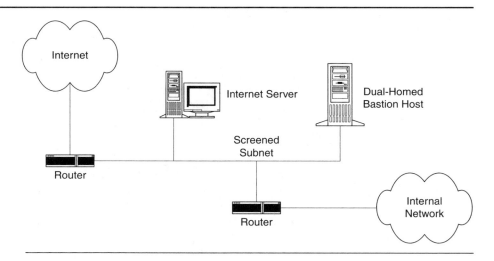

connected to other networks, such as the Internet. The difference between proxy servers and firewalls is often confusing. Proxies may be services that run on a firewall, where the firewall is a physical device, that sits between the internal network and the Internet. **Proxy services** run on the firewall at the application level (OSI Layer 7) to provide a sophisticated traffic control system. A firewall runs proxy services for each different type of Internet application, such as HTTP and FTP, which needs to be controlled. The physical server device running the proxy service has two Network Interface Cards (NICs), one connected to the internal network, and the other connected to the Internet. This is called a dual-homed or multi-homed system.

proxy services

A dual-homed proxy server does not perform routing functions between the NICs. Packets are only transmitted between networks after first passing up the protocol stack and across through the proxy services, but only if the services allow it. Figure 6–6 shows how the proxy server relates to the OSI Model protocol stack. If a proxy service does not exist for a particular application, no packets related to the application are allowed to pass through the proxy server. A system on the internal network must communicate with an external Internet system, and vice versa, through

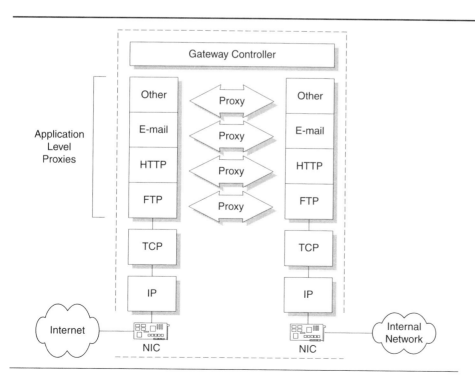

Figure 6–6
A Firewall/Proxy Server

the dual-homed host. No direct communication can occur and all communication may be monitored. When properly implemented, this prevents external hackers accessing internal systems. While there are advantages, a downside exists to using a dual-homed host—it is difficult to set up and administer.

Proxy services are usually one-way services that block Internet users from any access to the internal network. The services are designed for internal users only, and only packets that are in response to an internal user's requests are allowed back through the firewall. A proxy differs from a standard packet filter function since a packet filter only determines whether a packet can be forwarded based on its IP address or TCP port number.

For security reasons, incoming response packets may be inspected for viruses or possible alteration by an external hacker. The proxy server may also set up a secure encrypted session with the Web server to ensure that eavesdroppers cannot look at the packets and glean useful information.

Proxy servers also provide caching functions, where information is stored in memory in anticipation of the next request for information. Because they provide a centralized location where internal users access the Internet, the proxy server can cache frequently accessed documents from sites on the Internet and make them quickly available for other internal users. This can also save overall disk space and provide more efficient distribution of frequently requested documents.

Since a proxy server handles all packets for internal users, it is relatively easy to perform such activities as access control, content filtering, and virus scanning. Packets that contain undesirable content or words can be discarded.

Most proxy servers provide proxies for applications like FTP, HTTP, Telnet, and other Internet protocols. Some proxy servers use SOCKS, an authentication protocol, and require that clients have software that can negotiate with SOCKS. SOCKS will be discussed in greater detail in Chapter 7.

There are both advantages and disadvantages to using proxy servers for security. These are summarized as follows:

Advantages
- Users do not need to login or have accounts on a bastion host, which allows the bastion host to be kept simple and lean.
- The use of a proxy server allows audit trails of user activity. Detection of inappropriate access by internal users is possible.
- Proxies can hide internal IP addresses.
- Logging can include client IP address, date and time, URL, byte count, and HTTP transactions.

Disadvantages
- Most common Internet services have proxy services, however a particular service desired may not be available.
- Support for a particular protocol such as SNMP may not be available.

Routers and Gateways
Two devices deployed in today's Enterprise networks include routers and gateways. **Routers** are intelligent devices that connect like and unlike LANs, MANs, and WANs. They support such broadband technologies as X.25, Frame Relay, and Asynchronous Transfer Mode (ATM). Routers are protocol-sensitive, typically supporting multiple protocols. They commonly operate at the bottom three layers of the OSI model, and move network traffic based on a high level of internal intelligence. They can route traffic based on destination address, minimum route delay and distance, packet-priority level, route-congestion level, and community of interest. Because routers have the ability to consider an Enterprise network as comprising multiple physical subnetworks, they are able to confine data traffic within a particular subnet. This process is based on a policy-based routing table, which includes user privileges (Newton 1999).

routers

A router-based firewall is really a packet-filtering router. Because a router is a computer through which only authorized incoming and outgoing packets must pass, it is often considered a firewall. The router supports network logon as its base level of security for the network. For an additional level of security, they support access lists to protect the servers from unauthorized users, and route authentication to prevent other devices from acting as default gateways. A **default router** is the address of a router on the local network segment that a workstation uses to reach remote services, sometimes called a **default gateway**.

default router

default gateway

Since the router uses login on the network infrastructure, only designated individuals can log into them to change a configuration. Encryption is available on the router to provide data integrity on the network. Encryption can take place at the router and between the workstation and server. To provide the highest data integrity for the network infrastructure, encryption should be resident on all the routers in the network.

Servers and Workstations
Note that encryption of a router provides only partial data integrity on the campus network. When data is sent from a workstation to a server, or vice versa, the traffic often crosses two routers. With encryption at the router, the traffic is only encrypted when traversing the backbone. The backbone is the network (OSI Layer 3) domain responsible for transferring cross-campus traffic as quickly as possible without any processor-intensive operations.

If the data is accessed by another user in the LAN before it reaches the router, the data is readable. The LAN is the network domain that consists of the switching and routing devices required to connect users to the network. Even more critical, unauthorized users in the server farm can read the data after it passes through the router. Because most network traffic goes to the Enterprise servers, the server farm is a likely spot for a security breach and information theft. This situation is unacceptable for highly sensitive data. The server farm is the network domain that consists of the Enterprise servers in the network, and the switching and routing devices required for connection to the rest of the network.

Figure 6–7 depicts the relationship between the backbone, LAN, and server farms (Quin-Andry 1998). If the network carries extremely sensitive data, workstation and server encryption should be implemented along with router encryption. The level of encryption used at the workstation should be based on the level of confidentiality required for the message to be sent across the network. If the message requires complete confidentiality, then hash functions and asymmetric and symmetric encryption can all be used for data integrity. Note that encryption slows down the performance of workstations and some network devices.

gateway

A **gateway** is an entrance and exit into a communications network. In data networks, a gateway is typically a node on a network that communicates with an otherwise incompatible network, which often performs code and protocol conversion processes. They connect compatible networks, owned by different entities, such as X.25 networks linked by X.75 gateways. Gateways can eliminate duplicating wiring by giving all users on the network access to the mainframe without each having a direct,

Figure 6–7
Backbone, LAN, and Server Farm Infrastructure

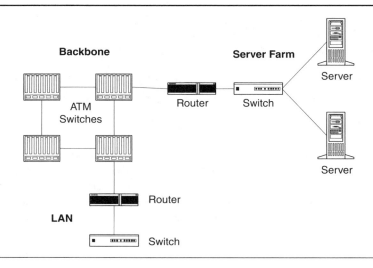

hard-wired connection. Gateways are commonly used to connect users on one network, like Ethernet or Token Ring, with those on a WAN.

According to the OSI model, a gateway is a device that provides mapping at all seven layers of the model. A gateway may be used to interface two incompatible e-mail systems or for transferring data files from one system to another. E-mail systems that sit on LANs often have gateways into larger e-mail systems, like the Internet (Newton 1999).

6.6 VPN HARDWARE AND SOFTWARE

After the user has a connection to the Internet or the public access network, the important network devices for a Virtual Private Network (VPN) are the ones that control access to the protected LAN from remote and external sources. The external source might be another of the corporate LANs, a mobile worker with a laptop, or a corporate partner. VPN access devices should be able to handle all of these situations; however, all are not equally adept at handling the different connectivity situations.

VPN hardware and software can be placed at various locations in the network. These include security gateways, policy servers, and certificate authority holders for preventing unauthorized intrusions. This section looks at firewalls and routers, which are implemented in the VPN to provide access control.

Firewalls and Routers

Firewalls have long been used to protect LANs from other parts of an IP internetwork by controlling access to resources on the basis of packet type, application type, and IP address. Figure 6–8 shows the locations of routers

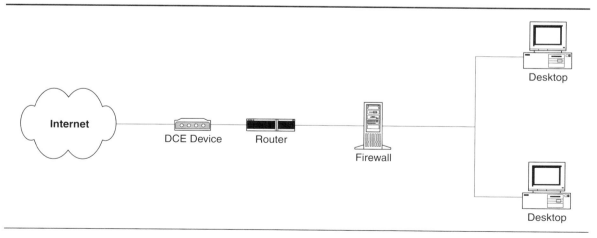

Figure 6–8
Firewall and Router Locations

and firewalls in the VPN. These devices can be placed anywhere between the DCE device and the LAN.

A firewall is a device acting as a network filter to restrict access to a private network from the outside, implementing access controls based on the contents of the packages of data that are transmitted between two parties or devices on the network. There are three main classes of firewalls: packet filters, application and circuit gateways (proxies), and stateful multi-layer inspection firewalls.

Packet Filtering Firewalls

Packet filtering firewalls were the first generation of firewalls. Packet filters track the source and destination address of IP packets, permitting packets to pass through the firewall based on rules set by the network manager. Two advantages of packet filter firewalls are they are fairly easy to implement and they are transparent to the end users. They can, however, be difficult to configure properly, particularly if a large number of rules have to be generated to handle a wide variety of applications and users.

Packet filtering often requires no separate firewall, because it is often included in most TCP/IP routers at no extra charge; but it is not the best firewall security that can be implemented. One of its deficiencies is that filters are based on IP addresses, which can be forged, and not authenticated user identification.

Attackers can determine the rules implemented on a packet-filtering firewall using the Firewalk tool to scan the target network. Defense against firewalking includes filtering outgoing ICMP Time Exceeded messages or using proxy-based firewalls (Skoudis 2002).

Packet filters can be used as part of the organization's VPN because they can limit the traffic that passes through a tunnel to another network, based on the protocol and the direction of traffic. For example, it is possible to configure a packet filter firewall to disallow File Transfer Protocol (FTP) traffic between two networks, while allowing Hypertext Transfer Protocol (HTTP) and Simple Mail Transfer Protocol (SMTP) traffic between the two, further refining the granularity of the control on protected traffic between sites.

Application and Circuit Gateways

Application and circuit gateways enable users to use a proxy server to communicate with secure systems, hiding valuable data and servers from potential attackers. The proxy accepts a connection from the other side and, if the connection is permitted, makes a second connection to the destination host on the other side. The client attempting the connection is never directly connected to the destination. Because proxies can act on different types of traffic or packets from different applications, a proxy firewall (server) is usually designed to use proxy agents. In this case, an agent is programmed to handle one specific type of transfer, such as FTP or TCP traffic. The more types of traffic that must pass through the proxy, the more proxy agents need to be loaded and running on the machine.

Circuit proxies focus on the TCP/IP layers using the network IP connection as a proxy. Circuit proxies are more secure than packet filters because computers on the external network never gain information about internal network IP addresses or ports. A circuit proxy is typically installed between the company's network router and the public network (that is, the Internet), communicating with the public network on behalf of the company's network. Real network addresses can be hidden because only the addresses of the proxy are transmitted to the public network.

Circuit proxies are slower than packet filters because they must reconstruct the IP header to each packet to its correct destination. Also, circuit proxies are not transparent to the end user, because they require modified client software.

Stateful Multilayer Inspection

A firewall technique called **Stateful Multilayer Inspection (SMLI)** was invented to make security tighter while making it easier and less expensive to use, all without slowing performance. SMLI is the foundation of a new generation of firewall products that can be applied across different kinds of protocol boundaries, with an abundance of easy-to-use features and advanced functions.

Stateful Multilayer Inspection (SMLI)

SMLI is similar to an application proxy in the sense that all levels of the OSI model are examined. Instead of using a proxy, SMLI uses traffic-screen algorithms optimized for high-throughput data parsing. With SMLI, each packet is examined and compared against known states of friendly packets.

One advantage of SMLI is the firewall closes all TCP ports and then dynamically opens ports when connections require them. Stateful inspection firewalls also provide features such as TCP sequence-number randomization and User Datagram Protocol (UDP) filtering. These firewalls, however, must be supplemented with proxies to support such other important functions as authentication.

Many firewall vendors include a tunnel capability in their products. Like routers, firewalls have to process all IP traffic. Because of all the processing performed by firewalls, they are ill-suited for tunneling on large networks with large amounts of traffic. Combining tunneling and encryption with firewalls is probably best done on small networks with low volumes of traffic. Also, like routers, firewalls can be a single point of failure for the VPN.

VPN Hardware Security

Because routers examine and process every packet that leaves the LAN, it is natural to include packet encryption on routers. Vendors of router-based VPN services usually offer two types of products, either add-on software or an additional feature of a coprocessor-based encryption engine. However, adding the encryption tasks to the same box as the router increases the risks of losing access to the VPN if the router has a failure. Virtual networks are discussed in Chapter 12.

Many of the requirements for an encrypting router are the same as those for firewalls. Encrypting routers are appropriate for VPNs if they have the following features:

- Supports both transport mode and tunnel mode IP Security (IPSec).
- Restricts access by operations personnel to keys.
- Supports a cryptographic key length that best matches the security needs.
- Includes separate network connections for encrypted and unencrypted traffic.
- Supports the default IPSec cryptographic algorithms.
- Supports automatic rekeying at regular periods or for each new connection.
- Logs failures when processing headers and issues alerts for repeated disallowed activities.

Because routers are generally designed to investigate packets at Layer 3 of the OSI model and not authenticate users, it will probably be necessary to add an authentication server in addition to the encrypting router to create a secure VPN.

Another possible VPN solution is to use special hardware that is designed for the task of tunneling and encryption. These devices usually operate as encrypting bridges that are typically placed between the network routers and the WAN links. Although most of these hardware devices are designed for LAN-to-LAN configurations, some products also support client-to-LAN tunneling.

A router (and firewall) feature, called an access list, provides the capability of permitting or denying access to network traffic based on source address, destination address, and protocol (Hudson 2000). This feature is available for IP, IPX, AppleTalk, and other protocols. Although access lists can help with network security, they do not take the place of more advanced security measures such as firewalls. Without significant advance planning, an access list can be very problematic, in that network access can be totally blocked with an incorrectly formulated configuration.

6.7 VOICE COMMUNICATION SYSTEMS

voice systems

Voice systems include PBXs, ACDs, key systems, and hybrid systems. These systems are computer based and are controlled by an operating system and a collection of voice system applications. They have similar vulnerabilities as client/server and mainframe installations. The computer telephony integration (CTI) application provides an interface between the PBX and a computer system. This provides an additional concern for the network, because it provides another avenue into the Enterprise resource that must be protected.

Private Branch Exchange

In the typical **private branch exchange (PBX)** environment, a T-1 digital trunk interface, which acts like a channel bank, resides in a shelf in the PBX system. As an integral part of the PBX (Figure 6–9), this T-1 interface is designed primarily around voice circuits; however, newer PBX generations use this interface to provide video and data services. The PBX offers some features that are not accessible to the multiplexer or network, including the following:

private branch exchange (PBX)

- Queuing a channel on a T-1
- Station Message Detail Recording (SMDR)
- Call forwarding
- Call transfer
- Conferencing
- Hunting
- Least-cost routing
- Redialing

Digital PBXs can switch data as easily as voice, but office automation does not function well in a circuit switched environment. The bandwidth required is often greater than a PBX supports.

Thieves and hackers use telephone and computer networks to commit fraud by stealing telephone services. Employees, historically, committed this theft by making long distance calls during business hours. With the increased sophistication of PBXs and the external access afforded to users, intruding into a PBX system is easier and has also become a security and integrity issue. Phone hackers use computers with auto-dialing modems to break security passwords and gain access into PBX systems, where they can use or sell long-distance services at the expense of the PBX owner. Organizations must take preventive measures such as blocking outgoing international telephone calls after-hours. Other efforts include blocking remote access features of PBXs such as direct inward dial access (DISA) and changing access codes and passwords frequently.

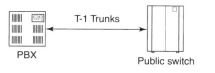

Figure 6–9
PBX T.1 Connectivity to Public Telephone Network

Automatic Call Distributors

An **automatic call distributor (ACD)** is a specialized telephone system that was designed originally for processing many incoming calls, but now is increasingly used for outgoing calls. An ACD can be integrated into a PBX for multi-functionality support. ACDs are historically used for processing many calls in industries such as airlines, rental car agencies, and mail-order houses.

    Computer-telephony integration (CTI) is used with the PBX, ACD, and computer system to provide a link for passing information to a database system. The primary CTI service is screen synchronization or "screen pop." In this application, the host computer identifies the caller by calling identification digits passed from the ACD or by prompting the caller for an account number. The computer directs the ACD as to which call-center agent should receive the call and simultaneously transmits the customer's record to the agent's workstation display.

Key Systems

Many **key systems** are designated by the capacity in central office lines and stations. For example, an 824 system could accommodate a maximum of eight lines and twenty-four stations. In a key system each line is terminated via a button on a telephone instrument. The number of buttons that can be physically terminated on the telephone tends to put an upper limit on the feasible size of a key system. At some point it is awkward to select lines manually for outgoing calls and to use the intercom for announcing incoming calls. A hybrid, in which the attendant transfers calls and the user dials an outgoing call access code, is usually required. Many key systems require proprietary telephone sets, and may not use the standard 2500 telephone set. Because key system attendants must announce each call to the called party, most modern systems have an internal paging feature that allows the attendant to hold a two-way conversation over a speaker/microphone that is built into the telephone set.

Hybrid Systems

Private branch exchanges are rarely economical in small-line sizes, and large organizations require features and capacity that key systems lack. For companies in between, the market offers a combination PBX and key system that is called a hybrid. A **hybrid system** has characteristics of both the PBX and key system and can usually be configured as either.

Voice Security Issues

Dial-up connections to voice systems still exist today. Dial-up with a modem is the normal method for a vendor to access the voice system remotely for upgrades or maintenance. Dial-up connections that are available continuously invite an intruder into the network. Attackers can use the war dialing method to hack the voice system and then proceed through the voice system to any network resources that might be attached.

Hacking **voicemail** systems provides another opportunity to access the Enterprise voice system. These instances may occur when there is a layoff of employees or there is a company merger. Often mailboxes remain active after employees depart, which provides for voicemail hacking using a brute force attack. Countermeasures include the deactivation of any mailbox not currently assigned to an employee. Passwords must be used and regularly changed. A lockout mechanism limiting attempts can be installed to counter a brute force attack.

There are general security issues that must be addressed to maintain a secure voice system. The telephone system and PBX may be part of the network administrator's responsibility. Console operator training and staff awareness of security issues must be addressed. The following questions need to be answered:

- Are passwords to administrative terminals maintained?
- Are telephone records limited to authorized individuals?
- Are equipment rooms and circuit demarcations kept locked?
- Are cable routes and vaults protected from unauthorized entry?
- Are sensitive communications carried over secure circuits?
- Does the network have virus protection software?
- Does the system allow for dial-in access?
- Are telephone facilities shared within a building?
- How are wireless transmissions controlled?
- Who reconciles the telephone bills to actual usage?
- Is there a system for monitoring and logging long distance calls?
- Do the employees know how to avoid toll fraud situations?

A checklist for auditing a voice communication system is presented in Appendix B.

6.8 INTERNET AND WEB RESOURCES

CERN proxy server site	www.w3.org/Daemon
Les Carleton's list	www.zeuros.co.uk/firewall/proxy.htm
Yahoo's Proxies links	www.yahoo.com
Microsoft's Proxy Server site	www.microsoft.com/proxy
Netscape's Proxy Server site	www.netscape.com/comprod/server_central/ Product/proxy/index.html

■ SUMMARY

The Enterprise organization uses a number of hardware devices to accomplish its mission. These include computer systems, client/server systems, network systems, and voice communications systems.

A computer system consists of a mainframe, front-end processor, controllers, storage devices, peripherals, and workstations. These devices are subject to intrusion and disruption by both internal and external sources.

Client/server environments include hardware, database, and software components that can be subjected to malicious attacks. Integrity, security, and access control must be provided for these systems.

Storage management for computer, client/server, and network systems must receive a high priority from security systems. Often, these data and database assets are the basis of an organization's existence.

Voice communications systems include PBXs, key systems, ACDs, and hybrid systems. These systems play an important part in the accessibility of a number of Enterprise resources.

A number of security solutions are available to protect the Enterprise hardware components. These include firewalls, gateways, and routers. These devices, in association with sophisticated software, are the first lines of defense against intruders and attackers.

Key Terms

Access lists
Application gateway
Automatic call distributor (ACD)
Bastion host
Circuit filtering
Client/server
Computer system
Computer-telephony integration (CTI)
Controller
Default gateway
Default router
Demilitarized zone (DMZ)
Disk mirroring
Disk storage
Disk striping
Duplexing

Fibre Channel
Filters
Firewall
Front-end processor (FEP)
Gateway
Hybrid system
Key system
Mainframe
Operating System (OS)
Packet filtering
Private Branch Exchange (PBX)
Proxy
Proxy server
Proxy services
Redundant Array of Inexpensive Disk (RAID)
Router

Screened subnet
SCSI Enclosure Standard (SES)
Server
Simple Network Management Protocol (SNMP)
Spoofing
Stateful Multi-layer Inspection (SMLI)
Voice systems
Voicemail

REVIEW EXERCISES

Questions and Analysis

Provide a short answer or in-depth analysis as required.

1. Identify the components of a computer system.
2. What is a mainframe?
3. What is a client/server?
4. What is the function of an operating system?
5. Discuss the uses of disk technology.
6. What is a RAID?
7. What is the function of a FEP?
8. Describe the Fibre Channel system.
9. What are the responsibilities of storage management?
10. Describe the client/server environment.
11. Identify server security issues.
12. What are the methods to secure a server?
13. Describe the issues associated with passwords.
14. What are the most common methods to protect mechanized transactions?
15. What are typical functions of a firewall?
16. Describe the DMZ.
17. What is an access control list?
18. What are the three classifications of firewalls that provide for traffic filtering?
19. Describe circuit filtering.
20. Describe the function of an application gateway.
21. Describe packet filtering.
22. What are the characteristics of traffic control?
23. What is spoofing?
24. What are the limitations to packet filtering?
25. Describe the function of a bastion host.
26. What are the components of the bastion host environment?
27. What is the function of a proxy?
28. What is a proxy server?
29. What are the advantages and disadvantages of a proxy server?
30. How are routers used in the computer network?
31. Explain how servers and workstations can contribute to security flaws in the computer network.
32. Describe the functions of a gateway.
33. How is hardware used in the VPN?
34. Draw a router/firewall network segment that connects to the Internet.
35. How is a packet-filtering firewall used in a VPN?
36. Explain the use of application and circuit gateways.
37. What is Stateful Multilayer Inspection?
38. How is hardware used in VPN security?
39. What systems comprise the voice communications configurations?
40. Describe the functions of a PBX.
41. How is an ACD related to the CTI application?
42. What is a key system?
43. What is a hybrid system?

44. Why are modems a threat to voice systems?

45. How are voicemail systems protected?

46. List ten voice security issues.

Definitions and Terms

Provide the most correct answer from the list of key terms.

1. A type of firewall that bases access decisions on the nature of the application's communications.

2. A computer system that interacts directly with an untrusted network.

3. A perimeter network that adds an extra layer of protection by establishing a server network that exists between the protected network and the external network.

4. In data networks, a node on the network that connects two otherwise incompatible networks.

5. The limiting of TCP/IP traffic, based on the type of traffic, source and destination IP address, and so on.

6. An application running of a gateway that relays packets between a trusted client and an untrusted host.

7. Like a firewall, it is designed to protect internal resources on networks that are connected to other networks such as the Internet.

8. Defines techniques for combining disk drives into arrays.

9. An OSI Layer 3 intelligent device that connects like and unlike LANs and WANs.

10. A network protected by a filtering router or other network device.

True/False Statements

Indicate whether the statement is true or false.

1. The client/server environment is immune to security violations.

2. RAID devices are used to improve performance and to provide for system recovery in the event of a failure.

3. Fibre Channel does not have any software provisions for Enterprise system management.

4. A server consists of both hardware and software components, which means that security issues must be addressed for both.

5. The most common ways to secure a server is through access control, however complex security systems may defeat the purpose of such a system.

6. Common mechanisms for securing mechanized transactions include SSL, TCP/IP, HTTP, and PCT.

7. A firewall is a hardware device that only limits access into a corporate resource.

8. The purpose of a DMZ is to keep unauthorized users from an untrusted network from entering a trusted network.

9. Access lists perform three functions: traffic flow control, traffic forwarding and blocking, and data integrity.

10. Three classifications of firewalls include application gateways, circuit filtering, and packet filtering.

11. Spoofing is the process of trying many different password combinations to illegally access a network.

12. If a proxy is the authority to act for another, then a proxy server acts on the network user's behalf.

13. A gateway possesses additional functions than those of a router, because it operates at a lower level of the OSI reference model.

14. A firewall technique called SMLI was invented to make security tighter while making it easier and less expensive to use.

15. Voice systems such as PBXs, ACDs, and key systems are not as vulnerable to attacks as are client/server systems.

Activities and Research

1. Research and identify hardware and software products that can be used to protect the Enterprise's computer, client/server, and network components.

2. Research the product offerings of storage technology devices. Produce a matrix that shows the device, the provider, the function, and the cost of each product.

3. Research and identify the providers of client/server security products and services. Produce a matrix of these providers and the offering, function, and cost of each.

4. Research and identify the various security products that can be used to secure mechanized computer network transactions. Prepare an overview of each product.

5. Compare and contrast the functions of firewalls, gateways, routers, proxy servers, and bastion hosts. Provide feature comparisons.

6. Compare and contrast the functionality of circuit filtering, application gateways, and packet filtering.

7. Identify the security provisions for major voice system providers. Show how they would enhance Enterprise security.

8. Schedule a field trip to a Network Operations Center that supports an Enterprise computer network.

9. Using Appendix B as a guide, prepare a list of questions that would be answered by an Enterprise security manager to ensure the hardware assets are being protected.

CHAPTER

7

Software and Database Systems

OBJECTIVES

Material presented in this chapter should enable the reader to

- Become familiar with the various software and database systems that are part of the Enterprise computer and networking environment.
- Learn the various terms and acronyms associated with application, database, and operating systems.
- Identify various methods and techniques for securing and protecting Enterprise data and database resources.
- Look at the various options for backup and restoring of database and software resources.
- Explore the alternatives available for ensuring the integrity and security of Enterprise software in the wireless and distributed environment.
- Identify the various software applications that are used to provide integrity, control, and security for Enterprise resources.
- Understand network management systems and how they relate to Enterprise security.
- Look at the various protocols and standards that contain provisions for the software security environment.

■ INTRODUCTION

In today's computer and internetworking environment, software and database security is a critical consideration because of the variety of access methods available. Several levels of security must be scrutinized if these resources and assets are to be made secure from internal as well as external forces. To provide this safe and secure environment, access must be restricted and monitored on all levels (Gelber 1997).

System maintenance programs must be password protected to ensure that only authorized personnel have access. Passwords must be frequently changed as well as managed. Policies and procedures must be established and managed so security maintenance is performed on a regular basis.

User access security is an on-going issue and techniques must be in place to ensure that unauthorized access is denied. Both system and application software must incorporate multiple levels of password protection. This can be accomplished by implementing multiple levels of security, including specific application access, and read, write, and update access, not only on software, but also on databases.

Data security is primarily a data processing issue, but it can cause concern in the Enterprise environment. Data transported in plaintext can be accessed via taps on the transmission facility. Using one of the many data compression or encryption security techniques can protect against this type of intrusion. As described in Chapter 3, these techniques operate by scrambling the bit patterns on the sending side and reassembling them on the receiving side.

It is essential that data and databases remain safe and secure from intruders, however they can be compromised by authorized internal personnel as well. A process must exist that can be used to promptly and effectively restore damaged database resources, no matter the cause.

7.1 PROTECTING THE DATA AND DATABASE ASSET

Can the organization survive if the database asset is destroyed? How much would it cost to recreate the asset? What are the repercussions if it cannot be recreated? These and other questions are pertinent if the organization uses a database to conduct its day-to-day activities. The issue is not *if the database will be corrupted*, but *when*? The organization must take proactive steps to ensure that a complete recovery can be effected in a reasonable time period.

Planning for the Inevitable

It is essential for the administrator to identify the possible causes of a database disaster and the steps that will be required to recover it completely. Enterprise network security is not the same for each file, data item, or database. This means that potential incidents are identified, the ramifications explored, and corrective measures identified to be taken for each occurrence.

Data and Database Integrity and Security

What is **data** and what can be done to ensure its integrity and security? Data is basically any material that is represented in a formalized manner so that it can be stored, manipulated, and transmitted by a computer. Data are stored in a **database** according to a predefined format and an established set of rules, called a **schema**. A database that consolidates an organization's data is called a **data warehouse**. **Data warehousing** is a software strategy in which data are extracted from large transactional databases and other sources and stored in smaller databases, which makes analysis of the data easier.

Two issues that must be addressed include data integrity and data security. The possibility of errors in data transmission exists in all telecommunication networks. Correcting and preventing these errors is called maintaining data integrity. Even though accurate transmission of data is essential, data security may be even more important.

Data integrity is addressed by using a number of error-control techniques. These include parity checking, cyclical parity checking, hamming codes, Checksums, and cyclical redundancy checks (Beyda 2000). Data security can be assured by implementing the five basic goals of the **National Institute of Standards and Technology (NIST)**. These goals state that secure data messages should be sealed, sequenced, kept secret, signed, and stamped.

NIST Security Goals

The NIST goals that should be considered when safeguarding data includes the following (Beyda 2000):

- An unauthorized party cannot modify a sealed message.
- A sequenced message is protected against undetected loss or repetition.
- An unauthorized party cannot understand a secret message.
- A signed message includes proof of the sender's identity.
- A stamped message guarantees receipt by the correct party.

Backup and Data Archiving

There are immense amounts of data collected and information generated continuously on computer and client/server systems. A **backup** is described as a copy of online storage that provides fault protection, and an **archive** is a historical backup. Hopefully, administrators are aware that some type of backup or data archiving is required in the event of some catastrophic event. There are a number of techniques to accomplish this task, including copying data to magnetic media, or copying or replicating data to other systems. **Replication** is the process by which a file, a database, or some other form of computer data in one location is created exactly the same way on another computing device in another location. This process should provide for user data access in both locations (Sheldon 1998).

backup

archive

replication

There are three basic types of backup—normal, incremental, and differential. Choosing a backup method depends on issues such as the volume of data, the frequency, and whether off-site storage is used.

The normal process copies all elements selected to a backup device (e.g., magnetic tape, jukebox, or Zip disk) and marks the original elements with a flag to indicate the activity. This method is the easiest to use and understand, since the most recent backup media is the most current backup. This can become prohibitive because of the volume of media required for the backup process. These backups must be stored somewhere and a log must be maintained showing location and content.

The incremental process only copies elements that have been created or changed since the last backup. The elements are marked with an archive flag so that only those changed or added are copied to the backup device. This process requires that a normal backup is conducted according to a scheduled time frame, because it will be used with the incremental backup to recreate the original database if there is a problem. These incremental updates do not occupy as many backup media as the normal process, however they must be logged and stored somewhere.

With the differential method, the elements that were created or changed since the last normal or incremental process are backed up. This method does not mark the elements with an archive flag, which means they are included in a normal backup process. A normal backup must be created on a regular basis. If there is a need to restore the database, first restore the normal backup, and then restore the last differential backup. It is essential that these backup media are properly labeled and stored.

These three backup procedures assume that data and files are being backed up one at a time, which is called a file-by-file backup. An alternative method is the image backup, which streams all the elements onto a disk without regard to the file structure. The advantage of this method is speed, however the entire disk volume must be backed up at the same time and the restoration must be done on a disk that is physically the same as the original. For proper security, a copy of this disk must be part of the off-site storage process.

With any backup system, a restoration test must be performed to ensure that the data and database can be restored successfully. These tests must be performed on a regularly scheduled basis. It is useful and prudent to have duplicate hardware to test the backup and restoration processes.

Data and Database Restorations

In many installations, data items change frequently, while others seldom change. When reconstructing a file, variations in the number and frequency of changes become important issues. A backup is a copy or a part of a file used to assist in reestablishing a lost file. In professional computing and networking systems, periodic backups are performed. Everything on the system is copied, including system files, user files, scratch files, and directories, so that the system can be regenerated after an incident. This type of backup is called a complete backup and is usually conducted during off-hours, like three A.M. Sunday.

Major installations may perform revolving backups, in which the last several backups are kept. Each time a backup is performed, the oldest backup is replaced. Many organizations keep three generations of backups. Another form of backup is a selective backup, in which only files that have been changed or created since the last backup are saved. In this case, fewer files must be saved, which means the backup can be completed more quickly. A selective backup combined with an earlier complete backup give the effect of a complete backup in the time needed for only a selective backup. Associated with performing a backup is saving the means to move from the backup forward to the point of failure. In critical transaction systems, keeping a complete record of changes since the last backup solves this problem.

Personal computer users often do not appreciate the need for regular backups. Minor events, such as a failed hardware component, can seriously affect personal computer users. With a backup, users can simply change to a similar machine and continue work.

A backup copy is useless if it is destroyed in the disaster. Many major computing and networking organizations rent space some distance from the primary location. As backups are completed, they are transported to the backup location, and the oldest copies are returned to the primary location. It is essential that these backups are labeled both internally and externally, and an inventory system is used to manage the backup storage.

Personal computer users concerned with integrity and safety can take copies of files home for protection. If both secrecy and integrity are important, a bank vault or other secure place in another location can be used. The worst place to store a backup copy is next to the computing device.

7.2 DATABASE MANAGEMENT SYSTEM

A **database management system (DBMS)** is a software program that operates on a mainframe system or database server to manage data, accept queries from users, and respond to those queries. Security and integrity is fast becoming an issue since databases are being organized so they can be accessed remotely. Most databases are operating in "real time," which means that transactions to the database occur instantly when an action is made against the data. In the near past, transactions were collected throughout the business day and batch processed at night. There were opportunities to correct problems using the batch processing method; however, this is not true with online real time processing. A DBMS may possess the following features:

- A data dictionary that describes the structure of the database, related files, and record information.
- Input and storage method for the data.
- A query language for data manipulation.
- Multiuser access and user access restrictions.
- Data integrity of the data from multiple modifications.
- Security of the data to authorized users.

Database Security

There are a number of malicious attacks that might occur on a computer database system. The results of three database attacks that can impact stored information are as follows (Amoroso 1994):

- Sensitive information is disclosed to an unauthorized individual.
- Information is altered in an unacceptable manner.
- Information is made inaccessible to an unauthorized individual.

These situations correspond to the threat categories of disclosure, integrity, and denial of service.

A technique for mitigating database attacks is to consider the architecture of the target database system. Most databases are configured as applications running on some underlying operating system. As a result, database attackers can either attack the database directly or attack the underlying operating system. Several different means exist for unauthorized individuals to obtain information from a database. Several problems discussed include the **inference problem** and the **aggregation problem**.

Inference Problem

There are subtle methods for an intruder to obtain information from a database. One method, called the inference problem, involves a malicious attacker combining information that is available from the database with a suitable analysis to infer information that is presumably hidden. This means that individual data elements, viewed separately, reveal confidential information when viewed collectively. It is, therefore, possible to have public elements in a database reveal private information by inference from this backdoor approach.

Aggregation Problem

The database disclosure problem called aggregation occurs when pieces of information, which are not sensitive in isolation, become sensitive when viewed as a whole. This is a common problem in military applications where several related secret documents might be viewed collectively as top secret. Databases could deal with this situation by allowing pieces of data to be labeled as nonsensitive until a collection or aggregate is created via some query, at which time the database could upgrade the mode of this aggregate to sensitive.

Security Approaches

Several security approaches have been proposed to deal with the previous two problems. These include **polyinstantiation** and database porting.

Polyinstantiation means there are several views of a database object existing so that a user's view of the database is determined by that user's security attributes. It addresses the aggregation problem by providing a means for labeling different aggregations of data separately. It also addresses the inference problem by providing a means of hiding certain information that might be used to make inferences.

The other approach is to port a database application to a secure operating system that counters these malicious attacks. Since most database applications rely on the underlying services of an operating system, it can be used to increase the security of the database. In a scheme in which the underlying operating system provides security services to the database application, the file system protection within the operating system is commonly used.

7.3 OPERATING SYSTEM

The **operating system (OS)** is the software that controls the execution of programs on a processor and manages the processor's resources. Functions performed by the OS include process and task scheduling and memory management. The OS determines which process should run at a given time. This is accomplished by an interrupt process, which allows the sharing of processor time among numerous processes and users. The process of memory management allows multiple users and processes to share a limited amount of computer main storage (Stallings 2000).

It should be noted that a computer system is a set of resources for the storage, processing, and movement of data, and the control of these functions. The operating system, therefore, is responsible for managing these resources. It is a sophisticated computer program. There are two types of operating systems—batch and interactive. In an interactive system the user communicates with the computer through a terminal device to request the execution of a program. The results often are immediate. In a batch system, the user's program is "batched" together with programs from other users and submitted by a computer operator for subsequent execution. The results from batch processing can be delayed for some time. Controls must be in place for both interactive and batch-processing OSs to ensure integrity and security of the resources and the data involved.

Kernel

Kernel refers to the core components of most operating systems. It is the portion that manages memory, peripherals, files, and computer system resources. The kernel typically executes processes and provides interprocess communication among those processes. Core functions include the following:

kernel

- Communication among processes
- Scheduling and synchronization of events
- Memory management
- Management of processes
- Management of input and output routines

A kernel-based operating system implements a layered approach, where the kernel provides the interface between the system hardware and the operating system modules. A generic set of services can be installed on top of the kernel, and it runs on a variety of hardware platforms. Processes interact by passing messages that run as "threads" within the microkernel. Threads provide a way to divide a single task into multiple tasks (Sheldon 1998).

The kernel also allocates and manages the memory used by processes. Each process has it own limited view of memory, and the kernel prevents one process from accessing the memory used by another process. With this memory protection capability, a renegade process trying to read or over-write the memory of another process will be stopped by the kernel (Skoudis 2002).

Stack-based Buffer Overflow

Stack-based buffer overflow attacks are common today because of poorly written OS code. They allow the attacker to gain access and control over a vulnerable computer system. Stacks are data structures used to store information concerning the functions of the computer. They operate on a last-in, first-out basis, where the last instruction on the stack is the first

stack-based buffer overflow

one executed. This pushdown stack is similar to the stack of dishes at a cafeteria. Function calls use this stack to store instructions and data elements.

Buffers or storage areas are set aside for stacks. Overflowing these buffers and stacks can result in machine code being written in memory locations outside the confines of the stack. Pointers to the next instruction may be pointing to memory locations outside the stack. Attackers can exploit programs that exceed the bounds of the stack, if they can be identified. This subject is not for the faint of heart and a good technical presentation is found in Chapter 7 of the book *Counter Hack* (Skoudis 2002).

The defense for this situation is to ensure that software checks the length of input supplied by a user before it is entered into memory space on the stack. Without the proper bounds checking, a buffer overflow can result, allowing an attacker to exploit the system.

7.4 DIRECTORIES

directory

A **directory** is a structure containing one or more files and a description of these files. Most network operating systems will create a directory called Users, under which a personal directory is created for each user account. Users typically have full rights to their own directories, which means they can create new branching subdirectories and share files or directories with other users.

directory service

A **directory service** is the facility within networking software that provides information on resources available on the network, including files, printers, data sources, applications, users, and so on. It also provides users with access to resources and information on extended networks. A directory service is to the network what white pages are to the telephone system. Large organizations have a need for these services to identify distributed and dispersed operations. A universal directory is also a requirement for the Internet and other public service networks.

Common distributed networks have resources and users at many locations. If a user needs to send a message to someone or access a service at a remote location, directory services can help the user locate the intended recipient or service. Because directory services are essential to locating services on the network, they must be continuously available. Directory service information is typically stored in a directory database that can be partitioned and replicated to other locations. This benefits users at remote locations who can access the database locally rather than over wide area network links.

As organizations open up their internal applications to customers and suppliers, they must ensure that only authorized users access corporate data. Directories have emerged as the primary traffic cop connecting users to needed information. These directories often support digital certificates and electronic signatures that identify users to a network. To simplify this process, suppliers have been developing standards, which would allow robust, end-to-end security for enterprise applications.

X.500 is the ITU-T standard designation for a directory standard that permits applications such as electronic mail to access information that is located either at a central location or distributed remotely. The X.500 goal is to make all users on the global networks accessible using directory services. The Internet is an example of an international system where millions of users can exchange e-mail, and users can search for services that might be available on Internet-connected networks.

X.500

Many organizations are not ready to expose their internal employee and resource lists to outsiders. Intercompany directory services and the X.500 standard have limited acceptance due to possible security problems. Another aspect of directory service security is the ability of the service to authenticate users and grant them access to the directory. Many vendors are implementing X.509 public-key encryption standards, which can provide a single logon to a variety of network services after users have checked in with the directory. A unique session ID is created to track users during their online sessions. In addition, an encrypted channel can also be implemented to prevent hackers from making sense out of information that crosses the line (Sheldon 1998).

7.5 ACCESS RIGHTS

Access rights are the "keys" that define a user's ability to access resources on a network. A network administrator or a departmental supervisor usually assigns them. Access rights affect the ability to access directories and files located on the computer system. Typical access rights or permissions include the following:

access rights

- no permission
- execute-only
- write-only
- write/execute
- read-only
- read/execute
- read/write
- read/write/execute

Access rights also control the resources accessible by the user, which includes communications services, printers, and fax machines. They can also dictate the time a user can login or the specific computers available to the user. Users who have rights to access files and directories are usually called **trustees** of those files and directories.

trustees

Microsoft Windows NT provides definitions for both rights and permissions (Sheldon 1998):

- **Rights** control the actions a user can perform on the system, such as logging on from the network, backing up files, and managing printers.
- **Permissions** give users access to directories, files, and resources such as printers. Permissions include the ability to read, write, execute, and delete files, among other things. There are also shared permissions, which give users the ability to access shared files over the network.

7.6 APPLICATION PROGRAMMING INTERFACE

Application programming interfaces (APIs) are the language and messaging formats that define how programs interact with an operating system, with functions in other programs, and with communications systems. APIs are available that interface network services for delivering data across communications systems. In database systems, APIs bind the user applications with the DBMS.

APIs are often called hooks. Programmers see APIs as software routines or modules that can be used to quickly build programs for specific systems. A cross-platform API provides an interface for building applications or products that work across multiple operating systems or computer platforms. There are three types of APIs for communications in a network or the Internet. These are the conversational, remote procedure call (RPC), and message APIs. These APIs can be used to provide efficient routing, guaranteed message delivery, priority-based messaging, and security (Sheldon 1998).

Windows NT

Windows NT includes two versions of the operating system—**Windows NT Workstation** and **Windows NT Server**. The Windows NT Server operating system was primarily designed to manage file and printing services for a variety of client configurations. It also behaves as an application server for both local and global organizations. The Windows NT Workstation operating system is a powerful desktop, with performance, reliability, and security. It can be used in the home and business environment. Common features include the following (Hudson 2002):

- Flexibility to run on many hardware platforms
- Security, auditing, and protection of files
- Multitasking of several different applications
- Networking capabilities
- Virtual memory management and protection

- Portable applications
- Memory protection for application reliability

The main difference between Windows NT Server and Windows NT Workstation is their functionality and implementation. NT Workstation is designed as a desktop operating system, whereas NT Server has additional features that allow powerful networking for server-based applications, such as Web servers.

Windows NT Security

Security is a significant feature of the Windows NT operating system. It requires that the user logs on and is authenticated by submitting a username and password to gain access to the computer. The strong authentication is also integrated into the sharing of resources. Every resource can be protected in many ways, based on the access rights and groups assigned to users. Windows NT uses the username as a key into a database to identify which permissions a particular user has for a unique resource (Badgett 1999).

For proper implementation of Windows NT security, it is essential that management makes decisions or is involved in making them. Management experience and security experience, however, is often two different skill sets. It is necessary that business decisions be balanced with security and integrity requirements. If the organization is not-for-profit, planned or otherwise, the issues are still relevant.

The results that an organization receives using a Windows NT application translates into some meaningful attribute; either a profit, recognition, or prestige. The importance of Windows NT security must be stressed. A model to accomplish Windows NT security is presented in Appendix B of *Windows NT Administration and Security* (Hudson 2002).

An important concept in Windows security administration is the domain. Microsoft defines a Windows NT domain to be a logical grouping of network servers and other computers that share common security and user account information to control access to domain resources. A domain can be of any physical size and can be connected by LANs, WANs, or the Internet. Authenticated users can log on to the domain and thereby access all domain resources. The domain can be monitored and managed from any domain computer. Domain security policies and the components of Windows NT security is presented in *A Guide to TCP/IP on Microsoft Windows NT 4.0* (Burke 1999).

Windows 2000

One of the major advantages of an operating system (OS) such as Windows 2000 is its built-in security. It uses multi-user security to control which users have access to system resources. Multiple users can logon the system at the same time, while the OS maintains a list of permissions that are assigned to each user and group on the system. This allows the administrator to customize how the users access system resources.

Many encryption tools are available, however Windows 2000 uses a native encryption capability in its New Technology File System (NTFS). This feature allows the user, or those that share the private key, to view a document. Windows 2000 encryption tools and techniques are presented in *A Guide to Microsoft Windows 2000 Core Technologies* (Tittel 2000).

Windows 2000 Security
Microsoft IIS has multiple security options for keeping a server and its data from possible intruders and hackers. These methods include IP access, user authentication, Web permissions, and NT File Security (NTFS). These allow the administrator to configure the Web server to prevent specific computers, a group of computers, or an entire network from accessing the Web server content.

Users can be required to authenticate by providing a valid Windows username and password before accessing any information on the server. For ISS, authentication can be set at the Web site, directory, or file level. IIS provides the following WWW and FTP authentication methods for controlling access to the server content (Regan 2000):

- Anonymous authentication
- Basic authentication
- Certificate authentication
- Digest authentication
- Integrated Windows authentication

In most situations, users accessing a Web server will logon as an anonymous user. Anonymous authentication gives the user access to public areas of the Web or FTP site without prompting for a username or password.

Windows XP
Windows XP is based on Windows 2000 technology and includes a refinement of the user interface and features designed to support home users. Visual themes have improved usability by stepping users through common tasks. Additionally, a remote access capability is provided.

Windows XP Security
Windows XP has several new and enhanced features for protecting files and providing overall security to the organization's data. Two important features include file encryption using Encrypting File System (EFS) and file permissions using access control lists (ACL). These provide easy ways for securing files and folders in the Windows environment.

Microsoft Authenticode

Microsoft's **Authenticode** is part of its Internet Security Framework. It addresses the issue of how users can trust software code that is published on the Internet. It provides a method to *sign* code so users know that programs obtained from the Internet are legitimate. This would be similar to the shrink-wrap that seals software packages available off the shelf. Authenticode provides for authenticity, which means the publisher is known, and integrity, which ensures tampering has not occurred. Authenticode also provides a legitimate and safe method to exchange programs over the Internet.

Authenticode

UNIX

UNIX is a multiuser, multitasking operating system with built-in networking functions. As a multiuser system, it allows many users to simultaneously access and share the resources of a server computer. The multitasking capability allows a user to execute more than one program at a time. Users must logon by entering a username and password before being allowed to use the system. This validation procedure protects each user's privacy and safeguards the system against unauthorized use.

UNIX

UNIX is also a portable operating system, which means that it can be used in a variety of computing environments. It runs on a wider variety of computers than any other operating system (Dent 2000). It also runs on the Internet, regulating utility programs such as FTP and Telnet.

UNIX Security

One aspect of UNIX system administration is that the operating system should keep track of authorized users and their passwords. To accomplish this, the system administrator uses the operating system to maintain a file named "/etc/passwd." Each time a user is added to the system, an additional line is added to this file. This security entry includes the user's login name, an encrypted password, and user identification. The /etc/passwd file also contains additional information about each user, however, only the login name and password are required to access the computer.

One way to thwart a password attack is to deny the opponent access to the password file. If the encrypted password portion of the file is accessible only by a privileged user, then the opponent cannot read it without already knowing the password of a privileged user. An effective strategy to keep the network secure is to force users to select passwords that are difficult to guess.

The security model in the UNIX operating system makes it desirable for providing many Internet functions, such as Internet server and firewall (Badgett 1999). It is possible to turn services on and off as required, and it is also possible to execute services so they do not result in security issues for other services on the computer.

Of particular interest to the UNIX administrator is the issue of backdoors. Trojan horse backdoors using such tools as **Netcat** and **RootKits** can be used to compromise the UNIX system.

Novell NetWare

Novell created the **NetWare** 4.11 network operating system in 1993. Its most important feature is **NetWare Directory Services (NDS)**, which enables network administrators to organize users and network resources such as servers and printers into a logical structure. Novell markets NDS for use on a number of non-Novell operating systems and the operating systems fully support the TCP/IP suite and Web technologies (Sheldon 2001).

Novell NetWare provides a wide range of built-in services, including the following:

- Directory services
- Security
- Routing
- Messaging
- Management
- File and print services
- TCP/IP support
- Symmetric multiprocessing

It also provides a platform for expanded services such as Internet and intranet browsing and publishing, multimedia services, and telephony.

NetWare Security

NetWare (sometimes called **IntranetWare**) provides five different security levels on the network. These security levels include login, file system, server, network printing, and NDS security. Login security is further divided into three categories—account restrictions, authentication, and intruder detection (Awwad 2000).

NetWare uses a highly secure logon sequence that implements public- and private-key authentication schemes and an auditing feature that monitors and records network-wide events. The security features are processed through the NDS directory services tree. The first level of security is the login name and password level, where the messages "Access denied" or "Intruder Detection Lockout" will be displayed if either the login name or password is incorrect. NetWare 4.1x operations are presented in *Introduction to Networking Using NetWare 4.1x*, (Velasco 2001).

Accounting features allow administrators to track user access to resources and charge users for use of those resources. Users login once to access resources everywhere, as opposed to logging in for each different resource.

Java

Java is a programming language invented in 1995 by Sun Microsystems for Web sites. Java is used to create Java applets and Java applications. A Java applet can run on any computer that supports the Java runtime, with the downloaded code being checked automatically by the Java security subsystem.

Java

The security features of Java are critical to its success as a platform for Internet application development. Java applet security features the security manager, which is an abstract class that limits the functions of the applets. There are a number of Java API classes that check with the security manager before executing a process.

There are two actions not covered in these API classes that could potentially be a security issue. These unsafe conditions include the allocation of memory and the invocation of threads. (Note that a thread is a sequence of computing instructions that make up a program.) Currently, a hostile applet can crash a user's browser by allocating memory until it runs out and firing off threads until everything slows to a crawl, resulting in a denial of service attack.

7.7 INTERNET SEARCH TOOLS

A variety of tools have been developed to help with navigation of the Internet. Although many of these tools have rendered themselves obsolete because of the advent of Web browsers, they are still extensively used as reference tools by the Web. These tools include the FTP, Anonymous FTP, Telnet, Gopher, Veronica, Archie, and Wide Area Information Servers (WAISs).

Browsers

Navigating the Web requires the use of a program called a client. The client software provides the communications protocol to interface with the various host computers on the network. The more common name for client software is called a Web **browser**. A Web browser is a program that handles most of the details of document access and display and permits the user to navigate the Web by accessing Web documents that have been coded in a language called HTML.

browser

In addition to text, an HTML document contains tags that specify document layout and formatting. Some tags cause an immediate change; others are used in pairs to apply an action to multiple items. To make document retrieval efficient, a browser uses a **cache**. The browser places a copy of each document or image the user views on the local disk. When a document is needed, the browser checks the cache before requesting the document from a server on the network. The latest Web browser products display all reachable resources, from the local PC to the worldwide Internet, on a single hierarchical file tree display.

cache

HTML was invented at CERN by Tim Berners-Lee. Documents written in HTML are plain-text ASCII files that can be generated with a text editor. They reside on the Web server and can be identified with an ".html" or ".htm" extension. Several Web browsers are used to navigate the Web (e.g., Netscape Navigator and Internet Explorer). Each browser must contain an HTML interpreter to display documents. One of the most important functions in an HTML interpreter involves selectable items. The interpreter must store information about the relationship between positions on the display and anchored items in the HTML document. When the user selects an item with the mouse, the browser uses the current cursor position and the stored position information to determine which item the user has selected.

One thing to remember about Web browsers and their ability to search the Web for a particular Web page or topic is that Web browsers still require a master Web index service or home page. The Lycos home page at Carnegie-Mellon University is an example of such a master Web indexing service.

File Transfer Protocol

To download or transfer information back to their client PCs, users would access another TCP/IP protocol called **file transfer protocol (FTP)** or anonymous FTP servers. Users can access FTP servers directly or through Telnet sessions (explained shortly). The difficulty with searching for information in this manner is that users must know the Internet address of the specific information server that they wish to access.

FTP makes it possible to move files across the Internet and handles some of the details involved in moving text and other forms of data between different types of computers. Although FTP is not a highly graphic application, it remains an important tool for individuals and organizations that must exchange files containing data or documents.

Trivial File Transfer Protocol

Trivial File Transfer Protocol (TFTP) is a simplified version of FTP that transfers files but does not provide password protection or user directory capability (Forouzan 2000). It depends on the connectionless (UDP) datagram delivery service and is associated with the TCP/IP family of protocols. Precautions must be taken to prevent hackers from accessing files; therefore, it is a good idea to limit access of TFTP to noncritical files. A remote intruder could easily use TFTP to obtain copies of password, system, or user files.

Another way to add security to TFTP is to first access Telnet, which checks whether the user has the right to access the system and the corresponding file. It then calls the TFTP client and passes the file name to the client. The client then makes the TFTP connection to the TFTP server at the user site.

Gopher

Gopher is a menu-based client/server that features search engines, which comb through all the information in all the information servers looking for a user's specific request. Gopher provides a way to index and organize all kinds of different collections of textual data, as well as other kinds of documents. Before the introduction of the Web, Gopher was the premier tool for browsing the Internet to look for information. Gopher was named after the mascot of the University of Minnesota where the system was developed.

Gopher client software is most often installed on a client PC and interacts with software running on a particular Gopher server. The server transparently searches multiple FTP sites for requested information and delivers that information to the Gopher client. Gopher users do not need to know the exact Internet address of the information servers that they wish to access.

Gopher

Newsgroups

Based on a TCP/IP service known as **USENET**, **newsgroups** provide a way for individuals to exchange information on specific, identifiable topics or areas of interest. This technology lets users read information on a variety of subtopics that are pertinent to a newsgroup's focus. Over 10,000 newsgroups covering selected topics are available. USENET servers update each other on a regular basis with news items that are pertinent to the newsgroups housed on a particular server. For technical matters, this is an especially useful way to exchange opinions and information on a broad range of topics.

USENET newsgroups

Telnet

Text-based information stored in Internet-connected servers can be accessed by remote users logging into these servers via a TCP/IP known as **telnet**. Telnet permits a user on one computer to establish a session on another computer elsewhere on the Internet, as if the local computer was directly attached to the remote computer. Given the proper access to remote machines, this program lets users achieve many tasks remotely that they might ordinarily only accomplish locally. Telnet is an application of choice for configuring all kinds of networking equipment, especially routers and hubs.

telnet

The administrator must ensure that only legitimate users can access the various devices via a telnet session. The main defense is in the use of passwords. According to the following checklist, a telnet password

- cannot be shared
- should be encrypted
- is restricted by device location
- contains at least eight characters
- must be changed every six months

- requires callback for remote access
- is access logged
- profiles legitimate users
- audits the use

Internet Explorer Security

Microsoft's Web browser is called **Internet Explorer**. Despite its near universal acceptance, it has a large number of security holes that can leave a user's PC open to the exploitation of hackers. This section will look at these security flaws and the efforts by Microsoft to correct them (Palmer 2001).

Even though Internet Explorer has been subject to numerous complaints about security, there are those who argue that the extent of these problems have been inflated by critics. Proponents suggest that Microsoft is simply a victim of its own success, since hackers tend to concentrate their efforts on crashing, or otherwise compromising, the security protocols of popular applications. Further, as Internet Explorer resides in the public domain, any security problems will naturally be aggravated.

While Microsoft has been diligent in providing timely security patches, many of the problems are attributable to design shortcomings. Hackers would not experience such great success but for the flaws inherent in the application itself. Regardless of which version one uses to access the Internet, security issues are rampant. Some experts contend that one of the reasons behind Microsoft's apparent disregard for tightening up its browser's security is the very nature of the browser market. There is little incentive to invest large sums of time and money into perfecting a product that is offered for free. The trend toward free browsers has been an impediment to the browser's evolution into a more reliable, robust application.

Microsoft seems to be implementing cosmetic changes in an attempt to deflect negative publicity. While the 128-bit encryption it implemented into Internet Explorer 5.5 is useful, Microsoft's browser is still being criticized for its numerous other problems.

Microsoft has addressed four security issues that impact Internet Explorer. These include Secure Socket Layer (SSL), Private Communication Technology (PCT), CryptoAPI, and 128-bit encryption modifications. While these measures have not been sufficient to protect Internet Explorer from hacker attacks, it should be noted that all browsers have vulnerabilities.

7.8 STANDARDS AND PROTOCOLS

Open Systems Interconnection (OSI) is the internationally accepted framework of standards for communications between different systems made by different vendors. This seven-layer standard allows for an open systems environment so that any vendor's device can communicate with any other vendor's equipment. Many vendors manufacture their products to conform to layer one and layer two of the OSI model. Appendix A provides an explanation of the standard.

Protocols

Protocols are sets of rules governing the format of message exchange. They include an orderly sequence of steps that two or more parties take to accomplish some task. Anyone communicating over the Enterprise network must use some protocol to successfully transmit data. It is essential that one be aware of the protocol issues when designing a system and making a proposal in this environment. There are numerous protocols, including those for file transfers (Kermit), data compression (V.42bis), error control (V.42), and modulation techniques (V.34).

The reason why different networks are able to communicate over different interfaces, using different software, is the establishment of protocols. Protocols provide the rules for how communicating hardware and software components negotiate interfaces or talk to one another. Protocols may be proprietary or open and may be officially sanctioned by standards organizations or market driven. For every potential hardware-to-hardware and hardware-to-software interface, there is likely to be one or more possible protocols supported. The sum of all protocols used in a particular computer is sometimes referred to as that computer's protocol stack.

Challenge/Response Protocol

A **challenge/response** is a security mechanism for verifying the identity of a user or a system that is attempting to connect to a secure system. When a user contacts a server, the server responds with a challenge. The user then takes this challenge and uses it to perform a cryptographic operation that produces some result that is transmitted back to the server. The server then performs a similar operation, and if the results match, the user is flagged as authentic. This process is depicted in Figure 7–1. This protocol assumes both the server and user have the user's password and both have the same encryption method.

This technique does not require the user send a password over the network to the server. The password is used as the key for encryption, so if both the client and server encrypt the same information with the same key, the results should match. Another important feature allows the server to issue a different challenge at every logon, which reduces the incidence of spoofing future sessions.

Point-to-Point Protocol

Point-to-Point Protocol (PPP) is used to dial into the Internet through a normal phone line and modem. PPP begins with a negotiation phase, during which two PPP processes can negotiate the transmission and security

Figure 7-1
Challenge/Response Mechanism

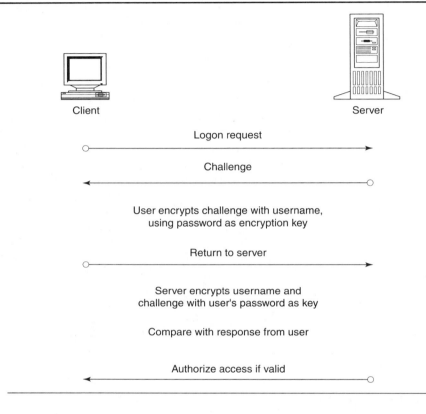

processes they will use. These two sides can decide to not use any authentication or can use one of the following protocols:

- Password Authentication Protocol (PAP)
- Challenge-Handshake Authentication Protocol (CHAP)
- Microsoft CHAP (M-CHAP)
- Extensible Authentication Protocol (EAP)

Challenge-Handshake Authentication Protocol

Challenge-Handshake Authentication Protocol (CHAP)

The **Challenge-Handshake Authentication Protocol (CHAP)** is an authentication method that can be used to connect to an ISP. It allows the user to login to the provider automatically, without the requirement for a terminal screen. With this protocol, the server sends a challenge message to the client, and the client sends back a response message that should authenticate the client to the server.

CHAP works on the basis of a shared secret. When the client receives the challenge message, it adds the shared secret to the challenge message and then hashes the combined stream using MD5 or another agreed-upon hashing algorithm. The client sends the resulting hash back to the server as its response message. The server also adds the shared secret to the challenge message, hashes the result, and compares the hashes. If they match, the client must know the shared secret and is then authenticated (Panko 2000).

Extensible Authentication Protocol

A general protocol for PPP authentication, **Extensible Authentication Protocol (EAP)** supports multiple authentication mechanisms. Instead of selecting a single authentication method for a connection, EAP can negotiate an authentication method at time of connection. This allows for methods such as secure access tokens or one-time passwords, and is supported by various EAP types. Windows 2000 currently supports EAP MD5-CHAP and EAP Transport Level Security (TLS) (Caudle 2001).

Extensible Authentication Protocol (EAP)

SOCKS

SOCKS is a proxy protocol that provides a secure channel between two TCP/IP systems. This is typically used where a Web client on an internal corporate network wants to access an outside Web server located on the Internet, another company's network, or on another part of an intranet. SOCKS provides firewall services, as well as auditing, fault tolerance, management, and other features. This allows an internal corporate network to be connected to the Internet and provides a safe method for internal users to access servers on the Internet.

SOCKS

SOCKS is a circuit-level proxy gateway that provides security based on connections at the TCP layer. A SOCKS configuration consists of SOCKS-enabled clients on the internal network, a SOCKS server, and a packet-filtering router acting as a firewall. It includes Web servers or other application servers on the Internet or any other external network. Figure 7–2 shows a SOCKS environment and component interaction.

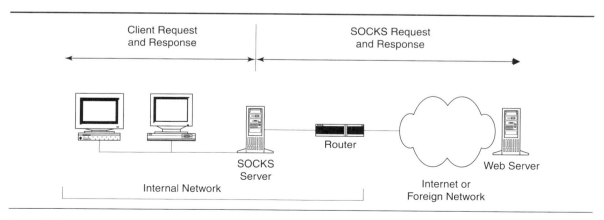

Figure 7–2
SOCKS Configuration

When a client needs to contact a Web server on the Internet, the process is as follows:

- The SOCKS server intercepts a request.
- It evaluates whether the request is allowed based on company policies.
- The SOCKS server relays this request to the target Web server.
- When the response arrives from the Web server, the SOCKS server evaluates the packet before relaying it to the internal client.
- The SOCKS server collects, audits, screens, filters, and controls data flowing into and out of the network.

According to NEC, a company that is a promoter of SOCKS, SOCKS is a "rich network application data warehouse." It provides the foundation for other critical networking services such as auditing, accounting, management, and security. **SOCKS5**, which is the latest version, adds authentication, allowing network administrators the ability to control user access through the SOCKS server. A variety of authentication methods are available, and the scheme is negotiable. SOCKS5 was created by a committee of industry supporters and submitted to the IETF for standardization (RFC 1928) (Sheldon 1998).

Common Data Security Architecture

Common Data Security Architecture (CDSA) is a specification of the Open Group that provides a set of APIs that independent software developers can use to embed security into desktop and network applications. It is designed for use in e-commerce and content delivery applications. An important feature of CDSA is its ability to provide the highest level of encryption allowed in the country that is using it. This means that international organizations can use CDSA to comply with the government encryption regulations of the countries they operate in. The architecture is designed around the use of digital certificates and portable digital tokens (Sheldon 2001).

The CDSA architecture supports data encryption, data integrity, authentication, and nonrepudiation. Support includes X.509 digital certificates, Digital Signature Algorithm, and Lightweight Directory Access Protocol (LDAP). This multilayered architecture consists of cryptographic service provisions, published APIs, and trust model, certificate, and data storage libraries.

CDSA was originally developed by Intel to provide application developers flexibility, consistency, and portability when implementing security applications. By using a common API, a software developer can easily add new and improved security features without rewriting entire programs. As vendors support and build products to the CDSA specification, a unified method of executing encryption and digital signature functions will emerge as well as a common method of executing digital certificate lookup, storage, and retrieval processes.

A major component of the CDSA architecture is the Common Security Services Manager (CSSM). The CSSM component provides a set of core services that are common to all categories of security services and includes the following modules (Traeger 2001):

- Cryptographic Service Provider
- Trust Policy
- Certificate Library
- Data Storage Library
- Authorization Computation
- Elective Modules

One common application for these services is to hide the addition of security capabilities to legacy services.

Crypto API
Crypto API (CAPI) is an application programming interface (API) from Microsoft that makes it easier for developers to create applications that contain encryption and digital signatures. RSA Data Security is working with CAPI, and Intel may layer CDSA on top of it. An API is a set of software calls and routines that can be referenced by an application program to access predefined network services.

Crypto API (CAPI)

Generic Security Services API
Generic Security Services API (GSS-API) consists of two specifications (RFC 1508 and RFC 2078), which provide a high-level interface that works to give applications an interface into security technologies such as private and public encryption schemes. It works in conjunction with CDSA.

Generic Security Services API (GSS-API)

7.9 WIRELESS TECHNOLOGIES

In addition to the previously discussed protocol standards, there exist new standards and systems for transporting wireless communications. These include Bluetooth, GPRS, 3G, EDGE, UMTS, CDMA2000, and WCDMA. Application of these technologies will be presented in Chapter 14.

Bluetooth
Bluetooth is a wireless personal area network specification, which is being developed by the Bluetooth Special Interest Group (SIG). It will enable electronic devices to spontaneously establish wireless networks within small local areas. Bluetooth is being designed for a number of devices, including notebook computers, inventory scanners, data/voice access devices, handheld and wearable devices, and telephones. It provides for connectivity to printers, keyboards, joysticks, fax machines, PDAs, and other digital devices (Sheldon 2001).

bluetooth

A typical mobile device can use Bluetooth to exchange information such as calendars, business cards, and phone book address information with other Bluetooth devices. Implementations are being developed for Windows and other popular operating systems.

The Bluetooth specification is considered a third-generation (3G) mobile technology, since its communicators are designed for speech, pictures, and video. A typical Bluetooth phone will include two radios—one for the metropolitan cellular system and one for the Bluetooth personal area network. These devices will include software to discover other devices, establish links with those devices, and exchange information. The Bluetooth software framework uses existing software specifications, including TCP/IP, Human Interface Device (HID), OBEX, and vCard/vCalendar.

Bluetooth users connect their computing and telecommunications devices using short-range radio links, and any peripheral devices can use the technology to gain connection to any computer device. It has built-in encryption and authentication to secure wireless signals. Bluetooth uses a data encryption scheme to obscure data being transmitted in the open. It uses a stream cipher that is well suited for a silicon implementation with secret key lengths of 0, 40, or 64 bits.

Bluetooth supports both point-to-point and point-to-multipoint connections. A collection of two or more connected Bluetooth devices is called a **piconet**. A **scatternet** is a collection of piconets that can connect to form a larger piconet. Each piconet in a scatternet environment is linked together via a different frequency-hopping sequence. These devices within each piconet synchronize with the unique sequence of the piconet of other devices. Data rates as high as 6 Mbps can be achieved in scatternet environments that include as many as ten fully loaded piconets.

GPRS

General Packet Radio Service (GPRS) is a standard for wireless communications. It operates at speeds of up to 150 kbps, compared with the current GSM Communications, which operates at 9.6 kbps. GPRS is an essential stepping stone to Third Generation (3G) personal multimedia services. The higher data rates will allow users to participate in video conferences and interact with multimedia Web sites and similar applications using mobile devices. GPRS will also complement Bluetooth and will support IP and X.25. GPRS is a step toward Enhanced Data GSM Evolution (EDGE) and Universal Mobile Telephone Service (UMTS). It is a way of giving handheld devices, mobile phones, or laptop computers an "always on" connection to the network, whether it is a corporate network or the Internet. Unlike the GSM network, which has to create a dedicated connection whenever exchanging data, GPRS sends bits of data only when needed. This results in a cheaper, more efficient technique of exchanging data; and, because GPRS uses existing mobile phone base stations, it is relatively inexpensive to install.

Third Generation

A new radio communications technology called **Third Generation (3G)** will provide high-speed access to Internet-based services. These include Internet videoconferencing and sharing of database information. Applications targeted include shopping, banking, reservations, computer games, and health care. Examples include videoconferencing from a limousine, a subway, or a train or providing on-the-spot vacation coverage to your friends. 3G is generally considered applicable to mobile wireless; however, it is also relevant to fixed wireless and portable devices. Proponents of 3G say that it will now be possible to provide connectivity to all users at all times in all places.

Third Generation (3G)

EDGE

Enhanced Data GSM Evolution (EDGE) is a fast version of GSM and can deliver data at rates of up to 384 kbps, it provides for multimedia and other broadband applications on mobile devices. EDGE is basically a new modulation scheme that is more bandwidth efficient than the scheme used in the GSM standard. The technology defines a new physical layer that has the potential of increasing the data rate of existing GSM systems by a factor of three.

Enhanced Data GSM Evolution (EDGE)

UMTS

Universal Mobile Telecommunications System (UMTS) is a 3G broadband, packet-based transmission of text, digitized voice, video, and multimedia at data rates up to 2 Mbps. It is based on the GSM communication standard. It will provide users with mobile access through a combination of terrestrial wireless and satellite facilities. The higher bandwidth of UMTS will provide for such services as videoconferencing.

Universal Mobile Telecommunications System (UMTS)

CDMA2000

CDMA2000, also called IMT-CDMA MultiCarrier, is a Code Division Multiple Access (CDMA) version of the IMT2000 standard that was developed by the ITU. It is a 3G mobile wireless technology. CDMA2000 can support mobile data communications at speeds ranging from 144 kbps to 2 Mbps. Deployment has been initiated by a number of mobile vendors.

CDMA2000

WCDMA

Wideband Code Division Multiple Access (WCDMA) is an ITU standard derived from Code-Division Multiple Access (CDMA). It is officially known as IMT2000 direct spread. WCDMA is a 3G mobile wireless technology that offers much higher data speeds to mobile and portable wireless devices than is presently available in today's market. WCDMA can support mobile voice, images, data, and video communications from 384 kbps (wide area) to 2 Mbps (local area). The input signals are digitized and transmitted in coded, spread-spectrum mode over a broad range

Wideband Code Division Multiple Access (WCDMA)

of frequencies. A 5 MHz carrier is used compared with a 200 kHz carrier for narrowband CDMA.

7.10 ENTERPRISE NETWORK MANAGEMENT

Any network, whether it is a LAN, MAN, or WAN, is really a collection of individual components working together. Network management helps maintain this harmony, ensuring consistent reliability and availability of the network, as well as timely transmission and routing of data. A **Network Management System (NMS)** is defined as the systems or actions that help maintain, characterize, or troubleshoot a network. The three primary objectives to network management are to support systems users, to keep the network operating efficiently, and to provide cost-effective solutions to an organization's telecommunications requirements. Network management and its use by Enterprise security administrators is presented in Chapter 5.

A large network cannot be engineered and managed by human effort alone. The complexity of such a system dictates the use of automated network management tools. The urgency of the need for such tools and the difficulty of supplying them increase if the network includes equipment from multiple vendors and manufacturers.

Network management can be accomplished by using dedicated devices, by host computers on the network, by people, or by some combination of these. No matter how network management is performed, it usually includes the following key functions:

- Network Control
- Network Monitoring
- Network Troubleshooting
- Network Statistical Reporting

These functions assume the role of network watchdog, boss, diagnostician, and statistician. These functions are closely interrelated and are often performed on the same device.

This section begins with an overview of network management, focusing on hardware and software tools, and organized systems of such tools, which aid the human network manager in this difficult task. Considerable information concerns the requirements for network management, the general architecture of a network management system, and the Simple Network Management Protocol (SNMP), which is the standardized software package for supporting network management.

Network Management
The term **network management** has traditionally been used to specify real-time network surveillance and control; network traffic management; functions necessary to plan, implement, operate, and maintain the network; and systems management.

Network management supports the users' needs for activities that enable managers to plan, organize, supervise, control, and account for the use of interconnection services. It also provides for the ability to respond to changing requirements, such as ensuring that facilities are available for predictable communications behavior and for providing for information protection, which includes the authentication of sources and destinations of transmitted data.

The task of network management involves setting up and running a network, monitoring network activities, controlling the network to provide acceptable performance, and assuring high availability and fast response time to the network users.

Network equipment manufacturers have developed an impressive array of network management products over the past several years. As with most telecommunications technologies, network management began as a proprietary system, but the real growth started when the Internet community introduced Simple Network Management Protocol (SNMP) in the late 1990s.

Network Management Systems

An effective network management system requires trained personnel to interpret the results. An effective network management system will include most of the following elements:

- An inventory of circuits and equipment
- A trouble report receiving and logging process
- A trouble history file
- A trouble diagnostic, testing, and isolation facility and procedures
- A hierarchy of trouble clearance and escalation procedures
- An activity log for retaining records of all major changes
- An alarm reporting and processing facility

Not every network management system will have all of these elements, but most systems will contain the above functions to some degree. The more complex the network, the more likely the functions will be automated on a mechanized system.

7.11 MANAGEMENT INFORMATION BASE (MIB)

A **Management Information Base (MIB)** is a collection of information that is organized hierarchically. MIBs are accessed using a network management protocol such as SNMP. They are comprised of managed objects and are identified by object identifiers. Table 7–1 provides information on the MIB variables.

Devices that support the MIB are manageable by SNMP. The names for all objects in the MIB are defined by either the Internet-standard MIB

Management Information Base (MIB)

Table 7–1
MIB Variables

MIB Variables	Purpose
Address translation group—AT	Converts network addresses to subnet or physical addresses.
Electronic gateway protocol group—EGP	Provides information about nodes on the same segment as the network agent.
Interfaces group—interface	Tracks the number of network NICs and the number of subnets.
Internet control message protocol group—ICMP	Gathers data on the number of messages sent and received through the agent.
Internet protocol group—IP	Tracks the number of input datagrams received and the number rejected.
SNMP group	Gathers data about communications with the MIB.
System group—system	Contains information about the network agent.
Transmission group	
Transmission control protocol group—TCP	Provides information about TCP connections on the network, including address and time-out information.
User datagram protocol group—UDP	Contains information about the listening agent that the NMS is currently contacting.

or by other documents that conform to SMI conventions. Each "instance" of an object type is defined by a unique variable name.

A managed object is one of any number of specific characteristics of a managed device. Managed objects consist of one or more object instances, which are essentially variables. The two types of managed objects are scalar and tabular. Scalar objects define a single object instance. Tabular objects define multiple related object instances that are grouped in MIB tables.

An object identifier uniquely identifies a managed object in the MIB hierarchy. The MIB hierarchy can be depicted as a tree with a nameless root, the levels of which are assigned by different organizations. Figure 7–3 depicts a portion of the **MIB tree**. Each branch of the tree consists of logical groupings used to generate unique object IDs. The branch is referred to as a node. A node can have both "parents" and "children." A node that does not have children is referred to as a leaf node. The leaf node is the actual object. Only leaf nodes return MIB values from agents or have their MIB values altered. A subtree is used to refer to all nodes and children under a branch of the tree.

The top-level object MIBs belong to different standards organizations, whereas lower-level object IDs are allocated by associated organizations. An object ID uniquely identifies a managed object in the MIB hierarchy. The MIB hierarchy can be depicted as a tree with a nameless root, the levels of which are assigned by different organizations.

An MIB object is named by concatenating the numerical names of each node when traversing the MIB tree from ISO (1) to the particular node. A full object ID name contains all the nodes, including the lead

MIB tree

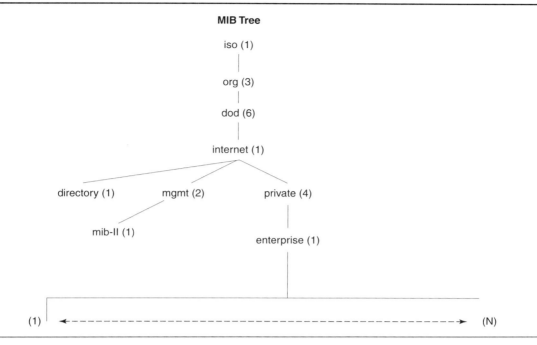

Figure 7–3
MIB tree

nodes. The nodes are concatenated and separated by periods. This MIB object notation follows the standard notation defined in ASN.1.

Vendors can define private branches that include managed objects for their own products. Table 7–2 shows an object associated with a Cisco device. The enterprise number associated with Cisco (9) is found in the SMI Network Management Private Enterprise Codes. This IEEE Organizationally Unique Identifier (OUI) assignment list can be found on the Internet or can be obtained from the IEEE Registration Authority.

Abstract Syntax Notation One

Managed objects are named and described using the abstract syntax notation (ASN.1). ASN.1 defines a powerful data description notation that allows the format and meaning of data structures to be defined without

Object identifier = 1.3.6.1.4.1.9.3
This decodes as:
iso.org.dod.internet.private.enterprise.cisco.temporary variables

Table 7–2
MIB Object Identifier Example

specifying how those data structures are represented in a computer or how they are encoded for transmission through a network. The advantage to using a notation like ASN.1 to define management information is it can be described independently of any particular form of information processing technology.

The objective of ASN.1 is the transfer of information from the internal representation of one machine to another machine's internal representation by way of a machine independent abstract syntax. Data types in ASN.1 are defined in a programming language similar to the Pascal programming language.

Table 7–3 provides a list of the terms that are relevant to ASN.1. ASN.1 is used to define the format of PDUs, the representation of distributed information, and operations performed on transmitted data.

Simple Network Management Protocol

Simple Network Management Protocol (SNMP) is the means by which the management station and the managed nodes exchange information. In 1988, the Internet Engineering Task Force (IETF) defined SNMP. Although SNMP has several deficiencies, it has become the de facto standard for the management of data networks. **SNMPv2** was produced in 1993 to tackle the deficiencies. Work is underway on the next generation, called **SNMPv3**.

SNMP was originally designed to provide an easy-to-implement but comprehensive approach to network management in the TCP/IP environment. Now, however, SNMP is being applied in networks that conform to many other network architectures in addition to TCP/IP. SNMP is an industry-standard protocol that is supported by most networking

Table 7–3 ASN.1 Notation

ASN.1 Term	Description
Abstract Syntax	Describes the generic structure of data independent of any encoding technique used to represent the data. The syntax allows data types to be defined and values of those types to be specified.
Data Type	A named set of values. A type may be simple, which is defined by specifying the set of its values, or structured, which is defined in terms of other types.
Encoding	The complete sequence of octets used to represent a data value.
Encoding Rules	A specification of the mapping from one syntax to another. Encoding rules determine algorithmically, for any set of data values defined in an abstract syntax, the representation of those values in a transfer syntax.
Transfer Syntax	The way in which data are actually represented in terms of bit patterns while in transit between presentation entities.

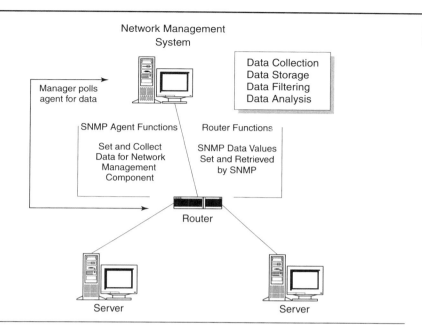

Figure 7-4
SNMP Environment

equipment manufacturers. In a network environment like the one depicted in Figure 7-4, software **agents** are loaded on each managed network device that will be using SNMP. Each agent monitors network traffic and device status and stores information in an MIB.

agents

To use the information gathered by the software agents, a computer with an SNMP network management program must be present on the network. This management station communicates with software agents and collects data stored in the MIB (RFC1157) component on the managed network devices. This information is then combined with the information obtained from all networking devices. Statistics and charts are then generated detailing current network conditions. With most SNMP managers, thresholds can be set and alert messages generated for network administrators when thresholds are reached or exceeded.

SNMP defines the formats of a set of network management messages and rules by which messages are exchanged. Many network components can be managed using SNMP. Through their software agents, it is possible to configure networking devices, and in some cases, reset them from the network management station. SNMP can manage network devices such as bridges, routers, and servers. An SNMP network management program can interrogate these devices and make configuration changes remotely to help managers control their networks from a single application from a central location.

The SNMP approach to network management was developed at a time when considerable work had already been done concerning the Common Management Information Protocol (CMIP) approach to network management. The developers of SNMP used the same basic concepts regarding how management information should be described and defined as those concepts that were developed for CMIP.

SNMP has a number of advantages over CMIP. An important one is it operates independently on the network, which means that it does not depend on a two-way connection at the protocol level with other network entities. This feature enables SNMP to analyze network activity without depending on faulty information from a failing node. SNMP also has an advantage in that management functions are carried out at a centralized network management station.

SNMP is based on the premise that all devices on a network are able to provide information about themselves. SNMP includes a **Structure of Management Information (SMI)** document that defines the allowable data types for the MIB. The MIB is a hierarchical structure of information relevant to the specific device and is defined in object-oriented terminology as a collection of objects, relations, and operations among objects. Each object has an **object identifier (OID)** to uniquely identify it among its siblings. Like ISO CMIP, SNMP uses **Abstract Syntax Notation One (ASN.1)** to define and identify objects in an MIB.

SNMP Components

SNMP provides a means to monitor and control network devices and to manage configurations, statistics collection, performance, and security. SNMP is an application layer protocol that facilitates the exchange of management information between network devices. Three versions of SNMP exist: SNMP, SNMPv2, and SNMPv3.

The original version of SNMP has some shortcomings that are addressed in the second version, which is called SNMPv2. One of the most important of these shortcomings is SNMP's lack of strict security measures. When SNMP is used, the community name is sent by the NMS without encryption. If captured, this password can be used to gain access to sensitive network management commands, providing the ability to remotely configure a managed network device, such as a router or gateway, comprising the security on a network. SNMPv2 provides an encrypted community name (password), improved error handling, and multiprotocol support. It also adds support for IPX and Appletalk, and provides the ability to retrieve more MIB information at one time.

Many of the deficiencies were corrected in SNMP with SNMPv2; however, additional enhancements were made in the latest version, SNMPv3. SNMPv3 was issued as a set of proposed standards in January 1998. This set of documents does not provide a complete SNMP capability, but rather defines an overall SNMP architecture and a set of security capabilities. These capabilities are intended for use with the existing SNMPv2.

SNMP and SNMPv2 can be used to monitor LANs, MANs, and WANs. An important SNMP-based tool used to monitor LANs connected through WANs is Remote Network Monitoring (RMON), an IETF standard developed in the early 1990s.

SNMP Security

SNMPv1 was not designed as a high-security protocol. Encryption was not used, therefore any nefarious user with physical access to the network could discover the community names (passwords) or addresses used and send a phony request to the SNMP agent bearing this information. Most SNMP MIBs are read-only, which prevents unauthorized modifications to the MIB parameters. To maximize the security available to SNMP, a unique community name should be used. Accomplish this by selecting the Send Authorization Trap and specify a Trap Destination, then specify Only Accept SNMP Packets from acceptable hosts (McLaren 1996).

SNMPv3 (RFC2570) assists in securing network devices, however its adoption is slow. Most of the devices on the current networks are probably the unsecure SNMPv1. It should be noted that attackers might easily guess SNMP community names, which could allow an intruder access to router options and addresses. So it is important to treat the community name as a password. SNMP is also often overlooked in security audits, which provides a gaping hole for attackers (McClure 2001).

One of the few countermeasures to sniffing SNMP traffic on the wire is to encrypt it. Both SNMPv2 and SNMPv3 have options for network encryption and uses DES to encrypt the sensitive information. Another alternative is to encrypt SNMP traffic through a point-to-point VPN tunnel. Using a VPN client ensures traffic will be encrypted from the client system to the end of the VPN tunnel.

7.12 VIRTUAL PRIVATE NETWORK SOFTWARE

To provide virtual private networking capabilities using the Internet as an Enterprise network backbone, specialized tunneling protocols were developed to establish private, secure channels between connected systems. Four VPN protocols that have been developed include the following:

- Microsoft's Point-to-Point Tunneling Protocol (PPTP)
- The Internet Engineering Task Force (IETF) Internet Protocol Security (IPSec)
- Cisco's Layer 2 Forwarding (L2F) Protocol
- A combination of the two called Layer 2 Tunneling Protocol (L2TP)

Figure 7–5 depicts the use of tunneling protocols to build a virtual private network across the Internet. Various tunneling protocols, including L2F, L2TP, and PPTP, are used to pass traffic across the VPN, a process

Figure 7–5
Utilizing Tunneling Protocols

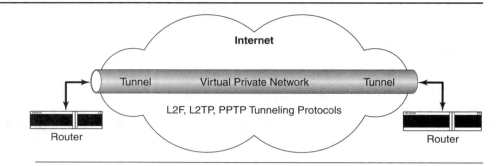

that is transparent to the end users. Virtual networks are described in Chapter 12.

Point-to-Point Tunneling Protocol

Point-to-Point Tunneling Protocol (PPTP)

The **Point-to-Point Tunneling Protocol (PPTP)** was first created by a group of companies calling themselves the PPTP Forum. The group consisted of 3Com, Ascend Communications, Microsoft, ECI Telematics, and U.S. Robotics. The basic idea behind PPTP was to split up the functions of remote access in such a way where individuals and corporations could take advantage of the Internet's infrastructure to provide secure connectivity between remote clients and private networks. Remote users would dial into the local number of the ISP and then securely tunnel into their corporate network.

The most commonly used protocol for dial-up access to the Internet is the Point-to-Point Protocol (PPP). The first connection is to the remote access host using PPP. A second connection is then made over the PPP connection to a PPTP server on a private LAN. A typical example of a PPTP installation is a client computer running PPP and PPTP to access an ISP. Figure 7–6 illustrates this connectivity. PPTP permits a user running Windows 95 to dial into a Windows NT Server running the Remote Access Service (RAS), and it supports the equivalent of a private, encrypted dial-

Figure 7–6
PPTP Environment

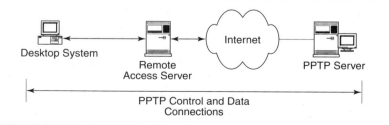

up session across the Internet. Similarly, a VPN could be established permanently across the Internet by leasing dedicated lines to an ISP at each end of a two-way link, and maintaining ongoing PPTP-based communications across that dedicated link.

Since it is based on PPP, PPTP is well suited to handling multiprotocol network traffic, particularly IP, IPX, and NetBEUI protocols. PPTP's design also makes it easier to outsource some of the support tasks to an ISP. By using RADIUS proxy servers, an ISP can authenticate dial-in users for corporate customers and create secure PPTP tunnels from the ISP's network access servers to the corporate PPTP servers. These PPTP servers then remove the PPTP encapsulation and forward the network packets to the appropriate destination on the private network. RADIUS is a popular remote-authentication dial-in user-service security system.

Layer 2 Tunneling Protocol

The **Layer 2 Tunneling Protocol (L2TP)** was created as the successor to two tunneling protocols, PPTP and L2F. Like PPTP, L2F was designed as a tunneling protocol, using its own definition of an encapsulation header for transmitting packets at Layer 2. One major difference between PPTP and L2F is that the L2F tunneling is not dependent on IP, enabling it to work with other physical media.

Layer 2 Tunneling Protocol (L2TP)

There are two levels of authentication of the user. Level one is by the ISP prior to setting up the tunnel and level two is when the connection is set up at the corporate gateway.

L2TP defines it's own tunneling protocol, based on the work of L2F. Work has continued on defining L2TP transport over a variety of packetized media such as X.25, Frame Relay, and ATM. Although many initial implementations of L2TP focus on using UDP on IP networks, it is possible to set up an L2TP system without using IP as a tunnel protocol entirely. A network using ATM or Frame Relay can also be deployed for L2TP tunnels. Figure 7–7 depicts the protocols involved in an L2TP connection between an access concentrator and a network server.

Because L2TP is a Layer 2 protocol, it offers users the same flexibility as PPTP for handling protocols other than IP, such as IPX and NetBEUI. Also, because it uses PPP for dial-up links, L2TP includes authentication mechanisms within PPP. The components of an L2TP system are essentially the same as those for PPTP: point-to-point protocols, tunnels, and authentication systems.

L2TP should be considered the next-generation VPN protocol, particularly for dial-in VPNs. Many vendors already have plans to supplant PPTP-based products with L2TP products. L2TP offers a number of the advantages of PPTP, particularly for handling multiple sessions over a single tunnel as well as assigning QoS parameters of different tunnels to the same site. In addition, L2TP's capability to run over such technologies as X.25, Frame Relay, and ATM while handling multiple network layer

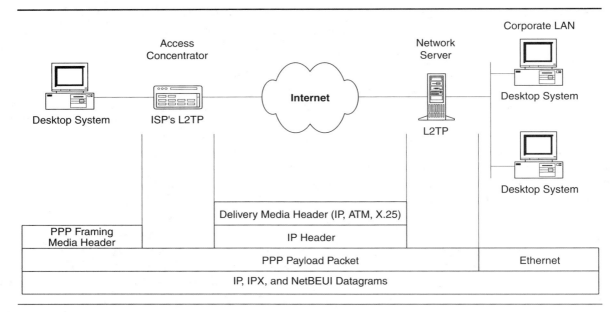

Figure 7–7
L2TP Connection Protocols

protocols, in addition to IP, affords users and ISPs a considerable amount of flexibility in designing VPNs. L2TP also provides stronger security for the corporate data, because it uses IPSec's Encapsulating Security Payload (ESP) for encrypting packets, even over a PPP link between the end-user and the ISP.

Layer 2 Forwarding Protocol

Layer 2 Forwarding (L2F) Protocol is a tunneling protocol developed by Cisco Systems, Inc. The key management requirements of service that are provided by Cisco's L2F implementation are as follows:

- Neither the remote end system nor its corporate hosts should require any special software to use this service in a secure manner.
- Authentication is provided by dial-up PPP and the various authentication protocols, as well as support for smart cards and one-time passwords; the user can manage the authentication independently of the ISP.
- Addressing will be as manageable as dedicated dial-up solutions; the address will be assigned by the remote user's respective corporation, and not by the ISP.
- Authorization will be managed by the corporation's remote users, as it would in a direct dial-up solution.
- Accounting will be performed by both the ISP and the user.

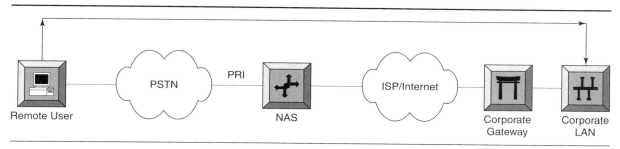

Figure 7–8
PPP Connection for Virtual Dial-up Topology

These requirements are primarily achieved based on the functionality provided by tunneling the remote user directly to the corporate location using the L2F protocol. In the case of PPP, all link control protocol and network control protocol negotiations take place at the remote user's corporate location. PPP is allowed to flow from the remote user and terminate at the corporate gateway. Figure 7–8 illustrates this process.

Internet Protocol Security

Internet Protocol Security (IPSec) is a standard created to add security to TCP/IP networking. It is a collection of security measures that addresses data privacy, integrity, authentication, and key management, in addition to tunneling. The IPSec system includes a considerable amount of flexibility in authentication and encryption algorithms, allowing it to meet the demands of both current and future networking situations.

The original TCP/IPs did not include any inherent security features. To address the issue of providing packet level security in IP, the IETF has been working on the IPSec protocols within their IP Security Working Group. The first protocols comprising IPSec, for authenticating and encrypting IP datagrams, were published by the IETF as RFCs 1825 to 1829 in 1995. Current specifications are included in RFCs 2401, 2402, 2406, 2408, and 2411.

These protocols set out the basics of the IPSec architecture, which include two different headers designed for IP packets. The IP **Authentication Header (AH)** is for authentication and the other, the **Encapsulating Security Payload (ESP)** is for encryption purposes. In addition, **Security Association (SA)**, which is a security agreement on key management, is an important concept in IPSec. An SA is a one-way relationship between a sender and a receiver that affords security services to the traffic carried on it.

The AH and ESP protocols can be applied either to authenticate or decrypt just the packet's payload or the entire IP header, including the IP

Internet Protocol Security (IPSec)

Authentication Header (AH)
Encapsulating Security Payload (ESP)
Security Association (SA)

addresses of the source and destination. Applying authentication and encryption in tunnel mode provides the greatest degree of security.

To enable secure communications between two parties, a system for exchanging keys is required. An IPSec Security Association is created between VPN sites to exchange keys and any pertinent details on the cryptographic algorithms that will be used for a session. Although manual exchanges of security associations and keys are possible for a small number of VPN sites, IPSec includes an involved, but workable framework for automatic key management called Internet Key Exchange (IKE) or ISAKMP/Oakley.

IPSec has multiple modes and services, including the following (Sheldon 2001):

- Connectionless integrity
- Confidentiality
- Data origin authentication
- Key management
- Replay protection

IPSec software can reside in stationary hosts, mobile clients, or security gateways. Only security gateways are needed if LAN-to-LAN tunnels are to be created. Mobile workers would require IPSec client software if they wanted to connect to a VPN site.

TCP/IP

TCP/IP consists of the Transmission Control Protocol, specified in RFC793, and the Internet Protocol, specified in RFC791 (*Guide to TCP/IP*, 1998). TCP operates at OSI Layer 4 and was designed for point-to-point communications between computers on the same network. IP operates at OSI Layer 3 and provides for communications between computers linked to different networks or to WANs. On a TCP/IP network, the packets of routed protocols contain internetwork layer addressing information, which allows user traffic to be directed from one network to another.

All legitimate TCP connections start with a **three-way handshake**, where the initiator sends a packet with the SYN code bit set, and the receiver responds with a packet with both the SYN and ACK code bits set. The initiator then completes the handshake by sending a packet with the ACK code bit set. This three-way handshake lets the two communicating systems agree on sequence numbers to use for the connection, which allows for proper transmission of packets in the correct sequence.

Address Resolution Protocol (ARP) and Reverse Address Resolution Protocol (RARP) are both Internet Protocol (IP) routed. ARP maps IP addresses to MAC addresses, which are used to send packets to their destinations. This means that ARP is used to obtain the physical address when only the logical address is known. The process includes two steps: (1) An

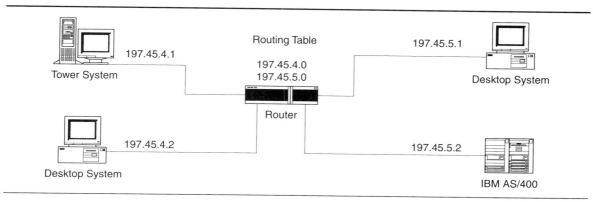

Figure 7-9
Router Configuration with Two Segments

ARP request with the IP address is broadcast onto the network; and (2) the node on which the IP address resides responds with the hardware address. RARP is similar to ARP in that it binds MAC addresses to IP addresses; however, it is used primarily for diskless workstations. The workstation uses the protocol to obtain its IP address from a server.

Routers connect two or more network segments and use the ARP in conjunction with their routing tables to transport data packets. A router requires an IP address for every network segment to which it is connected and a separate network interface for each network segment. Figure 7-9 provides an example of this router IP address and interface configuration. Note that each device on the network has both an IP and a MAC address. There are two segments in this example as indicated by the numbers 4 and 5 in the IP addresses.

When devices send packets to destinations not on their segment, the router connected to the segment on which the packet originated recognizes that the destination host is on a different subnetwork. The router's responsibility is to determine which network should receive the packet by referencing its routing table to determine which interface is connected to the destination network. If the destination address is not in the routing table, an ARP request must be issued to locate the path to the destination. This ARP request is forwarded through the network to as many routers as necessary to resolve the device address requested. This process continues until the hop count or time to live (TTL) reaches its maximum or the destination is located before discarding the packets.

When multiple routers exist on a network, they must be able to share routing information to update each other's routing tables. Routing protocols carry routing table information and do not transmit normal network traffic. Routers learn about paths to other networks in one of two ways: static or dynamic configurations. A static configuration is a manual process during which a network administrator must enter the paths used to traverse one segment to another into the router's routing table memory. With dynamic configurations, the routers communicate with each other and create and maintain routing tables automatically. Static routing configurations are time consuming and error prone, but in certain situations they are required. Following is a list of dynamic routing protocols that are used in the Enterprise networking infrastructure (Newman 2002):

- Routing Information Protocol (RIP)
- Border Gateway Protocol (BGP)
- Exterior Gateway Protocol (EGP)
- Interior Gateway Routing Protocol (IGRP)
- Enhanced Interior Gateway Routing Protocol (EIGRP)
- Open Shortest Path First Protocol (OSPF)

7.13 ELECTRONIC MAIL

electronic mail (e-mail)

Although **electronic mail (e-mail)** remains primarily character oriented, its ability to permit individuals to easily exchange information and files makes it the most popular networked application of any kind.

Millions of users are connected worldwide to the Internet via the global e-mail subsystem. From a business perspective, Internet e-mail offers one method of sending intercompany correspondence. Most companies have private networks that support e-mail transport to fellow employees, but not necessarily to employees of other companies. By adding Internet e-mail gateways to its private network, a company can send e-mail to users almost anywhere.

Internet electronic mail hosts exchange messages through the SMTP, which is in the TCP/IP architecture. E-mail hosts hold the mail until the subscriber is ready to read it. This allows the receiver to work on a client PC, which will often be turned off. E-mail hosts also transmit outgoing messages to other mail hosts. Together, mail hosts function as a network of electronic post offices.

E-mail Security

Millions of users rely on e-mail for a variety of purposes, including business correspondence and personal information. Many of these messages are sensitive, and users want to ensure their privacy is protected. The challenges for e-mail security include message confidentiality and encryption (Coleman 1997).

The contents of a message can be encrypted using a conventional encryption scheme such as DES. The most difficult technical challenge for such schemes is the secure exchange of encryption keys between pairs of correspondents. The goal is to prevent anyone but the intended recipient from reading the message.

Message authentication is often referred to as digital signature security. A digital signature, implemented using public-key encryption, makes use of two keys—a public key and a private key. If a block of data is encrypted with the sender's private key, any recipient may decrypt that block with the sender's public key. The recipient is assured the block must have originated from the alleged sender, since only the sender knows the private key.

Another development that might help in message security is the adoption of the Privacy Enhanced Mail (PEM) standard by e-mail vendors. This standard describes a common method of encapsulating encrypted messages. This IETF standard should help bring standard, secure e-mail to all users quicker.

7.14 REMOTE MONITORING

Remote monitoring (RMON) is a standard monitoring specification that enables various network monitors and network management systems to exchange network monitoring data. RMON provides network administrators with more freedom in selecting network-monitoring probes and network management stations with features that meet their particular networking needs.

remote monitoring (RMON)

The RMON specification defines a set of statistics and functions that can be exchanged between RMON-compliant console managers and network probes. RMON provides network administrators with comprehensive network-fault diagnosis, planning, and performance tuning information. The specifications are set out in RFC1757. This MIB complements the existing MIB-II (RFC1213). Special devices are attached to various subnets to collect information specific to the LAN. Examples of the types of data collected include

- Delivered packets per second
- Number of collisions
- Runt packet numbers (A runt is an Ethernet frame that is shorter than the valid minimum length, usually caused by a collision.)
- Giants (The opposite of a runt.)

Figure 7–10
RMON Agents

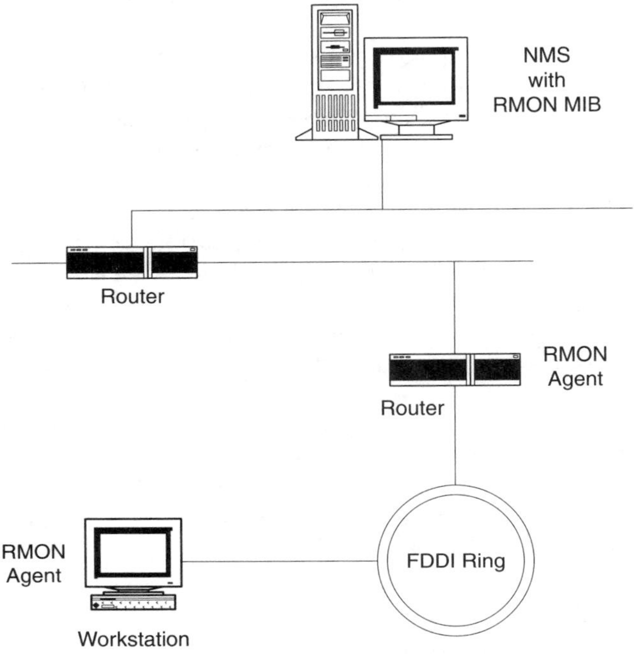

Each subnet is required to have an attached monitor collecting data. The devices can be servers, workstations, or routers. These remote monitors need to communicate with one or more network management stations. Figure 7–10 illustrates the components and connectivity to a RMON agent.

7.15 IPV4 AND IPV6 SECURITY

Internet Protocol Version 4 (IPv4)

Internet Protocol Version 6 (IPv6)

Internet Protocol Version 4 (IPv4) is the current version of the Internet Protocol, which is the fundamental protocol on which the Internet is based. While IPv4 has served its purpose for over 25 years, it has proven to be inadequate in terms of security and limitations of the address field. **Internet Protocol Version 6 (IPv6)** is the new proposed Internet protocol designed to replace and enhance IPv4. IPv6 has 128-bit addressing, autoconfiguration, new security features, and it supports real-time communications and multicasting (Newton 1999).

The Internet Architecture Board (IAB) issued a report in 1994 entitled *Security in the Internet Architecture*, identifying various security deficiencies in the Internet. The report stated the general consensus that the Internet needed more and better security and it identified key areas for security mechanisms. Among these were the need to secure the network infrastructure from unauthorized monitoring and control of network traffic and the need to secure end-user-to-end-user traffic using authentication and encryption mechanisms.

Many security incidents have occurred recently, impacting numerous network sites. The most serious types of attacks included IP spoofing, in which intruders create packets with false IP addresses and exploit applications that use authentication based on IP addresses. Incidents also included various forms of eavesdropping and packet sniffing in which attackers read transmitted information, including logon information and database contents (Stallings 2000).

A security mechanism employed by the U.S. government includes **security labels**. An optional IP header field that contains a security classification and handling label is used. This field allows hosts and routers to label IP traffic based on its security level. Other hosts and routers can then check these labels for arriving datagrams. Traffic can be accepted, rejected, or forwarded based on the labels. This labeling allows the implementation of government access control policies based on security clearances and information classifications. In response to these security issues, the IAB included authentication and encryption as necessary security features in the next-generation IP address, issued as IPv6.

security labels

7.16 CERN

CERN is the French acronym for the European Laboratory for Particle Physics, which is located in Geneva, Switzerland. The protocols that brought about the Web were created here. The protocols allowed users with browsers to access information on Web servers. CERN proxy services are a set of protocols that are standards for implementing proxy services on intranets and the Internet. Note that a proxy server is a device used for caching and security.

CERN

The proxy server can be viewed as a gateway between two networks, usually a private internal network and the Internet. Proxy servers and gateways are discussed in Chapter 6. With the proliferation of internal intranets, a proxy server may be used to control internal network traffic. CERN's proxy server provides application-aware proxy support for its HTTP Web servers. Microsoft and Novell implement CERN proxy services in their proxy server implementations.

7.17 INTERNET AND WEB RESOURCES

SOCKS Protocol V5	www.internic.net/index.html
NEC's SOCKS site	www.socks.nec.com
Aventail Corp. SOCKS	www.aventail.com/educate/security.html
RootKits	www.linuxsecurity.com
Packetstorm Security	www.packetstorm.security.org
Security Focus Bugtraq	www.securityfocus.com
IETF SNMPv3	www.ietf.cnri.reston.va.us/rfc/rfc2570.txt

■ SUMMARY

Protecting the data, database, application, and systems software are major goals and requirements of Enterprise management. In many cases, these assets represent the organization, which would fail if these resources were destroyed. Methods to prevent such an occurrence include backup and data archiving. If there is a disaster from internal or external forces, a process to restore these resources is possible through careful planning.

A database management system (DBMS) provides for data storage and access and manipulation of that data. It is essential that integrity and security of the DBMS receive high priority in the day-to-day Enterprise operation. It is essential that these systems be protected against attacks from both internal and external sources. These threats can include denial of service, modifications and deletions of data, and illegal dissemination of data.

Software components of the computer system include the operating system, the DBMS, software utilities, and application programs. The computer system uses directories to identify legitimate users and the files that are accessible. Directory services identify the resources, such as files or printers that are available to each legitimate user. The access rights of each user determine access to these resources.

An application program interface (API) defines how the programs interface with the various operating systems. Operating systems prevalent in today's Enterprise computer network include Windows NT, Windows NT Server, Windows 2000, Windows XP, UNIX, Novell NetWare, and others.

There are many software tools available in the computer network. These include Internet search tools such as browsers and file transfer software. Examples include FTP, TFTP, Gopher, Newsgroups, and telnet. Software is a critical component of virtual and distributed networks. Security protocols include PPTP, PPP, L2TP, L2F, IPSec, and TCP/IP.

Standards and protocols are important and essential parts of software components. The OSI specifications provide the guidelines and standards in the form of protocols. These protocols include various security mechanisms to protect the Enterprise from malicious attacks. These include challenge/response protocols such as PPP and CHAP, and authentication protocols such as SOCKS and EAP. Standards for wireless technologies include Bluetooth, 3G, GPRS, EDGE, UMTS, CDMS2000, and WCDMA.

Enterprise network management is a major component of the Enterprise operation. Network management helps to manage and control the reliability and availability between LANs, MANs, and WANs. It is used for real time network surveillance and network traffic management.

Key Terms

Abstract Syntax Notation One (ASN.1)
Access rights
Agent
Aggregation problem
Application programming interfaces (API)
Archive
Authentication Header (AH)
Authenticode
Backup
Bluetooth
Browser
Cache
CDMA2000
CERN
Challenge-Handshake Authentication Protocol (CHAP)
Challenge/response
Common Data Security Architecture (CDSA)
Crypto API (CAPI)
Data

Database
Database Management System (DBMS)
Data warehouse
Data warehousing
Directory
Directory service
Electronic Mail (e-mail)
Encapsulating Security Payload (ESP)
Enhanced Data GSM Evolution (EDGE)
Extensible Authentication Protocol (EAP)
File Transfer Protocol (FTP)
General Packet Radio Service (GPRS)
Generic Security Services API (GSS-API)
Gopher
Inference problem
Internet Explorer
Internet Protocol Security (IPSec)
Internet Protocol Version 4 (IPv4)
Internet Protocol Version 6 (IPv6)
Intranetware

Java
Kernel
Layer 2 Forwarding Protocol (L2P)
Layer 2 Tunneling Protocol (L2TP)
Management Information Base (MIB)
MIB tree
National Institute of Standards and Technology (NIST)
Netcat
NetWare
NetWare Directory Services (NDS)
Network Management
Network Management System (NMS)
Newsgroups
Object identifier (OID)
Operating System (OS)
Open Systems Interconnection (OSI)
Permissions
Piconet
Point-to-Point Protocol (PPP)
Point-to-Point Tunneling Protocol (PPTP)
Polyinstantiation
PEM
Protocol
Remote Monitoring (RMON)
Replication
Rights
RootKits
Scatternet
Schema
Security Association (SA)
Security labels
Simple Network Management Protocol (SNMP)
SNMPv2
SNMPv3
SOCKS
SOCKS5
Stack-based buffer overflow
Structure of Management Information (SMI)
TCP/IP
Telnet
Third Generation (3G)
Three-way handshake
Trivial File Transfer Protocol (TFTP)
Trustees
Universal Mobile Telecommunications System (UMTS)
UNIX
USENET
Wideband Code Division Multiple Access (WCDMA)
Windows NT Server
Windows NT Workstation
X.500

REVIEW EXERCISES

Questions and Analysis

Provide a short answer or in-depth analysis as required.

1. What are the basic goals of the NIST?
2. Describe backup and archiving procedures.
3. Why are backup and archiving important?
4. Describe a data and database restoration process.
5. What are the features of a DBMS?

6. Give three examples of malicious attacks that occur to a database.
7. What is the function of a directory?
8. What is the difference between rights and permissions?
9. Provide examples of operating systems. Discuss the features of the Windows' products.
10. How is security implemented in the Windows' products.
11. Explain UNIX security measures.
12. Identify the services of the NetWare system.
13. How is NetWare security assured?
14. Identify the various Internet search tools.
15. What is the function of FTP?
16. How is telnet used?
17. What are the security provisions of CHAP?
18. What are the security provisions of EAP?
19. Discuss CDSA. Describe its architecture.
20. Why is it important to understand the wireless technologies?
21. Provide an overview of a network management system.
22. What are the elements in a network management system?
23. What are the functions of SNMP?
24. What are the relationships between a MIB, SMI, OID, and ASN-1?
25. Describe the concepts of a managed device and an agent.
26. What protocols are used in a VPN?
27. Describe the security capabilities of PPP and PPTP.
28. What are the security provisions of L2TP?
29. What are the security provisions of L2F?
30. Describe the IPSec standard. Identify its major components.
31. Describe the functionality of TCP/IP.
32. Discuss e-mail security.
33. How is remote monitoring used?
34. What is the difference between IPv4 and IPv6?

Definitions and Terms

Provide the most correct answer from the list of key terms.

1. Part of the SNMP structure that is loaded into each managed device to be monitored.
2. A header included in IPSec, which provides authentication and integrity of the IP packet.
3. The process of copying data to magnetic tape or optical disks or copying or replicating information to other systems.
4. Software that looks at various types of Internet resources; also called a WWW client.
5. A process where a unique value or identification is generated and presented as a challenge that is then subjected to a defined algorithmic process to reproduce a response.
6. A subject-oriented, integrated, time variant, nonvolatile collection of data used primarily to support organizational decisions.
7. A TCP/IP service that permits the transfer of files between systems.
8. The core of UNIX and Windows NT/2000 OS, responsible for sharing the processor among running processes.
9. A collection of objects that can be accessed via a network management protocol.
10. A protocol that allows a computer to use a telephone line and a modem to connect directly to the Internet.

11. A formal description of messages to be exchanged and rules to follow for two or more systems to exchange information.

12. The rules used to define the objects that can be accessed via a network management protocol.

13. A standard Internet service that allows a user terminal access to a remote computer over a network.

14. A user or group of users who have specific access rights to work with a particular directory, file, or object.

15. A worldwide system of thousands of discussion areas, called newsgroups.

True/False Statements

Indicate whether the statement is true or false.

1. A database backup, archive, and replication all accomplish the same task.

2. Malicious attacks that can impact a database include disclosure, alteration, and inaccessibility of data.

3. Access rights are the keys that define a user's ability to enter a network facility.

4. TFTP is more secure than FTP.

5. The main defense for a telnet session is to provide a high level of password protection.

6. The challenge/response process includes a cryptographic operation to ensure privacy.

7. Authentication methods available for the connection to an ISP include EAP, CHAP, PAP, and RSA.

8. CDSA is a specification that provides a set of APIs that independent software developers can use to embed security into desktop and network applications.

9. Bluetooth is the code name for a transmission process that uses laser techniques to broadcast signals in a small area.

10. Third Generation (3G) technology may provide connectivity to all users at all times and in all places.

11. An NMS is defined as the systems or actions that help maintain, characterize, or troubleshoot a network.

12. SNMP is the means by which the network management station and a managed router exchange system traps.

13. Virtual private networking is possible because of PPTP, IPSec, L2F, and T2TP tunneling protocols.

14. The three main issues with e-mail security include confidentiality, integrity, and quality.

15. IPv4 was designed to provide a higher level of security than that offered by the older IPv6 protocol.

Activities and Research

1. Research and produce a list of database management systems. Show providers, major features, and benefits for each product.

2. Research software systems that provide support for computer network applications. Identify features and ease of use.

3. Produce an overview of the responsibilities of the NIST.

4. Produce a matrix comparing the various computer operating systems. Include Windows, UNIX, and NetWare products.

5. Identify the currently available browsers.

6. Use FTP or TFTP to download a file.

7. Access a newsgroup and produce an overview of some current topic.

8. Use the telnet function to access some network device. Show the steps and the processes required to accomplish this task.

9. Produce a comparison of the various protocols used in securing Enterprise networks.

10. Identify the currently available APIs. Show the application of each.

11. Provide a comparison of the various wireless protocols.

12. Draw a network that includes multiple locations that have managed devices with SNMP capabilities. Show how network management traffic might flow back to a centralized network management system.

13. Produce a drawing of a virtual network that uses some Internet security protocol. Show the connectivity between the locations.

14. Research dynamic routing protocols and produce a comparison matrix.

15. Research IPv4 and IPv6 and produce a matrix of features and differences.

CHAPTER 8

Attacks, Threats, and Viruses

OBJECTIVES

Material presented in this chapter should enable the reader to

- Identify the different types of attacks, threats, and intrusions that can occur in the Enterprise network.
- Learn the many terms, definitions, and topics that are part of security countermeasures for neutralizing computer networking attacks and intrusions.
- Become familiar with the virus and the various types of viruses that can invade a computer network.
- Recognize the various vulnerabilities that are inherent in the organization's resources.
- Identify a threat target or source and use a threat assessment process to determine the correct response to neutralize it.
- Look at the issues associated with cybercrime and cyberterrorism.
- Become familiar with the issues of ethics and social engineering in the Enterprise environment.
- Recognize a threat and employ a countermeasure to eliminate or minimize its impact on the Enterprise computer and networking asset.
- Understand the need for a gap analysis to identify holes in the security shield.
- Look at the issue of ethics in the Enterprise computer network environment.

■ INTRODUCTION

A **threat** is described as an expression of intention to inflict some evil, an indication of impending danger or harm, or someone that is regarded as a possible danger or menace. A threat to the Enterprise network is described as any potential adverse occurrence that can do harm, interrupt the systems using the network, or cause a monetary loss to the organization (FitzGerald 1999). Threat categories include security and privacy threats, integrity threats, delay and denial threats, and intellectual property threats.

Security threats can be classified as active or passive. Active attacks include the modification of transmitted data and attempts to gain unauthorized access to computer and networking systems, whereas passive attacks include eavesdropping and monitoring transmissions (Stallings 2001).

Destructive attacks on network security can take the form of computer viruses, which can rapidly spread among computers attached to the network. Viruses can corrupt data, delete files, slow system operations by spawning spurious processes, and prevent applications from saving files. Viruses can spread through tainted software and also via infected word processing document files. Antivirus software recognizes known viruses and can remove them from infected computers. New viruses continue to appear and they are identified as soon as possible by vendors of antivirus software. This software is usually available for download from the software vendors.

Denial of service attacks disrupt network operation by flooding the network with spurious messages, usually through an Internet connection. Attackers prepare denial of service attacks by invading computers that can legitimately access a target network and surreptitiously setting up those computers for a concerted attack. Such attacks are difficult to anticipate and can be combated only when they are underway.

Not all breaks in security are malicious, however the results can be just as damaging. Some may stem from a purposeful interruption of a system's operation and some may be accidental, such as a hardware failure. Security breaches must be minimized, whether they are malicious or accidental. The overall goal is to protect the Enterprise network from any attack, and to prevent theft, destruction, and corruption of the organization's assets.

threat

Countermeasures are procedures, software, and hardware that can detect and prevent computer networking security threats. Various countermeasures, when used in a coordinated effort, can protect against secrecy, integrity, and delay and denial attacks. A gap analysis can be performed to identify gaps or holes in the Enterprise network security program.

A new threat has arisen with the advent of the cybercrime, information warfare, and cyberterrorism. The Federal government is actively engaged in identifying and defeating these types of threats.

8.1 ENTERPRISE THREATS

There are many opportunities for threats to occur in the Enterprise computer and networking environment. This is particularly true in the Internetworking environment, where attackers, crackers, and hackers abound. A threat from these individuals can have a potentially adverse effect on the organization's assets and resources. Threats can be listed in generic terms, however they usually involve fraud, theft of data, destruction of data, blockage of access, and so on.

As these people continue to create more ingenious methods for penetrating the Enterprise network, administrators must take a comprehensive approach to security. The use of antivirus programs, firewalls, and the triple A techniques are good starting places; however much more is required to ensure network security. Of major importance is continuous vigilance with a security package of intrusion detection techniques, and updating of software with the latest patches. When the network is hacked, the response should be quick, thorough, decisive, and effective.

It is essential to identify the various threats and rank them in order of importance and impact on the organization. These assignments can be made on the basis of dollar loss, embarrassment created, monetary liability, or probability of occurrence. The most common threats to an organization include the following (FitzGerald 1999):

- Virus
- Device failure
- Internal hacker
- Equipment theft
- External hacker
- Natural disaster
- Industrial espionage

Surveys show that the most common security problem to an organization is the virus. The relative importance of a threat to the Enterprise usually depends on the business being conducted. For example, a financial institution or e-commerce site might be a frequent victim of an attack, whereas a fast-food Web restaurant might be spared. The impact of a threat could have different ramifications between these two types of sites.

These security problems could possibly have originated with attackers gathering information concerning the Enterprise organization. If the administrator was not prepared and not vigilant, attacks may be forthcoming.

8.2 VIRUS THREATS

In the past a virus was something that would send numerous people scurrying for medical attention. In today's environment, a virus can bankrupt a company and destroy a wealth of information. What is this new threat and what can be done to negate its impact? A **virus** is a computer program that infects other programs via replication. It clones itself from disk-to-disk or from one system to the next over computer networks. A virus executes and accomplishes it's damage when the host program executes. Some new viruses attack macros in programs like Microsoft Word and do their damage when the macro is executed. (Note that a macro is a set of instructions in a computer source language, such as Assembler).

The modifications caused by viruses may be harmless, such as a birthday greeting on a certain day, however they often cause considerable damage by destroying boot records, file tables, and user data. To maintain a protective stance against viruses the user must keep abreast of the current virus threats. There are a number of Web sites that provide information on viruses, such as recent activities and hoaxes. The URLs for these Web sites are listed at the end of this chapter.

There are a number of virus classifications set forth by the **National Computer Security Association (NCSA)** and the **Computer Security Institute (CSI)**. A brief description of each virus category is as follows (Sheldon 1998):

- **Boot Sector Virus**—Infects the master boot record by overwriting the original boot code with an infected version. A boot virus is spread to the disk drive when the system is booted with an infected floppy disk.
- **File-infecting Virus**—Infects executable files with com, exe, and ovl extensions. Operating system files are targeted. The original program is often replaced with an infected program.
- **Macro Virus**—Includes executable programs that attach themselves to documents created in Microsoft Word and Excel. The virus executes and does its damage when the user receives a Word or Excel document and executes a macro.
- **Multipart Virus**—Infects boot sectors as well as executable files. They are a real problem because they use stealth and polymorphism to prevent detection.

Polymorphic Virus

Stealth Virus

- **Polymorphic Virus**—Changes its appearance to avoid detection by antivirus software. It encrypts itself with a special algorithm that changes every time an infection occurs.
- **Stealth Virus**—Attempts to hide itself from the operating system and antivirus software. They stay in memory to intercept attempts to use the operating system and hide changes made to file sizes.

8.3 COUNTERING THE VIRUS THREAT

Computer Emergency Response Team (CERT)
CERT Coordination Center (CCC)

The **Computer Emergency Response Team (CERT)** is an organization with teams around the world that recognizes and responds to computer attacks. The **CERT Coordination Center (CCC)** is responsible for studying Internet security and responding to security incidents reported to it. CERT publishes various security alerts and develops plans for individual computer sites to improve their security. CERT is part of the Networked Systems Survivability program in the Software Engineering Institute, which is a federally funded research and development center at Carnegie Mellon University. The program was established to provide the following:

- Study Internet security vulnerabilities
- Provide incident response services to victimized sites
- Publish a variety of security alerts
- Research security and survivability in WAN computing
- Develop information to help improve host security

A system becomes contaminated with a virus through file system activity. A contaminated file is either copied from a floppy disk or downloaded from an online service. Users transport viruses from home and work on their portable computers, which have access to the Internet and other network services.

When files are contaminated, they may increase in size, which makes it relatively easy for a virus detection program to report such problems. However, stealth viruses are able to spoof the preinfected file size of a document so it appears that nothing has changed. Active stealth viruses can fight back against virus detection programs by disabling their detection functions.

It is, therefore, smarter to protect against a virus instead of fixing the damage. It is easy to avoid viruses if the user is careful to never copy files from unknown or untrusted sources. This is easier said than done. The Internet and Web have created a whole new way to spread viruses. It is now possible to execute programs while browsing the Web without actually copying a file to the user's system disk. Java and ActiveX modules are automatically copied to the user's system while browsing the Web. Web security programs such as Norton's Internet Security can give the user tools to avoid this situation.

Trojan horse programs and worms (described later) are usually installed by employees or contractors inside the organization, who have specific intentions to capture data or damage some system. The best protection is to limit system availability and carefully monitor the activity of employees and contractors. This applies particularly to people who might be leaving the organization or who are suspected of being malicious.

Even after detecting and eliminating a virus infection, there is still a chance that the virus is lurking somewhere in the organization, ready to reinfect systems. If the virus is detectable, all system must be checked for its existence. It might be necessary to run a virus scan on all floppy diskettes and hard drives.

The administrator must develop a security program for addressing viruses and other destructive elements. The administrator should keep up with the latest virus information by reading weekly computer journals or joining organizations like the NCSA. Frequently checking the NCSA and CSI Web sites for virus information is also recommended. There are many antivirus products, such as McAfee Office and Norton AntiVirus that prevent the spread of viruses and clean specific ones off the system. A directory listing of such products is available at www.faulkner.com/products/faccts/00016757.htm. Access to Faulker resources and references will require a userid and password, which can be obtained by accessing www.faulkner.com.

Virus Control Policies

An effective antivirus strategy must include policy, procedures, and technology. The following six general policies can be established to assist in controlling viruses:

- Create an education program oriented toward virus protection.
- Post regular bulletins about virus problems.
- Never transfer files from an unknown or untrusted source unless the computer has an antivirus scan utility installed.
- Test new programs or open documents on a "quarantine computer" before introducing them to the production environment.
- Secure computers to prevent malicious people from infecting systems or installing Trojan horse programs.
- Use an operating system that uses a secure logon and authentication process.

As collaborative applications such as groupware have become more commonplace in organizations, a new method of virus infection and virus reinfection has emerged. Since groupware message and data are stored in a shared database and since documents can be distributed throughout the network for document conferencing or workflow automation, viruses can be quickly spread throughout the network. Because groupware servers replicate their databases to other servers, a virus will continue to spread. Even if the virus is eliminated from the originating server, responses from still-infected servers will reinfect the original server. This infection/deinfection/reinfection cycle can continue until the virus is purged system-wide (Goldman 1998).

Software Detection

Computer viruses provide ever-changing means for hacking attacks on Enterprise networks, and antivirus measures are essential elements of defense against these attacks. Most networks employ antivirus software at numerous points throughout the network. Antivirus checking is most effective when it is performed automatically, which means it is always working and vigilant. The downside of antivirus software is a virus cannot be countered if it is new and unknown. Hackers are constantly inventing new viruses; therefore it is important to constantly update antivirus software with the latest versions.

There are a number of virus protection programs for desktop computers and for networks. Virus detectors attempt to detect known viruses that have infected files or memory locations. On workstations, interrupts can be monitored to detect and stop system subroutine calls that might indicate virus activity. Figure 8–1 depicts such an activity. Another technique is to look for unique identifiers that indicate a virus. These methods are called signature scanning. So that an antivirus program may detect the latest viruses, it must be periodically updated with the latest identifiers from the software vendor. The downside is that signature scanning is only as good as the most recent signature file. A representative sample of software detection programs include the following:

- BindView EMS
- LT Auditor +
- AuditTrack
- Kane Security Analyst

Additional tools that probe for system vulnerabilities include **Security Administrator Tool for Analyzing Networks (SATAN)**, and other scanning and intrusion detection packages.

Figure 8–1
Antivirus Software Detection

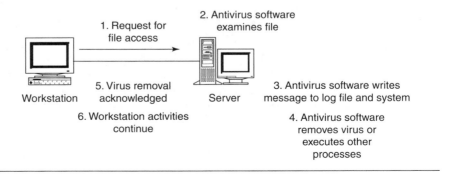

Table 8–1 Virus Vulnerabilities and Protective Measures

Element	Vulnerabilities	Protective Measures
Client PC	Infected diskettes	Strict diskette scanning policy
	Groupware conferences with infected documents	Autoscan at system startup
Remote access users	Frequent up/down loading of data	Strict diskette scanning policy
	Linking to customer sites	Strict policy about connection to corporate site after linking to other sites
Internet access	Downloaded viruses	Firewalls
	Downloaded hostile agents	User education concerning dangers of downloading
Server	Infected documents stored by attached clients	Autoscan run at least daily
	Infected documents replicated from other groupware servers	Active monitoring virus checking Rigorous backup procedures Audit logs to track down sources

Table 8–1 identifies various virus infection points of attack and protective measures that can be employed as countermeasures (Goldman 1998).

War Dialers

Before an attack can occur, the intruder must identify a target. One method is called a **War Dialer**, which is a program that tries a set of sequentially changing numbers to determine which ones respond positively. These are usually telephone numbers, however they can be passwords. There are many computer networks and voice systems that are accessible via some dial-up access. After these identities are established, the attacker can dial back the discovered listening modems and make attempts to enter the organization's network. It is essential that the administrator know the criteria being used by the hacker. A general list of factors has been developed that should be considered when employing a security shield against this type of threat. These are as follows (McClure 2001):

War Dialer

- The connection has a time-out or attempt-out threshold.
- Exceeding the threshold drops the connection.
- The connection is only allowed at certain times.
- Authentication levels required include userid, password, or both.
- The connection has a unique identification method, such as SecurID.
- The number of userid and password characters is deterministic.

- The makeup of the userid and password is deterministic.
- Will break characters produce any additional information?
- Are system banners present and what information is contained therein?

8.4 MALICIOUS ATTACKS

Security threats can be either active or passive, however both can have negative repercussions for the Enterprise network. There are a number of malicious attacks in addition to virus infestations that can impact the security of an organization's resources. Active threats include brute force, masquerading, address spoofing, session hijacking, replay, man-in-the-middle, and dictionary attacks. Passive threats include eavesdropping and monitoring. This section concentrates on providing descriptions of each and suggests methods to successfully counter them (Blacharski 1998).

port scanning

A prelude to these attacks might include active **port scanning** of the Enterprise's network devices. This is the process of connecting to TCP and UDP ports on the target systems to determine what services are running or in a listening state. Identifying listening state is critical to determining the type of operating system and applications in use.

socket

Both TCP and UDP use ports and port numbers to identify application protocols. A typical session involves sending packets from a source IP address and port to a destination address and port. The packet headers contain source and destination port address information and flow between the applications at either end of the connection. The endpoint of a connection is called a **socket**, which means that a socket identifies a specific application running on a specific computer. Port numbers are published in RFC 1700.

Brute Force Attack

brute force

Brute force is a type of attack under which every possible combination of a password is attempted. Forty-bit encryption is considered weak and vulnerable to attack. Although the 56-bit DES encryption standard is approved under certain circumstances, it is not necessarily hack proof. The 128-bit encryption is considered by most to be impregnable, however it may be possible, given the time and processing power, to even crack it.

cracker

Those who attack computer networks constantly seek out new methods while still hammering away at the old ones. Internal attacks and corporate espionage are real and can take place every day in many different organizations. As many different types of attacks exist as do attackers, and a standard approach is called brute force. A brute force attack is a cracker term for trying passwords on a system until it cracks. The attacker merely applies all possible combinations of a key to an algorithm until the message is deciphered. Remember from Chapter 1 a **cracker** is a person who "cracks" computer and telephone systems by gaining access to passwords, and a **hacker** is someone who "hacks" away at a program until it works.

hacker

Forty-bit encryption has long been considered weak and vulnerable to attack. Although the government allows the use of 56-bit encryption standards in certain circumstances, this standard has not proven to be hack-proof. Government experts had estimated that 56-bit encryption would take many years to crack, but clever crackers and hackers have proven otherwise. With today's large-scale computers, it is possible to try millions of combinations of passwords in a short period of time. Given unlimited time and access to a number of computers over the Internet, it is possible to crack most algorithms. A forty-bit encryption can be cracked in about an hour. The 56-bit encryption is many more times difficult, but still can be cracked. The 128-bit encryption is considered by most as impregnable and would require 4.7 trillion billion times as many calculations as would be required to crack 56-bit encryption (Blacharski 1998).

It is important to note that there is more to a key-search attack than simply trying all possible keys. Unless known plaintext is provided, the cryptanalyst must be able to recognize plaintext as plaintext. If the message is just plaintext in English, then the result is easily observable, although the task of recognizing English would have to be automated. If the text messages were compressed before encryption, then recognition would be more difficult. If the target contains compressed numerical data, it becomes more difficult to automate. To supplement the brute-force approach, some degree of knowledge about the plaintext is required, and some means of automatically distinguishing plain text from garble is necessary. If the only form of attack that could be made on an encryption algorithm is brute force, the corrective measure is obvious—use longer keys.

Masquerading

A **masquerade** takes place when an entity pretends to be a different entity. A masquerade attack usually includes one of the other forms of active attack, such as address spoofing or replaying. Authentication sequences can be captured and replayed after a valid authentication sequence has taken place. This enables an authorized entity with few privileges to obtain extra privileges by impersonating an entity that has those privileges.

masquerade

Eavesdropping

Eavesdropping, or snarfing, occurs when a host sets its network interface on promiscuous mode and copies packets that pass by for later analysis. It is possible to attach hardware and software, unknown to the legitimate users, and monitor and analyze all packets on that segment of the transmission media.

eavesdropping

A network protocol specifies how packets are identified and labeled, which enables a computer to determine its destination. Because the specifications for network protocols are widely published, a third party can easily interpret the network packets and develop a packet sniffer. A **packet sniffer** is a software application that uses a hardware adapter card in promiscuous mode to capture all network packets sent across a LAN segment (Cisco Press 2001). Because some network applications distribute network packets in plaintext, a packet sniffer can provide its user with meaningful and sensitive information, such as user account names and passwords.

Address Spoofing

Spoofing is a type of attack in which one computer disguises itself as another order to gain access to a system. This is generally done when an outside computer pretends to be a computer that exists on the legitimate network. If the local router is not configured to filter out incoming packets whose source address is internal, a spoofing attack can occur. Using IP spoofing, an attacker can easily masquerade as an authorized user. Because of this vulnerability, valuable information should be protected by more than a client's identity. Protection based solely on a client's identity should be limited to situations where the host computer is completely known and trusted. IP spoofing is facilitated by using an IP address that is within the range of the network or by using an authorized external IP address that is trusted for specific access to specific resources on the network.

An IP spoofing attack is normally limited to the injection of data or commands into an existing stream of data passed between a client and server application or a peer-to-peer network connection. To enable bidirectional communication, the attacker must change all routing tables to point to the spoofed IP address. If the attacker manages to change the routing tables, all network packets addressed to the spoofed IP address can be intercepted. The intruder can then respond to the messages as a trusted user (Cisco Press 2001).

A CERT advisory on IP spoofing reports that the CERT Coordination Center has received reports of attacks in which intruders create packets with spoofed source IP addresses (Blacharski 1998). This exploitation leads to user and root access on the target system. This means that the intruder can take over login connections and create havoc.

Session Hijacking

In **session hijacking**, instead of attempting to initiate a session via spoofing, the attacker attempts to take over an existing connection between two network computers. The first step in this attack is for the attacker to take control of a network device on the LAN, such as a firewall or another computer, so that the connection can be monitored. Monitoring the connection allows the attacker to determine the sequence numbers used by the sender and receiver.

After determining the sequence numbering, the attacker can generate traffic that appears to come from one of the communicating parties, stealing the session from one of the legitimate users. As in IP spoofing, the attacker would overload one of the communicating devices with excess packets so that it drops out of the communications session.

Session hijacking points out the need for reliable means of identifying the other party in a session. It is possible for the legitimate user who initially started a session to suddenly be replaced, maybe unknowingly, by an intruder for the remainder of a communications session. This calls for a scheme that authenticates the data's source throughout the transmission; however, even the strongest authentication methods are not always successful in preventing hijacking attacks, which means that all transmissions might need to be encrypted.

Replay Attack

Replay involves the passive capture of a data unit and its subsequent retransmission to produce an unauthorized effect. The receipt of duplicate, authenticated IP packets may disrupt service or have some other undesired consequence.

<small>replay</small>

Systems can be broken through replay attacks when old messages or parts of old messages are reused to deceive system users. This helps intruders to gain information that allows unauthorized access into a system. This is possible when partial encrypted or decrypted files are left on computer hard drives.

Man-in-the-middle Attack

A **man-in-the-middle attack** is a type of attack that takes advantage of the store-and-forward mechanism used by insecure networks such as the Internet. In this type of attack, an attacker gets between two parties and intercepts messages before transferring them onto their intended destination. Web spoofing is a type of man-in-the-middle attack in which the user believes a secured session exists with a particular Web server. The user then can be duped into supplying the attacker with passwords, credit card information, and other private information.

<small>man-in-the-middle attack</small>

To use encryption, the user must first exchange encryption keys. Exchanging unprotected keys over the network could easily defeat the whole purpose of the system, because those keys could be intercepted. A sophisticated attacker employing spoofing, session hijacking, and eavesdropping (sniffing) could actually intercept a key exchange in an insecure network. The attacker's key could be planted early in the process so that this intruder would appear to the sender as a legitimate recipient, when in fact it was a man-in-the-middle (Kosiur 1998).

The possible uses of such attacks are theft of information and data, denial of service, corruption of transmitted data, gaining access into the organization's internal computer and network resource, and introduction of new information into network sessions (Cisco Press 2001).

Dictionary Attack

The **dictionary attack** is a simple attack that illustrates the need for carefully chosen passwords. A simple cracker program takes all the words from a dictionary file and attempts to gain entry by entering each one as a password. End users often select a common dictionary word as a password, which is poor practice. A nonsensical combination of letters and numbers should be selected instead, and they should not reflect any personal information about the user. Additional types of programs that are destructive to systems include logic bombs, Trojan horses, worms, and bacteria.

Logic Bomb

A **logic bomb** is basically a Trojan horse with a timing device. It initiates some destructive mechanism based on a clock. As an example, a disgruntled employee may create a bomb to execute after leaving the company.

Trojan Horse

A **Trojan Horse** is a program that some unscrupulous person installs on a computer to capture passwords when entered by an unsuspecting user. These programs may perform an ostensibly useful function, but they contain a hidden function that compromises the host system's security. These programs can be installed on a computer when the user leaves the computer logged in when unattended.

Individuals may also attempt to capture passwords that are transmitted across communication channels. Some programs and network operating systems (NOSs) send unencrypted passwords that are easy to capture and reuse. Several methods that can thwart this situation are token-based authentication or challenge/response schemes. These schemes are discussed later.

The Trojan horse program is the most widely used class of computer attack methods. An attacker must trick the user into running a Trojan horse program by making it appear attractive and disguising its true nature. Some Trojan horse programs are merely designed to destroy data or crash systems, while others allow attackers to steal data or even remotely control computer systems. Another attacker method closely allied with the Trojan horse is called the backdoor, which is discussed later in the chapter.

After installing a Trojan horse backdoor, attackers often attempt to cover their tracks by manipulating the computer system files. This can be accomplished by deleting or modifying the computer system logs that show all of the processor activity. System log management is important to defend against log-editing attacks.

Worm

A **worm** is often mistaken for a virus. It is a single destructive program on a single system, often planted by someone who has direct access to the system. It is a program that duplicates itself repeatedly, potentially worming its way through an entire network. The Internet worm was the most famous; it successfully and accidentally duplicated itself on many systems across the Internet.

worm

Bacteria

Bacteria are programs that do not explicitly damage any files. Their sole purpose is replication. A typical bacteria program may do nothing more than execute multiple copies of itself simultaneously on a multiprogramming system. It may create multiple new files, each of which is a copy of the original source file of the bacteria program. Both of these programs may replicate twice, ad infinitum. Bacteria reproduce exponentially, eventually taking up all the processor capacity, memory and disk space, which denies users to those resources (Stallings 2000).

bacteria

8.5 OTHER SECURITY BREACHES

Not all breaks in security are malicious. Although some result from a purposeful disruption of system operations, others are accidentally caused by hardware and software failures. It matters not whether they are accidental or malicious; a security break damages an organization's credibility. Activities that can cause a **security breach** are as follows (Flynn 2001):

security breach

- Denial of Service
- Distributed Denial of Service
- Browsing
- Wire tapping
- Incorrect data encoding
- Accidental data modifications
- Backdoor

Denial Of Service

Denial of Service (DoS) attacks result in the blocking of a resource to legitimate users. They are synchronized attempts to deny service by causing a computer to perform an unproductive task over and over, making the system unavailable to perform legitimate operations. When a disk fills up, an account is locked out, a computer crashes, or a CPU slows down, service is denied. The attacks come in many forms and different levels of severity and can cost an organization millions of dollars in lost revenue.

Denial of Service (DoS)

Two basic types of denial of service attacks include logic and flooding. Logic attacks exploit software flaws to crash or seriously degrade the performance of remote servers. Many of these attacks can be prevented by fixing the faulty software or by filtering particular packet sequences. Flooding attacks overwhelm the victim's CPU, memory, or network resources by sending large numbers of spurious requests. Since there is no simple method to distinguish a good request from a bad one, these attacks are especially insidious.

When involving specific network server applications, such as an HTTP or FTP server, these attacks can focus on acquiring and keeping open all the available connections supported by that server, which effectively locks out valid users of the server or service. DoS attacks can also be implemented using common Internet protocols, such as TCP and ICMP. Most DoS attacks exploit weaknesses in the overall system architecture being attacked, rather than a software bug or security hole. A DoS attack can compromise the network by flooding it with undesired and useless packets and by providing false information about the status of network services (Cisco Press 2001).

One of the popular techniques for launching a packet flood is called a **SYN flood**. The SYN is a TCP control bit used to synchronize sequence numbers. Another popular technique is called **smurfing**. The **smurf attack** uses a directed broadcast to create a flood of network traffic for the victim computer.

Both insiders and outsiders can cause DoS attacks, however most are perpetuated by anonymous outsiders. Network intrusion detection is usually effective at detecting these attacks because it is designed to quickly identify them (Proctor 2001). RFC 2827 is a useful source of information for the security administrator.

Distributed Denial of Service

A **distributed denial of service (DDoS)** attack involves flooding one or more target computers with false or spurious requests, which overloads the computer and denies service to legitimate users. Researchers at the University of California at San Diego have estimated that more than 4,000 DDoS attacks are issued each week (Barr 2001). The threat of DDoS is so serious that it is an Enterprise priority to prevent or mitigate such attacks.

The term "distributed denial of service" describes the technique by which hackers hijack hundreds of thousands of Internet computers and plant "time bombs" on the victim systems. The hackers then instruct these time bombs to bombard the target site with forged messages, overloading the site and effectively blocking legitimate traffic. Products that are currently in the design stage will assist organizations in countering this type of attack.

Browsing

Browsing is the process of unauthorized users searching through files or storage directories for data and information that they are not privileged to read. The data and information could be stored in sections of main computer memory or to unallocated space on disks and other magnetic media. Browsing may occur after a legitimate process has completed because data still resides on these devices and is accessible to a knowledgeable intruder.

Intellectual property threats are a large problem because of the Internet and the relative ease with which a user can access existing material without the owner's permission (Perry 1999). Actual monetary damage resulting from a copyright violation is more difficult to measure than damage from secrecy, integrity, or delay and denial threats. Copyright laws were enacted before the creation of the Internet, which has complicated their enforcement by publishers. Many cases of copyright infringement on the Web occur because of ignorance of what cannot be copied. The most misunderstood part of the U.S. copyright law is that a work is protected when it is created and does not require the copyright notice.

intellectual property

Wire Tapping

Telephone lines and data communication lines can be tapped. Wire tapping can be either active, where modifications occur, or passive, where an unauthorized user is just listening to the transmission and not changing the contents. Passive intrusion can include the copying of data for a subsequent active attack.

Two methods of active wire tapping are *between the lines* and *piggyback entry*. Between lines does not alter the messages sent by the legitimate user, but inserts additional messages into the communication line while the legitimate user is pausing. Piggyback entry intercepts and modifies the original messages by breaking the communications line and routing the message to another computer that acts as a host.

Data Modifications

Problems with data integrity can cause a security breach. These problems include accidental incomplete modification of data and data values being incorrectly encoded. An incomplete modification occurs when nonsynchronized processes access data records and modify part of a record's field. When a record field is not large enough to hold the complete data item, some bits may be truncated and the value distorted. This situation can occur with most programming languages and may be difficult to detect; however, the results can be significant.

Backdoor

Backdoor software tools allow an intruder to access a computer using an alternate entry method. Whereas legitimate users login through front doors using a userid and password, attackers use backdoors to bypass these normal access controls. Once attackers have a backdoor installed on a computer the system can be accessed without using passwords, encryption, or other security controls. This backdoor can be used to bypass controls that

backdoor

the administrator has implemented to protect the computer system (Skoudis 2002). **Netcat** is one of the most popular backdoor tools in use today.

Another form of a backdoor tool is called **RootKits**. The traditional RootKits replaces critical operating system executables to let an attacker have backdoor access and hide on the host system. By replacing system software components, RootKits can be more powerful than application-level Trojan horse backdoors. These tools are primarily available on the UNIX platforms.

One of the most frequently discovered vulnerabilities that encourage the use of the backdoor is the default username and password. Almost every network vendor ships devices with a default username and password that must be changed when setting up a new system.

Finding and removing backdoors from the computer system is next to impossible, since there are many methods of creating them. The recourse for recovery is to restore the operating system from the original media. This would require the task of rebuilding and restoring user and application data from backups, assuming these exist. This becomes more difficult if the system has not been completely documented.

8.6 SOFTWARE THREATS

Software developed internally for day-to-day operations can be a threat source. Development of Web pages using HTML and **Common Gateway Interface (CGI)** scripts can unknowingly provide valuable information to hackers and even provide an access point into the Enterprise network.

Many Web pages display documents and hyperlink them to other Web pages or sites. Some have search engines that allow for site searches for particular information, which are accomplished through CGI scripts. Hackers can modify these CGI scripts for illegitimate uses. CGI scripts are normally used within the Web-server boundary; however, they can be modified to search outside of this server area. Applying a low privilege level to the scripts can reduce this problem. Any CGI script installed on the local server may contain bugs, and every bug can potentially cause a security hole (Gonclaves 2000).

8.7 SOCIAL ENGINEERING

A common and successful method of network intrusion is called **social engineering**, which occurs when an attacker uses deception to gain access to a resource. The easiest way to do this is by obtaining a password by lying and impersonating some authorized user. The person that would do this usually has a certain amount of self-confidence, is a fast and smooth talker, and possesses "quick thinking" abilities. The objective of this person is to test the computer and network resource security without getting caught.

The main objective of social engineering is to place the human element in the "network breaching" loop and use it as a weapon (Gallo 2002). A forged or stolen vendor ID or employee ID could allow access to a secure location, where the intruder could obtain access to sensitive assets. If this intruder was carrying network hardware or test equipment, an employee could easily escort this intruder to an equipment room, where an attack could occur. By appealing to the staff member's natural instinct to help a technician or contractor, it becomes easy to breach the perimeter of an organization and gain access to sensitive resources.

Responding to Social Engineering

Thoughtless users and employees contribute to the problems associated with social engineering. The following list shows examples of ways to reduce these occurrences (McClure 2001):

- Limit data leakage by restricting the detail of information published in directories, Yellow Pages, Web sites, and public databases.
- Enforce a strict policy for internal and external technical support procedures. Require some sort of user ID or customer ID.
- Be especially careful about using remote access. Validate the destination.
- Ensure both outbound and inbound firewall and router access controls are in place for all systems.
- Be careful when using e-mail. Learn how to trace messages via mail headers.
- Ensure that employees are educated on the basics of a secure environment. Use RFCs 2196 and 2504 to develop a security policy.
- Remove the opportunity of "dumpster diving" in corporate trash.

8.8 THREAT TARGETS

One can find information on the Internet on how to attack almost any type of protocol, operating system, or hardware environment. From the previous discussions it should be obvious that securing a system against threats requires a considerable amount of vigilance. After identifying the various threats, the next step is to identify the various computer and networking components that compose the threat environment. These include any hardware device or software component that might be subjected to the threats listed in Section 8.1. Potential candidates include the following:

- Computers, servers, PCs, administrative workstations
- Communication circuits
- Routers, gateways, switches

- Hubs, MAUs, repeaters, bridges
- Modems, DSUs, NT1s
- Front-end processors, communication controllers, multiplexers
- Network and Operating System software
- Application software
- Power and air conditioning systems

From this list, one can see boundless opportunities for an attacker to wreak havoc on the Enterprise's assets and resources. There are a number of resources and tools that an organization can use to mitigate the effect of the various threats.

8.9 RESPONDING TO HACKER ATTACKS

There are no simple means to protect the Enterprise organization against hacker attacks, so the organization must defend itself effectively and respond to any vulnerability that is either currently present or may emerge in the future. This is not an easy task, however the alternatives are unpleasant.

Although hackers continue to invent new methods of attacking Enterprise network resources, many of these are well known and can be defeated with a variety of tools currently available. Network and security administrators can respond to attacks by providing for rapid restoral of computer and network resources, closing holes in the organization's defenses, and obtaining evidence for prosecution of offenders.

Response to hacker attacks is a combination of planning and policy, detective work, and legal prowess. The lessons learned from a hacker attack should be used to further secure the network against similar attacks on the same vulnerability in the future. Also, there are law enforcement agencies, forensic experts, and independent response teams available to assist the organization in responding to a security incident and prosecuting the offender.

Hacker attacks are a cost of doing business in the information technology field. It is, therefore, essential that responses are both aggressively proactive and reactive, looking to prevent attackers as much as responding to them.

There are a number of books devoted to the subject of hacking that provide suggestions for countering the intruders. *Counter Hack* (Skoudis 2002) provides a step-by-step guide to computer attacks and effective defenses.

8.10 THREAT ASSESSMENT

A challenge for the administrator is to determine the proper level of security required to protect the organization's assets. This concept, called **threat assessment**, identifies the organization's assets and resources and the accessing entities (Gallo 2002). Organizations can best assess computer and network threats by using a structured approach. These threats can be directed at hardware, software, data and information, systems, or security measures. After the administrator has assessed the various vulnerabilities, the organization will be in a good position to protect its assets and resources. There are five requirements in this threat assessment process: identifying critical assets, identifying possible sources of the threat, developing the risk analysis, prototyping the solution, and documenting the results of the threat assessment.

Identifying Critical Assets

The first step in threat assessment is to decide which assets are critical. A number of organizations maintain a significant amount of sensitive and critical information that is vital to their survival. This information could include customer order information, sales figures, supplier contracts, customer lists, and many other proprietary assets. If this database was compromised, the business or organization might fail, or there could be other serious ramifications that could even affect national security. Therefore considerable attention should be devoted to determining which elements are crucial and how security lapses would damage the organization.

Identifying Threat Sources

After determining the critical elements of the organization, the next step is to identify any possible threat sources. Threats can originate from inside and outside the organization; from competitors, terrorists, and the hacker community. The object of this analysis is to determine who would benefit from the use of your resources and assets. Note that some threat sources are not after financial gain. A good starting point is to list all the possible assets and create a matching list of every entity that might receive some benefit from accessing into the asset. Determining who might benefit from the asset theft helps identify potential threat candidates and locations of attack.

Historically, break-ins occur from inside the computer network. Employees and contractors have a ready access to the organization's resources, both physically and through the network access. Most organizations do not take sufficient precautions to protect their assets from internal personnel and contractors. A list of potential insider actions that should be considered threats includes the following:

- to hamper corporate operations
- to sell to a competitor
- to hurt customer relations
- to cause internal strife
- to cause asset loss
- to create physical damage
- to cause internal outage of resources
- to compromise customer information
- to reduce the organization's ability to function

These activities are not difficult to initiate, given access to the appropriate resources. It should be noted that this list would apply not only to internal users, but also to anyone who could access the computer and networking asset. If the administrator can proactively identify the sources of potential threats, actions can be taken to limit the possibilities of a threat occurring.

Risk Analysis

Risk analysis is the process of understanding how the loss of an asset will impact an organization. This process can be formidable if the resources are significant, however performing some simple calculations will produce a ballpark figure. A look at such elements as replacing, repairing, upgrading, and managing a threat will assist in this endeavor. It is often cheaper to proactively fix a potential problem than to respond to a threat. After-the-fact problems can include negative publicity, a reduction in sales, loss of customers, and loss of credibility, particularly if this is an e-commerce operation. Therefore even a minor security breach can cause major repercussions that would be expensive to correct. Funds might be well spent to properly plan for all identified threats.

Prototyping

Prototyping involves developing a scenario concerning a security violation and determining the effect on the organization. This is similar to role-playing where someone tries to breach security and another group tries to detect and defeat the breach before any harm is done. Role-playing in security terms is essential to understanding the threat environment and making sound decisions on what can and cannot be protected (Gallo 2002). In a computer and network attack, efforts must be made to identify all the potential points of attack and counter measures. These measures include methods to identify, correct, and handle the repercussions of the various attacks. Actions for a proper defense include the following elements:

- Setting up information defenses
- Monitoring for information attack
- Delaying the attack until assessment and reinforcements are available
- Counterattacking
- Capturing and destroying the intruder
- Cleaning up problems discovered

By deploying some role-playing or simulation in the environment, threats can be identified and proper defenses planned.

Documentation

Documentation is the final phase of threat assessment. Without documentation, many of the efforts previously discussed will be in vain. Not only should these threat methods and procedures be documented, but up-

Table 8–2 Vulnerabilities and Protective Measures

Vulnerability	Protective Measure
Address spoofing	Firewalls
Dictionary attack	Strong passwords
Denial of Service Attack	Authentication, service filtering
Eavesdropping	Encryption
Masquerading	Authentication
Man-in-the-middle attack	Digital certificates
Replay attack	Sequence numbering or time stamping
Trojan horse attack	Firewalls
Virus attack	Virus management policy

dates must follow that represent the current situation. Documentation can take a considerable number of hours to complete, however this time and effort is an essential part of any plan for threat management.

Table 8–2 provides a summary of system vulnerabilities and associated protective measures that can be employed in the computer and networking environment (Goldman 1998).

8.11 GAP ANALYSIS

A logical follow-up to threat assessment is called **gap analysis**. All of the known threats and problems have been identified, and suggested solutions and countermeasures have been identified—or have they? Ensuring confidentiality, integrity, and availability of information are major goals of a computer system. Conducting regular reviews of day-to-day practices and comparing them with threat assessment documentation can reveal gaps in security precautions.

gap analysis

This gap analysis is an effective methodology for testing the authenticity, integrity, and confidentiality of an organization's hardware and software systems, and it can provide an assurance that security implementations are consistent with the real requirements. Not all gap analyses are the same. Factors that influence an analysis include the size of the organization, the industry, the cost involved, efforts involved, and the depth of the analysis. The following common steps are employed when conducting a gap analysis (Ferraiola 2001):

- Identify the applicable elements of the organizational standards.
- Collect methods and procedures and policy documents.
- Assess the implementation of the policies and procedures.
- Conduct a physical inventory of all computer and network hardware and software components.
- Interview users to determine compliance levels of policies.
- Compare current security environment to policies.

- Prioritize the identified versus actual gaps in policies.
- Implement the remedies to conform to policies.

Analysts may frequently overlook the human elements of security, however most security gaps are caused by people.

8.12 ETHICS

Ethics are the rules or standards governing the conduct of the members of a profession. The computer and networking industry has been associated with a lack of ethical behavior in the past. This can be attributed to the lag and lack of protective legislation with respect to electronic data and privacy issues. As a result, encouragement of ethical behavior throughout the industry has been inconsistent (Flynn 2001).

Ethics are not often addressed in computer and telecommunication's classes, however the implications are too significant to ignore. There are a number of conflicting issues: the user's need for privacy, the public's right to know, and the organization's necessity to protect proprietary information. There are a number of consequences for a system's ethical, integrity, and security lapses. The following significant incidents must be addressed:

- Plagiarism of copyrighted work.
- Eavesdropping on e-mail and other communications.
- Cracking or hacking a computer system.
- Unauthorized access to private or protected systems.
- Using illegally copied software.

There are a number of initiatives that can mitigate these situations. These include awareness training sessions and communications to employees and users of a system regarding the issue of ethics. The following specific activities can be employed by the organization to address this issue:

- Develop and publish policies that outline acceptable behavior and repercussions for failure to abide by them.
- Develop and schedule regular training sessions for users and employees on the subject of ethics in the organization's environment.
- Conduct normal business activities in an ethical way, providing an example to employees and users.

An example of a code of ethics can be viewed at www.acm.org/constitution/p98-anderson.pdf.

8.13 CYBERCRIMES

Criminal activity has increased in the area of networking with the advent of worldwide access to the Internet. This in turn has increased the vulnerability of the computer assets that are accessible through an organization's networks. The United States Department of Justice provides guidance to investigators and prosecutors in the areas of **cybercrime** and intellectual property matters. This support is provided under the **Computer Crime and Intellectual Property Section (CCIPS)** umbrella. Many documents containing computer crime guidance and case studies are available at the Web site www.cybercrime.gov.

cybercrime

Computer Crime and Intellectual Property Section (CCIPS)

There are many categories of cybercrime and security policy issues that must be addressed in today's Enterprise environment. The major issues can be summarized as follows:

- E-commerce legal issues
- Encryption and computer crime
- Federal code related to cybercrime
- Intellectual property crime
- International aspects of computer crime
- High-tech privacy issues
- Prosecuting crimes facilitated by computers and by the Internet
- Protecting critical infrastructures
- Searching and seizing computers and obtaining evidence in criminal investigations
- High-tech speech issues

There are a number of statutes used to prosecute computer and networking crimes. A sample of these include the following:

- Title 18 USC 1029. Fraud and Related Activity in Connection with Access Devices.
- Title 18 USC 1030. Fraud and Related Activity in Connection with Computers.
- Title 18 USC 1362. Communications lines, stations, or systems.
- Title 18 USC 2511. Interception and Disclosure of Wire, Oral, or Electronic Communications Prohibited.
- Title 18 USC 2701. Unlawful Access to Stored Communications.
- Title 18 USC 2702. Disclosure of Contents.
- Title 18 USC 2703. Requirements for Governmental Access.

These statutes are normally associated with domestic computer and networking crimes, however these crimes are starting to spread into the military and international environments. These new categories of crime are called **information warfare** and **cyberterrorism**.

information warfare
cyberterrorism

Information Warfare

This new category of crime includes a number of unique terms, including the following:

- Carnivore—A FBI system to monitor e-mail and other traffic through ISPs.
- Copernicus—The Navy plan to reformulate its command and control structure in response to the realization that information can be a weapon.
- Defense Information Structure (DIS)—The worldwide shared or interconnected system of computers, communications, data, applications, security, personnel, training, and other support structures serving the military's information needs.
- Defense Information Security Administration (DISA)—The military organization charged responsible for providing information systems to fighting units.
- van Eck Monitoring—Monitoring the security of a computer or other electronic equipment by detecting low levels of electromagnetic emissions from the device.
- Electromagnetic Pulse (EMP)—A pulse of electromagnetic energy capable of disrupting computers, computer networks, and other forms of telecommunications equipment.
- High Energy Radio Frequency (HERF)—A device that can disrupt the normal operation of digital equipment such as computers and navigational equipment.
- Information Security (INFOSEC)—Protection of classified information that is stored on computers or transmitted by radio, telephone, or other means.

Note that some of these systems relate to advanced forms of security attacks. Appendix D contains an expanded list of security acronyms, including both information warfare and cyberterrorism entries.

Cyberterrorism

cyberspace

Cyberspace is defined as the global network of interconnected computers and communication systems. Cyberterrorism is the convergence of terrorism and cyberspace (Denning 2000). It is usually understood to mean unlawful attacks and threats against computers, networks, and databases when done to intimidate or coerce a government or its people to further some political or social objective. Cyberterrorism includes attacks that result in violence against infrastructures, persons, and property, and often leads to death or bodily injury, explosions, plane crashes, water contamination, or severe economic damage.

Cyberspace is constantly under assault since cyberspies, thieves, saboteurs, and thrill-seekers break into computer and networking systems.

They steal personal data and trade secrets, vandalize Web sites, disrupt service, sabotage data and systems, launch computer viruses and worms, conduct fraudulent transactions, and harass individuals and organizations. These attacks are facilitated with increasingly powerful and easy-to-use software tools, readily available on the Internet.

The next generation of terrorists will grow up in a digital world, where more powerful and easy-to-use hacking tools will be available. They might see greater potential for cyberterrorism than the terrorists of today, and their level of knowledge and skill relating to attacking Enterprise resources will be greater. Hackers and insiders might be recruited by terrorists or self-recruiting cyberterrorists. Some might be moved to action by cyber policy issues, making cyberspace an attractive venue for carrying out an attack. Cyberterrorism could also become more attractive as the real and virtual worlds become more closely coupled, with a greater number of physical devices attached to the Internet.

8.14 SECURITY TOOLS

Properly securing a network is less expensive than rebuilding a corrupted one. There are several types of network security tools that can protect the organization's networks from unauthorized users and hackers. They include access control systems, data encryption, digital certificates, and vulnerability scanners. When selecting network security tools, it is important to determine which are best suited for the network being protected. Information on access control mechanisms is presented in Chapter 2.

Encryption is ideal for protecting credit card numbers and other data. Firewalls provide good protection from the outside, but will provide little protection from internal threats. Access control helps to secure a network from the inside. Because one tool will rarely get the job done, a combination that includes all three can provide an acceptable level of security.

There are a wealth of products and services available for protecting an organization's computer and networking assets. Functions provided by these include the following:

- Access control
- Antihacker
- Audit trail
- Authentication
- Auto logoff
- Cryptographics
- Data encryption
- Data integrity
- Encryption

- Filtering
- Monitoring and reporting
- Password protection
- Operating System security
- Risk analysis management
- User account restrictions
- Virus protection

Security tools are also available in hardware instruments. A network-connected device operating in **promiscuous mode** captures all frames on a network, not just frames addressed directly to it. A network analyzer operates in this mode to capture network traffic for evaluation and to measure traffic for statistical analysis. A hacker may also use such a device to capture network traffic for unscrupulous activities, so network traffic can be encrypted to protect against such eavesdropping.

Different types of surveillance and test equipment can be used in this war against the various threats. These devices include network analyzers, monitors, and protocol analyzers.

Network Analyzers

A **network analyzer** is a monitoring, testing, and troubleshooting device used by network administrators and technicians. This device is attached to a network for the purpose of capturing network traffic. Captured frames can be displayed in raw or filtered form. These devices are normally portable and can be inserted in many places in the network. The data obtained depends on the location of insertion in the network.

Network analyzers operate in promiscuous mode, as previously stated, and listen to all traffic on that particular portion of the configuration. The user can choose to capture frames transmitted by a particular network computer or frames that carry data or information for a particular application or service. The captured traffic is then monitored to evaluate network performance, locate bottlenecks, or track security breaches.

Analyzers are available in a number of configurations and possess many different features. Low-end analyzers are usually designed for traditional ethernet and token ring LANs. High-end devices are capable of addressing a variety of network types, including Frame Relay, ATM, and Fast Ethernet networks. WAN **protocol analyzers** are designed to handle a variety of link configurations and OSI Layer 2 data link protocols. Note that an analyzer product can consist of either hardware or software or both.

Some of the high-end network analyzers could cost tens of thousands of dollars. Software-based analyzers, however, are available as freeware.

Many network analyzers are made available on the Internet by the hacker community for the express purpose of capturing sensitive information on networks, such as passwords sent in the clear without encryption. Legitimate uses for network analyzers are presented in Chapter 5.

8.15 INTERNET AND WEB RESOURCES

NCSA	www.ncsa.uiuc.edu
CSI	www.csi.com
Data Rescue's virus encyclopedia	www.datarescue.com
Trend Micro	www.antivirus.com
Symantec Corp	www.symantec.com
McAfee	www.mcafee.com
USDOJ Computer Crime	www.cybercrime.gov
Terrorism	www.terrorism.com
Faulkner Information Services	www.faulkner.com
Smurf defense	www.pentics.net/denial-of-service/white-papers/surf.cgi
Smurf amplifier	www.powertech.no/smurf/
Security Focus	www.securityfocus.com
Security Portal	www.securityportal.com
Stake Security News	www.atstake.com/security_news
Packetstorm	www.packetstormsecurity.com
2600 Magazine	www.2600.com
White Hats	www.whitehats.com
Attrition	www.attrition.org

■ SUMMARY

Common threats to an Enterprise include computer viruses, resource theft, asset destruction, natural disasters, espionage, and computer hackers. Viruses are the most common security problem for an organization. Viruses come in all flavors and can attack a computer network at any time, almost with impunity. Considerable efforts are devoted to countering this threat.

The CERT Coordination Center is responsible for looking at Internet security issues. A number of virus control policies have been developed to counter the threat, as have a number of software packages.

Malicious attacks include brute force, masquerading, eavesdropping, spoofing, session hijacking, replay, man-in-the-middle, and dictionary attacks. Each of these has their own special techniques of compromising Enterprise network assets and resources.

Other security breaches include denial of service, distributed denial of service, browsing, wire tapping, and data modifications. These attacks might not destroy or damage network resources, however they may deny access to legitimate users.

An attacker or intruder can gain access to an Enterprise network by using social engineering techniques. The goal of the attacker is to identify passwords and other security protection mechanisms by posing as a legitimate user.

It is essential for the security administrator to identify potential threat targets, including hardware, software, circuits, and facilities. The administrator must determine the proper level of security to apply to mitigate possible threats. Threat assessment is used to identify critical assets and possible threat sources. This is coupled with risk analysis, prototyping, and a documentation process. These processes are followed up with a gap analysis to ensure all security issues have been covered.

Cybercrimes have increased with the advent of universal Internet access. Computer crime is of particular interest to the e-commerce community. The Federal government is addressing the issues relating to information warfare and cyberterrorism.

A number of hardware and software tools are being developed to counter the issue of computer network threats. Security products and services are available from many different suppliers.

Key Terms

Backdoor
Bacteria
Boot Sector virus
Browsing
Brute force
CERT Coordination Center (CCC)
Common Gateway Interface (CGI)
Computer Crime and Intellectual Property Section (CCIPS)
Computer Emergency Response Team (CERT)
Computer Security Institute (CSI)
Cracker
Cybercrime
Cyberspace

Cyberterrorism
Denial of Service (DoS)
Dictionary attack
Distributed Denial of Service (DDoS)
Eavesdropping
Ethics
File-infecting virus
Gap analysis
Hacker
Information warfare
Intellectual property
Logic bomb
Macrovirus

Man-in-the-middle attack
Masquerade
Multipart virus
National Computer Security Association (NCSA)
Netcat
Network analyzer
Packet sniffer
Polymorphic virus
Port scanning
Promiscuous mode
Protocol analyzer
Prototyping
Replay
Risk analysis
RootKits
Security Administrator Tool for Analyzing Networks (SATAN)
Security breach
Session hijacking
Smurf attack
Smurfing
Social engineering
Socket
Spoofing
Stealth virus
SYN flood
Threat
Threat assessment
Trojan horse
Virus
War Dialer
Worm

REVIEW EXERCISES

Questions and Analysis

Provide a short answer or in-depth analysis as required.

1. List the most common threats to an organization.
2. List the major virus categories.
3. Describe each of the major viruses.
4. What are the six general antivirus strategies?
5. Comment of virus software detection.
6. Identify software detection programs.
7. What are the points of attack for client and server devices?
8. What is the difference between a hacker and a cracker?
9. Describe masquerading and provide countermeasures.
10. Describe eavesdropping. Why is this activity a concern?
11. List activities that can occur during a security breach.
12. Define the denial of service (DoS) attack.
13. What are the two types of DoS attacks? What are the differences?
14. What is the difference between a DoS and a DDoS?
15. What is an intellectual property violation?
16. What is the function of a backdoor?
17. Why should the UNIX administrator be alert for RootKits?
18. Discuss social engineering. What are the countermeasures?
19. What are the common-threat targets?
20. Describe the threat assessment process.
21. What are some activities that can be considered an inside threat issue?

22. What elements are required when prototyping?
23. Discuss the solutions that can be employed to counter the various vulnerabilities discussed in this chapter.
24. Why is a gap analysis required? What are the steps?
25. Discuss ethics in the Enterprise. Provide examples.
26. Discuss the issue of cybercrimes.
27. What are the major issues that encompass cybercrime?
28. What are some functions provided by security tools?
29. What is the function of a network analyzer?
30. What is the difference between a protocol analyzer and a network analyzer?

Definitions and Terms

Provide the most correct answer from the list of key terms.

1. A type of attack that uses a large number of computers to simultaneously send packets to a target.
2. A form of warfare whose aims are to nullify or destroy the computing resources of the opponent.
3. A small set of instructions carried out by a certain program that implements destructive tasks.
4. A type of attack that takes advantage of the store-and-forward mechanism used by insecure networks such as the Internet.
5. An attack in which an attacker pretends to be someone else.
6. The process of identifying security risks, determining their impact, and identifying areas requiring protection.
7. A type of tool that allows an attacker to maintain super-user access on a computer.
8. A technique whereby an attacker steals an existing session established between a source and destination.
9. The collective name for techniques that deceive users into revealing sensitive information.
10. The combination of an IP address and port number.

True/False Statements

Indicate whether the statement is true or false.

1. Threat categories include integrity threats and intellectual property threats.
2. The most common threat to the Enterprise is from a natural disaster.
3. A virus is a computer program that infects the computer by creating unique software code in the processor's memory.
4. The CERT is an organization that recognizes and responds to computer attacks.
5. A War Dialer tries a random set of numbers in an attempt to learn the password of a computer system.
6. A brute force attack uses every possible combination of passwords in an effort to attack a system.
7. A hacker and a cracker can be the same person.
8. Eavesdropping is not considered an attack on a computer system.
9. Session hijacking and a replay attack are different names for the same attack.
10. The store-and-forward mechanism in a network device provides an opportunity for a man-in-the-middle attack.
11. A logic bomb has no relationship to a Trojan horse.
12. A denial of service attack does not negatively impact an organization.
13. A contractor can program a backdoor into an Enterprise system.
14. Social engineering is a technique that uses deception as the tool to gain access into an organization's resources.

15. Threat assessment includes identifying critical assets and documenting the results of the threats.

Activities and Research

1. Research and identify hardware and software security tools that are currently available for countering computer network threats.

2. Research and produce a list of all current computer viruses. Describe the threat caused by each virus.

3. Research and identify the different types of malicious attacks that have occurred within the past year. Identify the countermeasures required to defeat each attack.

4. Research the subject of computer crime and present an overview on an actual incident.

5. Look at the various Federal laws concerning computer crimes. Develop a presentation on one of these laws.

6. Develop an analysis on the migration of international computer crime into the domestic environment.

7. Prepare a document that shows how cyberterrorism can impact the computer and networking assets of some organization.

8. Research the topic of cybercrime legal and policy issues and prepare a paper on some aspect of the impact on the computer and networking environment.

CHAPTER 9

Building, Campus, and Facility Security

OBJECTIVES

Material included in this chapter should enable the reader to

- Understand the need to review and identify situations that might constitute a compromise to an organization's computer and network facilities and structures.
- Identify the internal and external threats that are pervasive in the physical facilities that house the computer and networking resources.
- Become familiar with the various solutions that can be used to counter physical security threats.
- Look at the physical security issues that arise from distributed and virtual networking systems.
- Learn how to develop a disaster recovery plan for the organization's computer and networking resources.
- Recognize potential security breaches and identify measures to proactively correct them before a disaster occurs.
- See how natural disasters can be as devastating to computer and networking assets as human-initiated events.
- Identify solutions that can mitigate both man-made and natural disasters.

INTRODUCTION

The primary emphasis of physical security is to prevent unauthorized access to the computing facility, which includes the computer room, the communications room, network control center, and network infrastructure. Efforts are directed at preventing malicious vandalism or subtle tapping of communications circuits. Physical security can be thought of as the lock and key part of security, since the primary protection mechanism is a lock of some type. Equipment rooms that house sensitive equipment are normally locked, workstations contain locking mechanisms, and doors into various sensitive areas are locked. In lieu of actual physical locks and keys, magnetically encoded cards are used to activate workstations and open doors. These methods of entry are also used as personal employee identification.

Physical security planning includes addressing a number of emergency situations, such as floods, storms, earthquakes, power failures, and fires. Disasters can occur from both fire and water damage originating from inside and outside sources. There are a number of proactive measures that can be taken to reduce the damage caused by these natural and man-made disasters.

Intrusion detection is the art of detecting and responding to Enterprise computer and network misuse. The benefits include attack anticipation, deterrence, detection, response, damage assessment, and prosecution support. The security administrator must evaluate the potential risks and implement the appropriate system for the maximum protection.

Auditing involves the process of collecting and monitoring all aspects of the computer network to identify potential holes and flaws in the systems. This information becomes part of contingency planning and disaster recovery efforts.

Security awareness programs are necessary to educate employees and users about the importance of physical security. Many users are not aware of the consequences of simple lapses in security, which could cost the organization millions of dollars in computer and network damages.

9.1 PHYSICAL SECURITY OVERVIEW

physical security

Physical security is the term used to describe protection provided outside the computer and networking system. Typical physical security components are guards, locks, fences, and cameras to deter direct attacks. Many physical security measures are the result of common-sense thinking and planning, and are often obvious solutions. Sources that can compromise an organization's ability to function include natural events, human vandals, power loss, fire, water, and heat and humidity levels. These physical vulnerabilities to security can and often do occur simultaneously. Issues to consider include the cost of replacing the resource, the speed with which equipment can be replaced, the need for available computing power, and the cost or difficulty of replacing operating system and application software (Pfleeger 1997).

The physical element of computer and network security involves the hardware, the facility, cable runs, and the software and database systems that reside in the physical location. An initial step when evaluating physical security is to conduct an audit that looks at all the elements that comprise the Enterprise environment that must be protected. The Enterprise Systems Security Review in Appendix B is a vehicle for accomplishing this task. Analyzing the results of this audit can reveal shortcomings in the organization's physical environment.

How easy is it for someone to access sensitive hardware components and software and database storage devices? It is essential to note that in today's distributed and e-commerce environments, a number of physical threats can be located at remote branches and locations. Devices that can be an access point for attackers include remote terminals, communications controllers, multiplexers, servers, printers, and PCs. Also included are cable runs and network demarcation points, which can be a source of eavesdropping and monitoring (Ratliff 1999).

9.2 PHYSICAL SECURITY CATEGORIES

Physical security issues are varied and include natural disasters, vandals and destructive individuals, fire and water damage, power loss, and heat and humidity (Figure 9–1). This section will provide an overview of each situation and discuss possible ways to mitigate the cost of each.

Natural Disasters

natural disasters

It is impossible to prevent **natural disasters**, however through careful planning it is possible to reduce the damage they cause. The Enterprise resources are subject to the same natural disasters that can occur to buildings, homes, and vehicles. They can be flooded, burned, and destroyed by earthquakes, storms, and tornadoes.

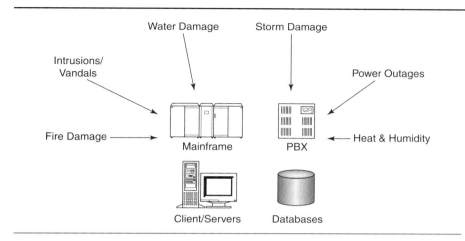

Figure 9–1
Physical Integrity and Security Impediments

Water from a natural flood comes from ground level, usually rising gradually, and bringing with it mud and debris. Of course the damage from a flood is greater if the site is below a dam break. There is usually enough time for an orderly shutdown of the systems, losing at worst some of the processing in progress. The hardware components may be damaged or destroyed, but most systems are insured and can be replaced by the vendor. Organizations with unique or irreplaceable equipment might have a duplicate, redundant system available in the event of some disruption.

Hardware is replaceable. The real concern is the data and programs that are stored locally on magnetic media. Tape libraries and disk packs usually are stored on shelves where they are readily accessible. Many of these media are probably not identified as to their importance. There could be thousands of volumes of tapes and hundreds of disk drives that need to be rescued in the event of a natural disaster. A simple scheme would be to color code the most important volumes so they could be removed quickly in the event of a disaster.

Vandals and Destructive Individuals

Because computers and their media are rather sensitive, a vandal could quickly inflict a considerable amount of destruction. **Human attackers** can be bored operators, disgruntled employees, saboteurs, ignorant employees, intruders, or thrill seekers. An unskilled vandal may use a brute-force frontal attack with an ax or brick, however these people will probably be noticed and stopped before any serious damage can occur. More skilled vandals can short-circuit a processor or disable or crash a disk drive with small tools, which are difficult to detect.

As distributed computing increases, protecting the system from outside access becomes both more difficult and more important. Protection

human attackers

mechanisms and policies are needed both to prevent unauthorized people from obtaining access to the systems and to verify the identity of accepted users. Physical access methods are addressed in a subsequent section.

Fire and Water Damage

Fire and water damage can occur from natural events, structural failures, or from human causes. Fires can be set by arsonists and internal water supplies can be broken, or attackers can activate sprinkler systems. All of these situations can be devastating to a computer system. Most computer rooms and electronic systems are protected against fire by something other than a sprinkler system. Sprinkler systems should not be installed over electronic systems. Most computer systems are installed over a raised floor, which allows for cabling to run under the devices and out of the way. Air conditioning is also supplied to the computing devices from under the raised floor up into the cabinet or chassis.

It should be obvious that water and all of these electrical connectors and cables do not mix very well. Water under a raised floor will quickly cause an electrical short and shut the computer system down. It is a good idea to keep a supply of plastic sheets available to cover the electronic components in the event of a water hazard from above.

Fire is a more serious issue since the fire itself, and the water used to put out the fire, can both cause damage. Once a fire starts, taking any time to protect the computing resources could put personnel in danger. A computing and network center should have a plan for shutting down the systems in an orderly fashion. Such a process can be accomplished quickly and makes recovery much easier and faster. This plan should include individual responsibilities in the event of a disaster, with backup assignments for those personnel who are not on duty. Every computer system installation should have a master power "kill switch" that shuts off all power to the computer system.

Power Loss

A constant source of predictable power is required to keep a computer system operating properly. For some time-critical applications, loss of service from a system is intolerable, and efforts must be made to provide some alternative power source if this occurs. Because of possible damage to media by sudden loss of power, many disk drives monitor the power level and quickly retract the recording head if power fails.

One protection against power loss is an **uninterruptible power supply (UPS)**. This device stores energy during normal operation so it can provide backup energy if power fails. One form of UPS uses batteries that are continually charged when the normal power is on and then provide power if the normal source fails (Figure 9–2). Several problems with batteries include heat, size, flammability, fumes, leakage, and insufficient output. Size

uninterruptible power supply (UPS)

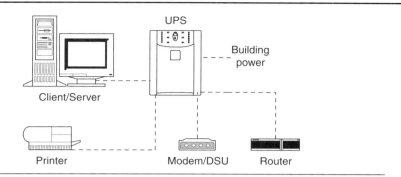

Figure 9–2
UPS for a Client/Server Configuration

and limited duration of energy output are limitations of the standard UPS system. Some UPS systems use outside generators and motor generators that provide support to the primary backup power supply, however this is a very expensive alternative. If backup power is to be provided for an extended period of time, the outside generator is a requirement. All telephone central offices use both UPS systems and outside generators, because the service they provide is expected to operate continuously and seamlessly.

Another issue with power is its quality. Most people are unaware that AC power fluctuates plus or minus ten percent, which means that the supply can be anywhere between 108 volts to 132 volts. A voltmeter can be used to measure this variation. Power cycling by a disk drive or air conditioner can cause a temporary drain on the system and cause the lights to dim. When a motor stops, a temporary surge can be sent across the system. It is also possible for lightning to send a momentary, large surge across the system. Instead of being constant, the power delivered along a power line shows many fluctuations, called drops, spikes, and surges. A drop is a momentary reduction in voltage, whereas a spike or surge is a rise. For computing equipment, a drop is less serious than a surge, however most electrical devices are tolerant of some fluctuations of current.

Voltage fluctuations can be destructive to sensitive electronic equipment. Simple devices such as surge suppressors filter spikes from a power source, clocking fluctuations that could affect the electronic devices. These surge suppressors are rated in joules—the higher the rating, the more protection afforded. These devices range in price from twenty dollars to several hundred dollars. Protection devices should be installed on every computing and networking device in the installation. This includes PCs, printers, servers, routers, and DCE devices such as modems and DSUs.

Another surge that can damage electronic equipment is a lightning strike. To increase protection, personal computers can be unplugged when not in use and especially during electrical storms. Lightning striking a telephone line and passing through the modem to the computer is another source of destruction. Disconnecting a modem during a storm is a wise preventive measure. A lightning strike on a telephone line can follow a path to the modem card or stand-alone modem and destroy the electronics. Additional lightning suppression devices may be available from the local telephone company.

Heat and Humidity

Excessive heat and high or low humidity can cause an electronic device to fail. Computer systems are sensitive to heat, and loss of cooling is common due to mechanical failure or electrical disruption. The normal response to this situation is to provide an alternative cooling source or shut down the system, since excessive heat can cause extreme damage to the electronic devices. There is usually adequate time to respond to this situation, assuming that someone is aware of the cooling failure.

Computing systems may perform normally even if the temperature exceeds the manufacturer's recommendations. However, as the temperature rises, components may perform unpredictably, sometimes appearing to work well but producing faulty results. It is quite serious when the system is sometimes correct/sometimes incorrect, because this uncertainty can corrupt the entire system while it appears to function properly.

High or low humidity is another issue that can negatively impact electronic components. When the humidity is extremely low, arcing of electronics components can occur, which can be devastating. High humidity causes moisture to form on the electronic components, which can cause shorts. Either high or low humidity must be corrected quickly, since the results of both can be circuit card failure. A monitor is available that charts humidity on a 24 hour basis.

Environmental Safeguards Recap

environmental safeguards

Adequate **environmental safeguards** must be installed and implemented to protect computer, database, and networking assets. The critical nature and sensitivity of each asset will determine the level of security required. The more critical the asset is for accomplishing the mission of the organization, the higher the level of security required. The following important safeguards should be considered:

- Fire prevention, detection, protection, and suppression
- Internal water hazard prevention, detection, and correction
- Electrical power supply provisions
- Temperature control
- Humidity control
- Natural disaster protection from floods, lightning, storms, and earthquakes

- Protection from electromagnetic interference (EMI) and radio frequency interference (RFI)
- Maintenance procedures for dust and dirt

9.3 PREVENTING DAMAGE TO PHYSICAL ASSETS

The organization cannot function without physical assets that include computer systems and networking systems. In many cases the database asset that resides on the computer system is the company or organization. Several good examples include the records that are located on the databases of the Social Security Administration, the Internal Revenue Service, stock exchanges, Internet Service Providers, and the airlines. This section suggests precautions that can be taken to prevent or minimize damage from the disasters discussed in the previous section. Appendix B provides a security review checklist that can be used as a basis for a facility audit (Green 1996).

Fire Hazards

Fire hazards can be both internal and external, with the former being the most controllable. An external hazard results when the building is vulnerable to combustion from a fire in the surrounding area or when the organization is a tenant in a larger building complex and a fire starts in some other part of the building. External fires can be started by hazards ranging from a smoker's carelessness to faulty wiring. The computer and telecommunication's systems are assets that must be protected with a fire suppression system. If an external hazard exists, the solution is to cure it or move away from it.

The most controllable fire is one that is localized in an equipment room or demarcation closet. Either removing the cause of combustion and combustible materials or using fire suppression systems can control hazards. The first line of attack is good housekeeping, which means removing debris or flammable liquids from the area. Ensure that cleaning fluids and flammable liquids are stored in fireproof cabinets outside the asset area.

There should be some form of fire suppression devices or a system that provides coverage for the computer and networking assets. Fire codes should be observed and monitored. Fire extinguishers must be located throughout the area in easily accessible reach. Personnel must be instructed on how to use the extinguishers.

The local fire department must be made aware of the critical nature of the computer and networking facilities. They must be informed of the location of the major electronic components so they can take precautions to prevent damage in the event of a fire, particularly water damage, which can be a disaster. All equipment rooms must be equipped with smoke detectors and fire alarms that can communicate with a central monitoring operation.

Water Damage

Two of the most common causes of **water damage** are from overhead sprinkler systems within the facility and runoff from sprinklers on other floors. Another common hazard is from ruptured pipes that can flood equipment rooms. Overhead sprinkler systems should not be installed over sensitive electronic equipment. Other alternatives for putting out fires should be pursued that do not cause damage to the circuit cards. If sprinklers are used for equipment room fire suppression, they should consist of high-temperature heads. A fire in another section of the building that poses no hazard to the equipment rooms should not activate sprinklers in these rooms.

Flooding is the other source of water damage to electronics. If possible, equipment rooms should not be located below ground level, even in areas that are not subject to flooding. Ruptured water mains and stopped-up drains can cause localized flooding. If the building is located in a flood plain, install the equipment above the high-water mark. Even though the building might be closed during flooding, the electronics can remain dry. There is a real-world example of this situation, in which a carrier's central office had equipment located below a high water mark and the unexpected happened—water backed up through a drain and flooded the electronics, which cost an enormous amount of money to repair.

Earthquake Damage

It is not possible to prevent an **earthquake**, so management must decide how great the risk is. If the facility is located in a high earthquake zone, damage can be limited if the building survives. If the building is earthquake proofed, the equipment should be braced to prevent tipping. It should also be bolted to some major support to reduce movement. There is also danger from secondary damage from fire and water during an earthquake. Water pipes should not be in the vicinity of equipment rooms in earthquake-prone areas.

Storm Damage

Hurricanes, tornadoes, snowstorms, and windstorms can cause severe **storm damage**, but usually only to outside facilities. Structures affected usually include antennas, dishes, towers, and overhead wiring. Companies with outside plant facilities are installing them underground, so this issue will be moot. Microwave and other antennas should be braced to withstand maximum expected wind velocities. They should also be equipped with deicing equipment when appropriate. Pedestals and wiring vaults can be affected by flooding and forest fires, however they must be located close to media aggregation.

Human-Initiated Damage

Humans cause a high percentage of outages and damage, which can be accidental or deliberate. The incidents range from simple work errors, sloppy work, or deliberate sabotage caused by disgruntled employees. **Vandals**, competitors, or real **saboteurs** can also cause destruction. All telecommunications areas, computer rooms, and networking facilities must be kept secure from unauthorized access. The most effective access control method is a card-operated lock that allows access to authorized people and maintains a record of the event. With this method it is easy to control access and to determine who was in the area at the time of failure. Good records are essential for recovering from damage. Equipment location, inventory, and configuration records should be kept where they are readily available to anyone restoring service.

vandals
saboteurs

9.4 PHYSICAL SECURITY CONTROLS

Physical security controls pertain to the physical infrastructure, physical device security, and physical access. The physical infrastructure includes the facilities that house the systems and the networks that are used to access these same assets. The physical network infrastructure encompasses both the selection of the appropriate media type and the path of the physical cabling. It is essential that no intruder is able to eavesdrop on the data traversing the network and that all critical systems have a high degree of availability (Kaeo 1999).

Network Access

Telecommunications access adds another dimension to security since data can be accessed from terminals outside the computer room. A number of questions arise concerning network access:

- How is the terminal user identified?
- Is this user authorized to access the computer?
- What operations can this user execute?
- Are the communication lines being monitored or sniffed?

Access control techniques, discussed in Chapter 2, include having a unique identification code for each user and a password to protect against these situations.

Users must be required to logon to the network and this information is automatically recorded in a central database so that a record can be maintained of all network activities. When someone tries to logon unsuccessfully a number of times, the user's ability to logon should be disabled. This logon failure should invoke a log entry and the notification of a security administrator. A supervisor, who could authorize a new password, may approve reactivation of this user.

Dial-up lines are especially vulnerable to unauthorized access. The most common security techniques used with dial-up circuits are callback and handshake. With callback the user dials the computer and provides identification, then the computer breaks the connection and dials the user back at a predetermined number that is obtained from the database. The disadvantage of callback is it does not work well with traveling employees who may not be at a predetermined location. The handshake technique requires a terminal with special hardware circuitry. The computer sends a special control sequence to the terminal, and the terminal identifies itself to the computer. This technique ensures only authorized terminals access the computer, however it does not regulate the user of the terminal. The provisions of network management, discussed in Chapter 5, should be used to ensure the highest level of network access.

Media Access

From a security point of view, the type of cable chosen for various parts of the network can depend on the sensitivity of the information that traverses it. The most common cable types used in networking infrastructures are coaxial cable (coax), twisted-pair copper, and optical fiber. Unlike copper or coax, optical fiber does not radiate any energy, so this lack of radiation provides a high degree of security against eavesdropping. Fiber is also much more difficult to tap into than copper or coaxial cables. The heat-fused cladding of multimode graded index and single-mode cable makes them practically impossible to tap. Even if a tap occurs, it can be detected, because the match on the fiber-optic cable must be perfect for light transmission without disruptions (Shelly 1998).

Wire taps can sometimes be detected by using tools to measure attenuation (reduction of signal strength) on the cable. A time domain reflectometer (TDR) is used with copper cable and an optical TDR (OTDR) is used with fiber optics cable to look for abnormalities. These devices are normally used to measure signal attenuation and installed cable length, however an experienced technician can use them to locate wiretaps. It should be noted that an expert intruder can insert a tap so that it not easily detectable by either device. It is a good practice to take an initial baseline signal level of the physical cable infrastructure and periodically verify the integrity of the entire physical cable plant. Creating a baseline is part of Enterprise security management, which is discussed in Chapter 5.

The physical path of the media, which is part of the network topography, is an issue that must be addressed to ensure availability of the network and its devices. It is imperative to have a structured cabling system that minimizes the risk of downtime. The cable infrastructure must be well secured to prevent unauthorized access. There are ITU recommendations that provide specifications and recommendations for the construction, installation, and protection of cable plants.

In the case of access into a controlled environment, such as a computer room or network wiring closet, a restrictive set of access rules should apply. Organizational site guidelines must be developed and implemented to restrict access for cabling to these areas by controlling access to cable trays and overhead cable support systems. Approved methods and access points must be established prior to cable routing and installation. The security system to the area can vary from simple lockable cable feeder openings to dedicated individual fiber and cable runs, each with alarm-equipped access covers and video surveillance.

The common areas of exposure when addressing cable security include (Herrick 2001)

- Cable trays in the crawl space under raised floors
- Wall mounted wiring access
- Overhead cable and wire trays

Chapter 10 provides details concerning wired LAN connectivity, where cabling would terminate in a centrally located wiring closet. There may be a patch panel in the wiring closet that provides for the interconnection between the WAN circuits and the LAN components. There may also be wiring between a terminal server or a communications server and the patch panel. The wiring closet is a major point of vulnerability and must be afforded a high level of security. Security of WAN circuits is discussed in Chapter 11.

Physical Device Security

Physical device security includes identifying the location of devices, limiting physical access, and the installation of environmental safeguards. All network infrastructure equipment should be physically located in restricted access areas. This means creating a secure space for wiring closets that contain switches, firewalls, modems, DSUs, and routers. It is essential that this equipment closet be well secured, however do not forget that intruders can enter through an opening in the ceiling. This network closet is sometimes called a demarcation point where all the telephone facilities enter the building. This is an excellent place for an intruder to place a monitor for eavesdropping on all of the organization's network traffic.

Any OSI Layer 1 LAN device is a potential candidate for illegal access. This could include a repeater, hub, or multistation access unit (MAU). The

physical device security

potential increases if shielded twisted pair (STP) or unshielded twisted pair (UTP) cabling is used. These elements are discussed in Chapter 10.

Infrastructure equipment includes more than networking equipment, which must be protected. Other sensitive components of the infrastructure include numerous servers, gateways, and administration terminals. Print servers, terminal servers, file servers, and domain name servers must all be afforded a high level of security and should not be available to the general staff. Consoles and administrative terminals must be secured from the general staff and all other unauthorized personnel because these devices provide access into the secured infrastructure of the organization.

The output from the office printer or fax machine is often overlooked. These devices are usually centrally located for the convenience of the staff. They are good places to look for confidential information, so do not overlook protecting them. It might be necessary to use a paper shredder and not allow any whole pages out of the facility. Do not forget that there are "dumpster divers" waiting for someone to dispose of confidential material that can be used for some illegal activity.

Physical Access Security

As mentioned earlier, wiring closets and rooms that house infrastructure devices and networking facilities must be secured from unauthorized people. These areas should be in the interior of an organization's operation. The exterior access must also be protected to control both nonemployees and employees alike. Many organizations use a guard station to control exterior access. In such cases, employees are required to show identification and be authorized for a specific level of access.

Part of physical security policies and procedures must address contract personnel, maintenance personnel, and others who do not possess unrestricted access. These people may be required to be in a controlled area, but still must be escorted by an authorized employee or must sign in before entering this area. Identification cards may be required if these are very sensitive areas. The larger the organization, the more important these measures become, since many employees may not even know who their coworkers are.

To ensure enforceable physical security, employees' work areas must mesh well with access restrictions. This includes the flow of personnel within an organization's premises. It is easy for an employee to leave a door propped open instead of locking and unlocking it many times a day. Installing an alarm that will sound when a door is left open can eliminate this type of situation. To avoid employee aggravation, a logical and workable security plan must be developed that considers the daily workflow.

Many organizations host both visitors and personnel from other departments. These people must be given access on a temporary basis, however they should be located in a controlled area away from sensitive operations. Logs for signing in and temporary identification should also be required of these visitors.

Personnel Security

It is essential that personnel in sensitive positions are screened to ensure their integrity. This process should not be an afterthought and must be part of corporate policy. **Personnel security** involves using one or more of the following techniques (Rowe 1999):

- Security and background screening of prospective employees before hiring
- An active security awareness program emphasizing security issues and ramifications
- Training of employees regarding security issues and their responsibilities in maintaining security
- Identification of employees, contractors, and maintenance workers
- Error prevention techniques to detect accidental mistakes and malicious activities

Customer Security

Physical computer and networking assets may be protected inside a building or they may be in a public area. These devices can include automated teller machines and cash dispersing machines. There has been a rising crime rate against both these devices and the customers that use them (Rowe 1999). Robberies can occur shortly after customers withdraw money from an ATM or criminals assault ATM customers and force them to withdraw funds. Physical violence is often part of these attacks. Banks must determine how far to go to protect ATM customers. The location of these devices and the time of day the transactions occur has an enormous impact on the security needed. Many banks provide video monitoring and lighting systems, however these initiatives are no solace to the customer being attacked. Banks do not accept responsibility for these crimes and say consumers bear part of the risk when they use these machines late at night. These situations exemplify the type of societal problems that can and will occur as communications systems become more widely used by the general public.

9.5 SECURITY FORENSICS

Data collected by intrusion detection systems, firewalls, and network devices such as switches, routers, gateways, and servers can be used to analyze and evaluate the extent to which a network has been compromised by an attack. Logs are an essential source of information for this type of forensic examination. In addition, if equipment that has been the target of an attack can be secured at a crime scene, photographs and fingerprints can be taken, and the equipment can be thoroughly examined. Documentation is an important element used in a court of law if an attacker is to be prosecuted. It is essential that the evidence trail be preserved and secured.

Intrusion Detection

Intrusion detection techniques and equipment are used to detect and respond to computer and networking misuse. Different intrusion detection techniques provide different benefits for different situations and environments; therefore, it is essential that the security administrator deploy the security system that best matches the need. Unnecessary costs can be incurred for nonessential detection devices. The key to selecting the right security detection system is to define the specific security requirements first and then implement a system based on those requirements (Proctor 2001).

Monitoring

Some firms deploy intrusion detection **sensors** at customer locations and provide security experts to monitor corporate resources such as routers and firewalls. Chapter 5 provides an in-depth analysis of network management systems. Other firms place probes on the customer's networks to collect audit data from network devices. The data is transmitted in encrypted form back to the central facility where it is monitored continuously.

In the security monitoring application, a video camera can be mounted on top of a building, in a doorway or corridor, or in the corner of a room viewing the area being monitored (Rowe 1999). Black and white television is most often used, with the picture being updated every 30 to 90 seconds. This periodic refreshing of the picture is called freeze-frame television. It requires a much slower, less expensive telecommunications line than one that is transmitting full-motion pictures, which is similar to commercial-quality television. Video cameras are not prevention devices, however they do provide significant value in monitoring intrusion events.

Some network-based intrusion detection systems consist of sensors deployed throughout a network that provide a data stream to a central console. Sensors may contain logic that collects network packets, searches for patterns of misuse, and then reports an alarm situation back to the console. Two types of sensor-based architectures include the traditional sensor design and the network node design. Sensor-based systems monitor whole network segments. They are not widely distributed because there are relatively few segments to monitor, as opposed to the sensor-based systems that are widely distributed onto every mission-critical target. Network-node systems place an agent on each managed network computer device in the network to monitor traffic bound only for that individual target (Proctor 2001).

9.6 DISASTER RECOVERY

Disaster recovery is an extension of planning for normal computer, database, and network systems outages. It is possible that a disaster has occurred where immediate repair is not possible and the computer facility and equipment is no longer usable. Having a computer in an alternate site and the ability to switch the telecommunications lines to this other site is a viable alternative in many situations. It may also be possible to quickly obtain a new system from a computer vendor, since many vendors have the capability to provide a new system and all its components on short notice. This solution requires a place to install this system that has suitable power and air conditioning. Of equal importance is the ability to switch the telecommunications network to the new computer location promptly. Telephone companies are developing techniques and facilities to allow network switching to alternate locations, however it is essential to determine if the broadband network in the alternate location supports the protocols used on the organization's networking system (Rowe 1999).

disaster recovery

It is beneficial to work out a mutual aid pact with another company, in which both organizations agree to provide backup in the event of a disaster. This assistance could be in the form of after-hours computer time or space to house a temporary operations center. A common problem with backup operations is the requirement of telecommunications facilities, which can be difficult to install on short notice. A backup location that is in close proximity could be problematic because both could suffer similar disasters simultaneously. An earthquake, hurricane, or tornado could destroy both locations and effectively neutralize any recovery plans.

There are several companies whose sole purpose is to provide disaster recovery plans and facilities. These companies make significant investments in computer hardware, which can be used in disaster situations. These services are not cheap, however it's "pay me now or pay me later." Subscribing to such a service is like buying insurance. Whatever plan is developed must provide for a specific disaster type, since the response to a fire is different than responding to a flood.

Some organizations develop a disaster recovery plan and never validate it for effectiveness and completeness. Rarely will all of the problems and issues of a real disaster be covered when a recovery plan is created. The recovery plan must be tested to identify weaknesses and a disaster recovery firm can provide this capability. Tests usually involve transferring software systems between computer systems and ensuring communications can be established in the event of a disaster.

As an organization becomes increasingly reliant on telecommunications and computer networks, it must constantly reassess how much downtime can be afforded. One way to accomplish this is to get users involved in assessing the impact of an extended outage caused by a disaster. This can result in building a case for additional funds for supporting and maintaining disaster recovery procedures.

9.7 AUDITING

auditing

Auditing is the collection and monitoring of events on servers and networks for the purpose of tracking security violations. It is also used to keep track of system utilization. A network auditing system also logs details of user transactions on the network so that malicious or unintended activities can be tracked. When auditing is implemented, vast amounts of data may be recorded and archived for future reference. Some audit systems provide event alarms to warn administrators when certain conditions or levels are met.

Auditing records can be viewed using special filters to produce reports that show specific activities. Filters can be applied to show specific date and time ranges, file and directory activities, and user events. Administrators should look for a large number of failed logon attempts and logins that take place at odd hours, which could indicate that an unauthorized person is attempting to access the system.

An auditor can use software to set up auditing features, view audit logs, and look at designated directories. A record can be kept for every activity designated for tracking. Events that can be tracked include the following (Sheldon 1998):

- Logon and logoff
- Connection establishment and termination
- Space restrictions
- Granting of trustee rights
- Dismounting volumes
- File creation and modification
- Directory creation and deletion
- Server commands and queue activities
- Changes in the security environment

An additional step taken by operations management is to monitor the activities of the network or systems administrator. The auditing system can be used to verify the integrity of this person, who basically has unlimited rights to the system.

9.8 DISASTER PLANNING

Disaster recovery occurs if the organization has developed a **disaster plan**. This plan should cover LAN, MAN, WAN, and PBX/ACD resources. It should also include computer systems, servers, and local network service. The plan should identify the most common hazards and assess their potential impact if there is a catastrophic event. The security section of the plan should describe how to limit access to critical facilities. Many organizations have developed formal disaster recovery plans to guide them through service restoration following a natural disaster or civil disturbance. The key to disaster planning is to prevent damage from occurring in the first place. A service loss can occur as a result of fire, water, snow, earthquake, landslide, lightning, or human activities. There are a number of precautions that can be taken to prevent or minimize damage from these events when damage does occur. Contingency planning can be used to address these issues.

disaster plan

Contingency Planning

A **contingency** is defined as an event that might occur but is not likely or intended, however there is a possibility that it may occur. The key to a successful recovery from a disaster is adequate preparation. Seldom does a crisis destroy irreplaceable hardware and software resources. Most of these computer and networking components are available from many suppliers and are readily available—at a charge. Databases, specialized software, and application software, however, is another issue. Locally developed software is more vulnerable because it cannot be quickly substituted from another source. Work must continue during and after a crisis (Pfleeger 1997). Data and database backup and restore processes are presented in Chapter 7.

contingency

■ SUMMARY

Physical security includes efforts to protect the Enterprise's assets from both man-made and natural disasters. Typical security components include physical guards, video monitors, locking mechanisms, and barriers. Physical destruction can result from such natural disasters as fire, water, power, heat, and humidity. These same disaster categories can also result from internal negligence, which can be just as destructive.

Environmental safeguards should include protection efforts to counter disasters caused by the following natural and man-induced factors:

- Fire
- Water

- Power
- Temperature
- Humidity
- Dirt
- Smoke
- Wind

The human is another potential destructive force that must be managed by the security administrator. Vandals and unscrupulous and disgruntled employees can exact a toll on the Enterprise's resources. Protection mechanisms and security initiatives must be part of an ongoing effort to protect the organization's assets.

A major part of physical security includes efforts to limit access to sensitive areas and computer network components. It is essential that the security administrator identify methods to restrict access to the network facility to only legitimate users. This also includes access to the computer and networking devices located in these facilities.

Attackers and malicious people can invade an organization's resources via physical media. Protection from wiretaps and eavesdropping is required for both internal and external users. The location of cabling should be considered when developing a wiring plan.

An ongoing program of security audits and monitoring must be part of the Enterprise's normal operating procedures. This is a top-down requirement supported by the organization's management and all personnel. Personnel must be screened before being placed in sensitive positions. The Enterprise must know which customers, contractors, partners, and other users have a need to access the organization's physical assets.

If a disaster occurs, a contingency plan must be in place for all possibilities, and a disaster recovery plan must be ready for the administrator to implement to bring the computer network back online quickly.

Key Terms

Auditing	**Flooding**
Contingency	**Human attackers**
Disaster plan	**Intrusion detection**
Disaster recovery	**Natural disasters**
Earthquake	**Personnel security**
Environmental safeguards	**Physical device security**
Fire hazards	**Physical security**

Saboteurs
Sensors
Storm damage

Uninterruptible power supply (UPS)
Vandals
Water damage

REVIEW EXERCISES

Questions and Analysis

Provide a short answer or in-depth analysis as required.

1. Provide an overview of the essence of physical security.
2. What are the physical security categories?
3. Describe a natural disaster.
4. Comment on the activities of vandals and destructive individuals.
5. Give examples of fire and water damage. Why does the administrator need to be concerned about fire and water in the facility?
6. Discuss the issue of power loss and recovery in the computer network environment.
7. What is the function of power supply systems? Why are they required?
8. How can heat and humidity effect the hardware components of a computer network?
9. Provide an overview of the safeguards that can be employed in the Enterprise physical environment.
10. What precautions must be undertaken to protect the Enterprise assets?
11. What are the natural disasters that can impact the operation of a computer network?
12. Discuss the issue of fire hazards and corporate assets.
13. How can water damage corporate hardware and physical assets?
14. Discuss the issue of flooding from both natural and internal causes.
15. How would the administrator prepare for an earthquake?
16. How does the Enterprise prepare for the various categories of storms?
17. What can be done about vandals and saboteurs?
18. Identify the physical security controls that can be employed to protect the corporate assets.
19. What can be done to limit network access to legitimate users?
20. Why is media access an issue? What can be done to improve media access security?
21. What are some common areas of media exposure?
22. Why is physical device security important?
23. Identify physical devices that need to be afforded some level of access security.
24. Why should access to corporate computer and network areas be controlled?
25. Give examples of methods for controlling physical access to corporate facilities.
26. How can the organization ensure personnel are not part of the security problem?
27. Discuss the issue of customer security.
28. What is security forensics?
29. Why is intrusion detection important?

30. How does monitoring assist in access control? What are sensors?
31. What is involved in disaster recovery?
32. Describe the auditing process.
33. What events must be tracked with an audit?
34. Discuss the disaster plan.
35. What is the difference between a disaster plan and contingency planning?

Definitions and Terms

Provide the most correct answer from the list of key terms.

1. The term used to describe protection provided outside the computer and networking system.
2. This device stores energy during normal operation so it can provide backup energy if power fails.
3. Data collected by intrusion detection systems, firewalls, and network devices such as switches, routers, gateways, and servers used to analyze and evaluate the extent to which a network has been compromised by an attack.
4. A person who willfully or maliciously defaces or destroys public or private property.
5. Person or persons who look through trash for sensitive proprietary information.
6. Techniques and equipment that are used to detect and respond to computer and networking misuse.
7. An event that might occur, but is not likely or intended, however there is a possibility that it may occur.
8. The activity that occurs if the organization has developed a disaster plan.
9. The collection and monitoring of events on servers and networks for the purpose of tracking security violations.
10. A device, such as a photoelectric cell, that receives and responds to a signal or stimulus.

True/False Statements

Indicate whether the statement is true or false.

1. Physical security components include guards, locks, cameras, and fences.
2. Physical security does not involve natural disasters.
3. Water from a natural flood is more damaging to tapes than a man-made flood.
4. Internal attacks can be as damaging as external ones.
5. An interruptible power supply is not needed where power fluctuations rarely occur.
6. Power fluctuations do not harm computer systems.
7. It is essential that heat and humidity be monitored in a computer center.
8. Dust and dirt are not environmental considerations.
9. Cleaning chemicals and their storage in proximity to electronic components is not an issue to the network manager.
10. It is possible to mitigate the effects of an earthquake on computer resources.
11. Saboteurs, vandals, and competitors may cause damage to an organization's assets.
12. Access control is not an issue if the physical security is sufficient.
13. Fiber, coax, and copper wiring are all impervious to eavesdropping techniques.
14. It is not necessary to screen employees who have access to sensitive corporate assets.
15. Disaster recovery is a normal extension of corporate planning efforts.

Activities and Research

1. Arrange for a field trip to an organization that has implemented a physical security system. Document the security efforts.

2. Arrange for a field trip to an organization that has little or no physical security provisions. Develop a program to make this organization secure from both internal and external users.

3. Research organizations that have experienced physical disasters and develop a proposal that would have mitigated the losses.

4. Research for products that can be installed and used to provide access protection to an organization's physical assets. Develop a cost/benefit analysis for each.

5. Identify organizations that have been the victim of natural disasters. Describe how damage could have been reduced.

6. Design a computer and network layout that shows how various devices and programs can be implemented to protect an organization's assets.

7. Research the topic of UPS systems. Prepare a matrix of different products showing capabilities and cost.

8. Research energy control systems. Prepare a cost/benefit analysis of using these systems.

9. Research fire monitoring and control systems. Create a matrix that depicts provider, model, description, features, and cost.

10. Create a matrix of the various types of physical media. Show the media type, attributes, use, and cost for each.

11. Use the audit presented in Appendix B to develop a specific checklist for either a hypothetical or real organization.

PART TWO

CHAPTER 10
Local and Metropolitan Area Network Security Environment

This chapter presents an overview of both Local Area Networks (LANs) and Metropolitan Area Networks (MANs). Addressed are topology issues of wiring and wireless transmissions in secure and nonsecure environments. Emphasis is placed on wireless LAN technologies and protocols that are emerging in the Enterprise environments. Also discussed are the security issues of intranets and extranets in the LAN and MAN operations. The chapter concludes with a discussion on standards that apply to LAN and MAN topologies.

CHAPTER 11
Wide Area Network Security

Considerable detail is furnished to provide the reader with enough information to understand the security issues that are prevalent in the wide area network (WAN). The various WAN components and their connectivity to the Enterprise computer networks are depicted. A number of sections are devoted to addressing the security issues of connecting the Enterprise network to the Internet. Details are presented that portray the threats that can emanate from the Internet and Web.

CHAPTER 12
Virtual Networks Security

The concept of a virtual network is presented along with a discussion of the Virtual Private Network (VPN). A look at the various VPN types includes details concerning access VPNs, intranet VPNs, and extranet VPNs. Tunneling, security, and connectivity issues are discussed in great detail. Details concerning VPN hardware and software are presented, including firewalls, gateways, and routers. The software section includes information concerning PPTP, L2F, L2TP, and IPSec. VPN implementation scenarios are presented along with VPN standards and applications. The chapter concludes with a discussion of security and integrity issues and problems with a virtual network.

CHAPTER 13
Distributed Systems Security

The concept of a distributed network is presented along with a discussion of the e-commerce environment. Various components are identified, including intranets and extranets, that go to make up an e-commerce environment. The various security techniques and protocols for protecting the e-commerce transactions are presented. The chapter concludes with a discussion on the various design issues for an e-commerce network.

CHAPTER 14
Wireless Networking Security

This chapter offers a fairly detailed overview of the components and interworkings of the wireless network. These include radio, microwave, satellite, and cellular technologies, and the various applications that are in place today. Personal communication services (PCSs) and the devices that are employed in the wireless networks are emphasized. The chapter concludes with a discussion on security and privacy issues in the wireless environment.

CHAPTER 10

Local and Metropolitan Area Network Security Environment

OBJECTIVES

Material included in this chapter should enable the reader to

- Learn the terms and definitions associated with the LAN and MAN environments.
- Understand the basic structures inherent in Local Area Networks.
- Understand where Metropolitan Area Networks fit in the Enterprise environment.
- Identify the security issues relevant to using EDI, CTI, and other computer networking technologies.
- Become familiar with the structures that form intranets and extranets.
- Look at the security issues that must be addressed with the various LAN technologies, wiring arrangements, and transmission techniques.
- Look at the special security issues associated with the wireless LAN configurations.
- Recommend countermeasures and programs that can provide protection from internal, external, active, and passive attacks.

INTRODUCTION

Organizations have expanded their use of personal computers, laptop computers, personal digital assistants (PDAs), and individual workstations to support the information needs of users throughout the Enterprise network. Today's organizations use small computers for word processing, order processing, messaging, sales reporting, financial analysis, engineering, and many other applications in support of the enterprise's mission. As the use of small computers has grown, so has the need for these computing devices to communicate, both with one another and with a centralized Enterprise mainframe computer or client/server system.

Small computing devices were often used in a stand-alone manner to support applications that were local in nature. The data were often stored on the device's hard drive or available on a removable storage medium. Typically, additional requirements arose for various levels of access that include the following:

- Provide access to existing data that might be stored on a computer in some other division or department
- Provide access to computer peripheral devices, such as color printers, that are too expensive to be used by a single user
- Give the user the ability to exchange electronic messages with each other
- Provide access to the Internet, a corporate intranet, or extranet

During the 1980s the proliferation of personal computers (PCs) in the office led to the introduction of local area networks (LANs). Once offices were computerized, companies began looking for effective ways to link the individual PCs, to allow for the sharing of documents and expensive peripherals. Today, LANs enable multiple-site connectivity and even telecommuting. Access to vital corporate information and e-mail is provided by file servers—computers connected to the LAN. Password features and firewalls restrict access to this information. LANs come primarily in two topologies: Ethernet/IEEE 802.3 or Token Ring. Several transmission speeds are available. Ethernet/802.3 operates at 10, 100, and 1000 megabits per second and Token Ring operates at either 4, 16, or 100 megabits per second (Mbps). Figure 10–1 depicts a logical representation of two different LANs connected to a common computer system. The computer system must provide support for the different **Ethernet** and **Token Ring** protocols.

Ethernet
Token Ring

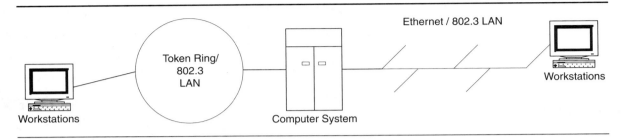

Figure 10–1
Local Area Networks

Many of today's LANs are used to interconnect PCs and workstations with centralized computers called servers. These topologies are called client/server systems. Personal computers, shared printers, shared databases, factory automation, and quality control systems connected in a local area data communications system are examples of LANs that can use a client/server system. Client/server environments have introduced a whole new set of security issues for the LAN administrator that must be addressed to ensure the integrity and security of the organization's resources.

With the introduction of wireless systems into the LAN environment, security of data, databases, hardware, software, and network components has emerged as a major challenge to the system administrator. Both virtual LAN and wireless LAN technologies and configurations are presented in this chapter.

Administration of a LAN and its derivations is a major undertaking for the management of Enterprise assets and resources. Some technical details are presented in this chapter to provide the reader with a basic understanding of the infrastructure of the LAN environment. Without a general understanding of the underlying structure and components of a LAN, it would be difficult to plan and implement a viable security system that would provide an adequate level of protection to the organization's resources.

10.1 LOCAL AREA NETWORK (LAN)

Local Area Network (LAN)

A **Local Area Network (LAN)** consists of a group of data devices, such as computers, printers, and scanners, which are linked with each other within a limited area, such as on the same floor or in the same building. There are two ways of looking at the LAN technology: (1) the LAN data link technology that is used to implement a computer network, and (2) the LAN networking software that is used to provide users with local area networking facilities. The LAN data link technology is used to implement a flexible, high-speed form of networking data link. The source system generates a message and uses the facilities of a LAN data link to deliver that message to the destination system. The most commonly used LAN data link technologies are as follows:

- Ethernet
- Token Ring
- Token Bus
- FDDI
- ARCnet
- LocalTalk

A typical network user views a LAN as a collection of computing systems that are capable of communicating with one another. A user primarily interacts with high-level networking software that allows the use of the networked computers. The following are examples of the most commonly used networking software systems:

- TCP/IP
- NetWare
- AppleTalk
- DECnet
- System Network Architecture (SNA)
- VINES
- LANtastic
- LAN Manager

10.2 WIRED LAN AND MAN CONNECTIVITY

To communicate with one another, each device must be connected to a LAN. Connections between devices may be any combination of twisted-pair copper, coaxial cable, fiber optic, or wireless media. Twisted-pair Category 4/5 cabling and coaxial cable most commonly connect devices to a LAN. Copper cabling is available in shielded (STP) or unshielded (UTP). Figure 10–2 depicts a simple physical Ethernet (802.3) LAN. The wiring method depicted is category 5 UTP cable. There are a number of opportunities for signals to be intercepted, modified, and deleted in the physical cabling plant environment such as this. Each of the LAN topologies present a different set of security issues that must be addressed. These topologies include a bus, ring, and star configurations.

Bus Topology

A **bus** is an electrical connection that allows two or more wires or lines to be connected together. All network interface cards (NICs) receive all the same information put on the bus, but only the card that is "addressed" will accept the information. With the bus topology, each system is directly attached to a common communication channel, where signals that are transmitted over the channel comprise the messages. As each message passes along the channel, each system receives it and examines the destination address that is contained in the message. If the destination address is for that system, the message is accepted and processed; otherwise it is ignored. Figure 10–3 illustrates a logical bus topology.

bus

Figure 10–2
Ethernet LAN Example

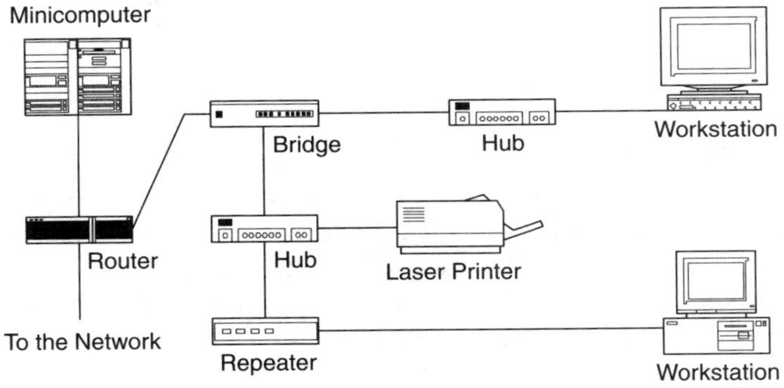

Figure 10–3
Bus Topology

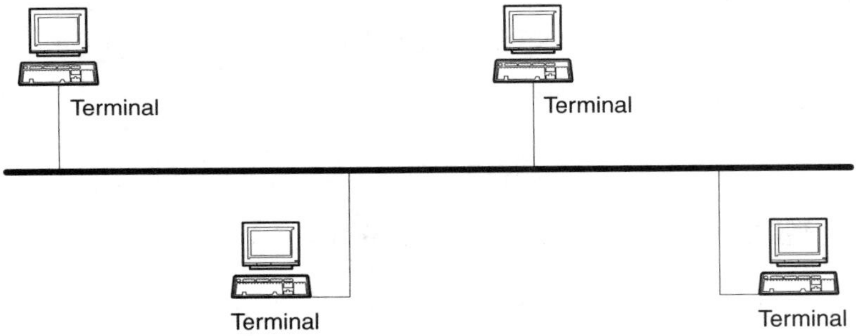

Ring Topology

ring

A **ring** is a LAN in which all the PCs are connected through a wiring loop from workstation to workstation forming a circle or ring. Data are sent around the ring to each workstation in the same direction. Each system acts as a repeater for all signals it receives and then retransmits them to the next system in the ring at their original signal strength. All messages transmitted by any system are received by all other systems, but not simultaneously. The system from which a message originates is usually responsible for determining that a message has made its way all the way around the ring and then removing it from the ring. Figure 10–4 shows a logical ring topology.

Star Topology

star

A **star** is a topology in which all phones or workstations are wired directly to a central service unit. The central service unit establishes, maintains, and breaks connections between workstations. With the star topology, all transmissions from one system to another pass through the central point, which may consist of a device that plays a role in managing and controlling communication. The device at the center of a star may act as a switching device. When one system wishes to communicate with another system,

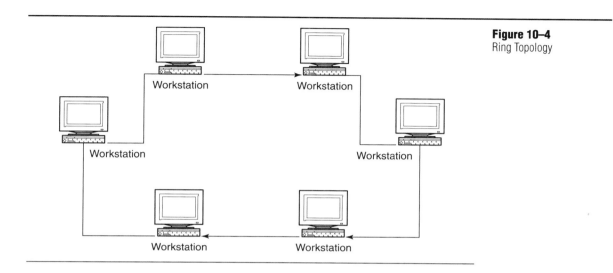

Figure 10–4
Ring Topology

the central switch may establish a path between the two systems that wish to communicate.

The star topology has been used for many years in dial-up telephone systems, in which individual telephone sets are the communicating systems and a private branch exchange (PBX) acts as the central controller. Figure 10–5 shows a logical star topology.

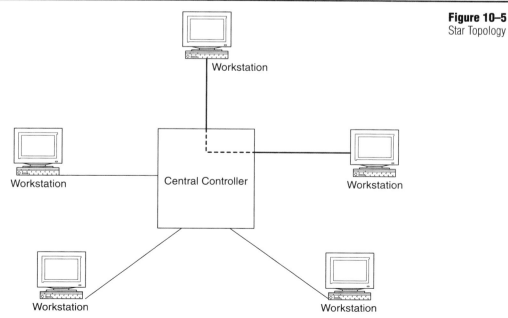

Figure 10–5
Star Topology

10.3 LAN COMPONENTS AND CONCEPTS

A common base of technology understanding is required so that the security administrator can communicate effectively with the user community and implement security procedures to protect the LAN assets. There are different and unique issues that must be addressed for each LAN component. The various elements and concepts are presented in alphabetical order. Sample configurations are presented where appropriate.

Client/Server

client/server

A **client/server** is a computer that sits on the customer site and splits computing operations and applications between the desktop and one or more networked PCs or a mainframe computer. This allows for distributed applications support and access to a central computer. In this environment, an application component called a client issues a request for services from another application component called a server. There may be multiple clients sharing the services of a single server, and the client applications do not need to be aware that processing is occurring remotely. This is possible because a communication network, such as a LAN, provides the means for transporting information back and forth between the client and server components.

The client/server computing environment allows many different types of server systems to be built. The following server systems are possible in the client-server environment:

- File server
- Print server
- Database server
- Communications server
- Applications server

Figure 10–6 depicts a print server that supports a number of different print image devices and is accessible from both the Enterprise network and the LAN.

CPE

Customer Premise Equipment (CPE)

Customer Premise Equipment (CPE), or Customer Provided Equipment, refers to such telecommunications equipment as key systems, answering machines, PBXs, routers, hubs, and switches that resides on the customer's premises. These devices are on the data terminal equipment (DTE) side of the DTE/DCE interface (Figure 10–7).

DCE

data communications equipment (DCE)

Data communications equipment (DCE) refers to devices that sit between the DTE and the communications line. Modems, network interface cards

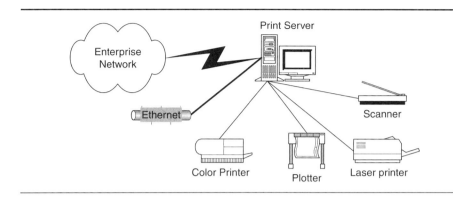

Figure 10–6
Print Server Available to Local LAN and Enterprise Network

(NICs), DSUs, and the interfaces on remote bridges, routers, and gateways are examples of DCE. In many cases, the long-distance carriers and local exchange carriers (LECs) will only work network problems up to the DCE interface. Usually the test to be conducted by the carrier is a remote loopback, which is used to show that the network has connectivity to the DCE device. The WAN connections to these devices are usually in some wiring closet or demarcation point.

DTE

Data terminal equipment (DTE) is the source or destination device that originates or receives data over communications networks. Dumb terminals, PCs, and fax machines are examples of DTE, as are devices that have interfaces to these devices such as routers, gateways, and remote bridges. The DTE/DCE interface is important because of the physical connectivity

data terminal equipment (DTE)

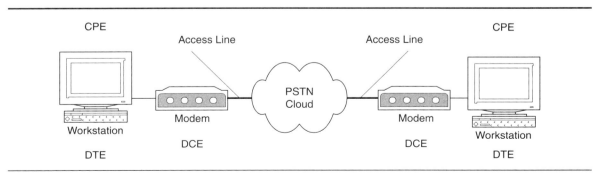

Figure 10–7
DTE/DCE Interface with Access Lines

requirements of cabling. The customer is usually responsible for network troubleshooting on the DTE side of the network interface.

Bridge

Bridges connect multiple LANs. They have less intelligence than routers because they use simple packet filtering to determine whether traffic stays on the LAN or is passed off to an adjacent LAN. Segmenting is a common function for a bridge. There are both local and remote bridges. See the following discussion on LAN switches for increased functionality.

A bridge forwards frames from one LAN segment, but is more flexible and intelligent than a repeater. A bridge interconnects separate LAN or WAN data links rather than only cable segments. Some bridges learn the addresses of the stations that can be reached over each data link they bridge, so they can selectively relay only traffic that needs to flow across each bridge. The bridge operates in the OSI Layer 2 Medium Access Control (MAC) sublayer, and is transparent to software operating in the layers above the MAC sublayer.

A bridge can interconnect networks that use different transmission techniques and/or different MAC methods. A bridge might be used to interconnect an Ethernet LAN with a Token Ring or an FDDI LAN. Examples of these multiple-protocol bridges are translational bridges and source-route transparent bridges. A pair of remote bridges with a telecommunications facility between them can be used to interconnect two LANs that are situated in different geographic locations. Figure 10–8 depicts two Ethernet LAN segments—one segment has a workstation and the other segment has a printer. Access to the hubbing devices is through the local bridge, whose access is based on the proper addressing scheme.

Data Service Unit

The primary function of a **data** (digital) **service unit (DSU)** is to convert a unipolar signal from the customer's DTE into a bipolar signal for the network (DCE) and vice versa. The channel and digital service units (CSU/DSUs) have evolved to the point of being combined, so they are not generally available separately. This combination allows for the transmission of high-speed data communications rates from 56 kbps to the T-3 rate of 44.736 Mbps, with additional capabilities for fractional T-1 (24 DS0s) connectivity. Some of the latest technology equipment, such as multiplexers and channels, integrates the functions of the CSU/DSU into the chassis.

Firewall

A **firewall** is a computer with special software installed between the Internet and a private network that prevents unauthorized access to the

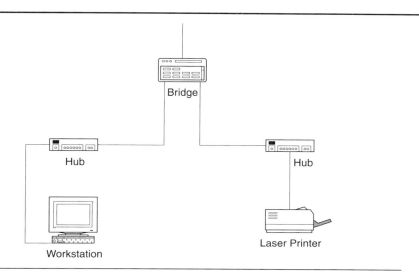

Figure 10–8
Local Bridge Configuration

private network. All network packets entering the firewall are filtered, or examined, to determine whether those users have authority to access requested files or services and whether the information contained within the message meets the criteria for forwarding over the internal network.

A filter is a program that examines the source address and destination address of every incoming packet to the firewall server. Routers are other network access devices that are capable of filtering data packets. Filter tables are lists of addresses whose data packets and embedded messages are either allowed or prohibited from proceeding through the firewall server into the corporate network. Filter tables can also limit the access of certain IP addresses to certain directories.

Attackers use network-mapping techniques to develop an inventory of the overall topology of the network being targeted. Using traceroute, an attacker can determine how systems, routers, and firewalls are connected. It is possible to defend against network mapping by blocking some of the ICMP messages used by the network.

Gateway

A **gateway** is both an entrance and exit into a communications network. Technically, a gateway is a translation device that converts electrical signals from one type of network to another, or converts code from one type of application to another. Gateways operate at all layers of the OSI model.

gateway

A gateway is a fundamentally different type of device than a bridge, repeater, router, or switch, and it can be used in conjunction with them. A gateway makes it possible to communicate with an application program that is running in a system conforming to some other network architecture.

A gateway performs its function in Layer 7 (Application) of the OSI model. The function of a gateway is to convert from one set of communication protocols to some other set of communication protocols. Protocol conversion includes message format conversion, address translation, and protocol conversion.

Internet gateways offer a LAN-attached link for client PCs to access a multitude of Internet-attached resources including e-mail, FTP/Telnet, news groups, Gopher, and the World Wide Web. The Internet gateway translates between the LAN's transport protocol and TCP/IP.

Access control can be performed by gateways that translate from one protocol to another. A gateway between a LAN running IPX and the Internet that uses IP can provide a security screen. IP traffic from the Internet can be blocked and the gateway can also hide local IP addresses and configurations from the Internet. Simultaneously, the IPX workstations on the LAN can run IP and access the Internet.

Port scanners can be used to determine which ports have listening services on a target network. By interacting with various ports on the target systems, a port scanner can be used to develop a list of running services. To defend against port scans, the administrator can harden the OS by shutting down all unneeded services and applying appropriate filtering (Skoudis 2002).

Hub

hub

A **hub** is the intelligent wiring delivery point to which printers, scanners, PCs, and so forth are connected within a segment of an Ethernet/802.3 LAN. Hubs enable LANs to be connected to twisted-pair (10BaseT) cabling instead of coaxial cable in the 10base2 topology. Since it operates in a half duplex mode, only one device at a time can transmit through a hub. Speed is usually 10/100 Mbps. A simple hub is an OSI Layer 1 device. Wiring hubs are useful for their centralized management capabilities and for their ability to isolate nodes from disruption. Figure 10–9 depicts a hub that is used to connect two printers, a desktop system, and a workstation with another LAN device. The desktop system and the workstation can access either printer.

A hub or a wiring hub can be a point of access for eavesdropping and monitoring. Unused ports can be an access point for a sniffer device, which could go undetected. An internal user or contractor would most likely initiate this type of intrusion, but it is easy to counter—keep hubbing devices in a secure location.

sniffer

An intruder can use **sniffer** software to access the network through a hub port. A sniffer is a program that gathers data-link layer packets from

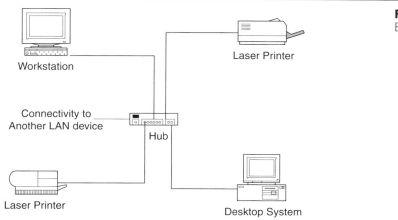

Figure 10–9
Ethernet Hub

the LAN. Using a sniffer (Network General)™, an attacker can read data real-time and store it for later use. It should be noted that the attacker needs an account on the LAN, which means this person could be an internal employee, a contractor, or an unauthorized user who has cracked the system.

Modem

A modulator/demodulator (**modem**) is a DCE device that modulates and demodulates signals. Modulation techniques include amplitude modulation, frequency modulation, and phase modulation. The modem provides an interface between digital devices and analog circuits and equipment. Modems are designed for different applications and are available in a number of configurations and speeds. Many PCs are currently being shipped with a 56 kbps modem card for Internet access. The use of modems should be regulated and monitored for potential security violations.

modem

Multistation Access Unit

A **multistation access unit (MAU)** is the intelligent wiring delivery point to which printers, scanners, PCs, and so forth, are connected within a segment of a Token Ring/802.5 LAN. Connectivity is usually via coaxial cable; however, shielded twisted-pair cable is currently being used. The transmission speed is usually 4/16 Mbps, with 100 Mbps possible.

multistation access unit (MAU)

An MAU is a small electronics device that usually contains eight ports for device connectivity plus a ring-in and a ring-out port for additional MAU devices that allow cascades to form larger rings. Figure 10–10 shows two MAUs cascaded into one ring topology. The server and the two computers can access the printer.

Figure 10-10
Token Ring MAU

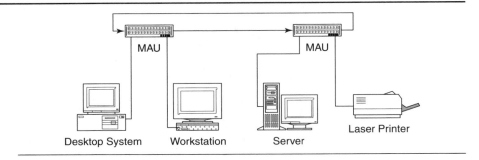

Network Interface Card

Network Interface Card (NIC)

A **Network Interface Card (NIC)** is an adapter that connects a device to the network. The NIC, usually a PC extension board, provides the functionality needed by the connected device to share a cable or medium with other stations. NICs are defined by physical and data link layer specifications. The physical specifications define mechanical and electrical interface specifications. The mechanical specifications define the physical connection methods for cable, whereas the electrical specifications define the framing methods used to transmit bit streams across the cable. Some types of computing devices that are designed for use on specific types of networks, such as a network printer, have the functions of an NIC integrated directly into the device.

Network Termination 1

network termination 1 (NT-1)

Network termination 1 (NT-1) provides functions relating to the physical and electrical termination of the local loop between the carrier and the user ISDN CPE. NT-1 functions are required for both BRI and PRI ISDN services. This DCE device usually resides in the wiring closet with modems and DSUs.

Repeater

repeater

The simplest facility used for network interconnection is the **repeater**. The major function of a repeater is to receive a signal from one LAN cable segment and to retransmit it, regenerating the signal at its original strength over one or more cable segments. A repeater operates in the OSI model Layer 1 (Physical) and is transparent to all the protocols operating in the layers above that layer.

Using a repeater between two or more LAN cable segments requires that the same physical layer protocols be used to send signals over all the cable segments. Repeaters are used in bus-structured LANs, such as Ethernet, to connect individual cable segments to form a larger LAN.

Repeaters are generally transparent to the networking software operating in the individual computers attached to a LAN. The main charac-

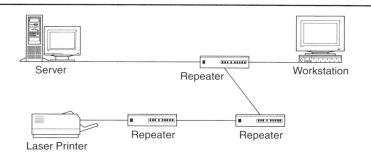

Figure 10–11
Ethernet LAN Repeater

teristic of a repeater is that all the signals that are generated on the cable segment on one side of the repeater are propagated to the cable segment or segments on the other side of the repeater. A repeater implements no form of filtering capability.

The 5-4-3 rule must be observed when deploying repeaters. This means that between two end devices on a five-segment LAN, four repeaters can be used, with three segments that have devices connected. Figure 10–11 shows three repeaters with devices attached in two Ethernet segments.

Router

Routers connect multiple LANs. They are more functional than bridges in that they can handle more protocols, offer traffic control capabilities, and perform better in highly meshed networks. Routers are the "central offices" for the Internet. Filtering and firewall functions are available in a router.

Routers provide the ability to route messages from one system to another where there may be multiple paths between them. A router performs its function in the OSI model Layer 3 (Network). Routers typically have more intelligence than bridges and can be used to construct Enterprise internetworks of almost arbitrary complexity. Interconnected routers in an internet all participate in a distributed algorithm to decide on the optimal path over which each message should travel from a source system to a destination system.

In the internetworking environment, a router is often called an intermediate system. By contrast, systems that originate data traffic and serve as the final destination for that traffic are called end systems. In general, a router performs the routing function by determining the next system to which a message should be sent. It then transmits the message to the next system over the appropriate link to bring the message closer to its final destination. Figure 10–12 illustrates two user networks that are connected through a network cloud using two DSUs.

Routers are often misconfigured or poorly designed, which provides a silent backdoor to the Enterprise LAN. Attackers may use port scanning

Figure 10–12
Router Connectivity

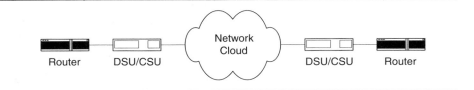

techniques, such as traceroute or tracert, to identify important routers. Always keep up with the latest version of the router's software. The newer releases can fix many recent DoS attacks. A router upgrade will sometimes translate into further expense in memory or firmware, however this may be a small price to pay for system security.

Other than updating the software, disabling remote management is often key to preventing both DoS and remote attacks. With a remote management port open, attackers have a way into the router. Some routers also fall victim to brute-force attacks against their administrative passwords.

Switch

In addition to repeaters, bridges, and routers, a variety of different types of switching facilities can be used in constructing internets. The main purpose of interconnecting LAN data link segments using interconnection devices is to allow a greater sharing of a communication medium. Network interconnection devices allow a larger number of different network devices to be attached to the same LAN.

A LAN switch functions like a bridge and acts as a backbone interface device. Like a bridge, a switch can be used to segment LANs to improve performance and enhance security. A **switch** provides switching through hardware, which provides improved performance over a bridge that switches through software. Speeds available are 10 Mbps and 100 Mbps. A switch operates at Layer 2 of the OSI model. Figure 10–13 depicts three switches that are used as backbone devices on three different locations in a building. These switches could be connected with multimode fiber, coaxial cable, UTP, or STP. LAN switches are being substituted for bridges.

Unlike hubs, switched Ethernet does not broadcast all information to all systems on the LAN. The switch looks at each frame's MAC address and sends data to only the required connection, which limits the data that a passive sniffer can gather. (Note: the MAC address is the physical hardware address on the Ethernet card). If an attacker activates Snort, Sniffit, tcpdump, or another passive sniffer, the only traffic available will be from a single network device, instead of the entire segment. There is, however, a tool called Dsniff, for capturing data from a LAN, that can overcome this limitation (Skoudis 2002).

switch

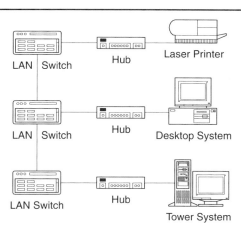

Figure 10–13
LAN Switch Configuration

Transceivers

A **transceiver** is the attachment hardware in 802.3 networks that connects the controller interface to the transmission cable. The transceiver contains the carrier-sense logic, the transmit-receive logic, and the collision-detect logic. It is called the attachment unit interface (AUI), however it is also called the medium attachment unit by some users. It provides MAC services such as jabber inhibit, and heartbeat for a LAN station. It provides connectivity for a number of different physical media, such as coax, twisted pair, and fiber cables.

transceiver

10.4 LAN MEDIA

The devices and hardware that comprise the physical design of a network vary greatly in function and cost. There are a number of different options concerning connectivity in the communications network, and thus numerous pitfalls to avoid. This section comments on the general categories of network components, including transport media, cabling, and general LAN/MAN devices.

Transport Media

Components of transport media include coaxial cable, copper wire, fiber, radio waves, and waveguide. Specifications for copper wire include 10Base2, 10BaseT, and 10Base5.

Copper Wire
10Base2 is the IEEE 802.3 Ethernet standard for data transmission over thin coaxial. It is commonly called thin Ethernet, or Thinnet, because the cable diameter is half the size of 10base5. 10base2 LANs run at 10 Mbps.

10baseT is known as twisted-pair Ethernet because it uses the same type of wire as is used for connecting telephone systems in an office building. It operates at 10 Mbps and 100 Mbps. It is available in both unshielded twisted pair (UTP) and shielded twisted pair (STP).

10base5 is a transmission specified by IEEE 802.3 that carries information at 10 Mbps using 50-ohm coaxial cable. It is sometimes called Thicknet Ethernet.

Coaxial Cable

Coaxial (coax) cable is composed of an insulated central conducting wire wrapped in another cylindrical conducting wire. This package is usually wrapped in another insulating layer and an outer protective layer. It is typically used on 10base2 Ethernet implementations and for connections to IBM 327x type terminals. It has a greater carrying capacity than UTP cable.

Fiber

Fiber is a shortened way of saying fiber optic. An optical fiber can be used to carry data signals in the form of modulated light beams. An optical fiber consists of a thin cylinder of glass, called the core, surrounded by a concentric layer of glass, called the cladding. This package is all wrapped in a protective sheath. Fiber is manufactured in single-mode and multimode versions. Fiber-optic cables have the potential for supporting high transmission rates. The emitter for single-mode fiber is a laser and the emitter for multimode is a light-emitting diode (LED). Multimode and single-mode fiber is constructed with ST (round) and SC (square) connectors.

10.5 WIRED LAN SECURITY

Providing appropriate security for the LAN, its data, and its users is the combined responsibility of management, the LAN administrator, and the LAN software. Management must establish the environment by providing security polices and by communicating the importance of security to all employees. They must also conduct regular audits of the LAN, which can be performed by both internal personnel and external auditors.

LAN hardware, such as servers and communications controllers, must not be accessible to unauthorized people. Someone, such as the LAN administrator, must assume responsibility for ensuring the security of the LAN components. An orderly process must be in place for establishing and maintaining access control tables in the various LAN devices. Changes must be approved by management and should not be made without proper approval.

There are a number of policies and procedures that must be enacted to ensure the proper level of security on the LAN. These items include, but are not limited to, the following (Rowe 1999):

- Use passwords for logging on to the LAN and frequently change them
- Sign-off when leaving a workstation
- Regularly backup disk systems
- Download data to the LAN from foreign systems
- Scan disks for viruses
- Provide dial-up access to the LAN
- Use legal, authorized software
- Encrypt data sent on the LAN or stored on the servers

Managing and providing a secure LAN requires careful attention at both the management and technical levels of an organization.

A passive attack such as eavesdropping or monitoring can be focused on e-mail, file transfer, and client/server exchanges in the wired LAN environment. As described in the previous sections, workstations are attached to some type of local wired topology. The user can reach other workstations, hosts, and servers directly on the LAN or other devices that interconnect over other LAN segments. This is an area of vulnerability and must be addressed with a combination of security procedures, including Triple A, firewalls, and routers.

It is possible that the eavesdropper may not be an employee, since it is possible to access the wired LAN through a dial-in capability. The security administrator must look at the possibility of internal and external attackers using a dial-in access line. Access to the outside, untrusted world is almost always available in the form of a bank of dial-out or dial-in modems, a terminal server, or a router (Stallings 2001). Solutions concerning WAN security are presented in Chapter 11.

10.6 WIRELESS LAN OVERVIEW

The emergence of **Wireless Local Area Network (WLAN)** technology is moving wireless connectivity from the back office to the front office. The advent of wireless resources requires security policies and technologies be responsive to this shared medium, where sensitive data and information may be transmitted beyond the organization's physical controls. Wireless technologies can increase flexibility and enhance the capabilities of the wired-LAN environment. The Enterprise can benefit through location independent access, increased accessibility of data, and reduced dependency on fixed wiring, which can improve employee productivity.

Wireless Local Area Network (WLAN)

There is, however, a price to pay for these benefits in the form of the unique security risks. An important step in implementing the use of wireless technology is a comprehensive review of the corporate information security policy, which is discussed in Chapter 1. The review should consider the types of information transmitted over the WLAN, privacy requirements of the information, and the identity of the recipients. This approach will assist in identifying the specific security measures required when implementing these technologies in the organization's computing infrastructure.

Wireless LAN Connectivity

In local area networks, wireless components act as part of an ordinary LAN, usually to provide connectivity for roving users or changing environments. They may provide connectivity across areas that may not otherwise be networkable, such as in older buildings where wiring would be impractical or across right-of-ways where wire runs might not be permitted.

The wireless components of most LANs behave like their wired counterparts, except for the media and related hardware involved in the physical connectivity. The operational principles are almost the same. It is still necessary to attach a network interface of some type to a computer device, but the interface attaches to an antenna and an emitter, rather than a wire or cable. Users still access the network just as if they were wired into it.

An additional item of equipment is required to link wireless users with wired users or resources. It is still necessary to install a transceiver or an **access point** that translates between the wired and wireless networks. This access point broadcasts messages in wireless format, which must be directed to wireless users, and relays messages sent by wireless users directed to resources or users on the wired side of its connection. An access point device includes an antenna and transmitter to send and receive wireless traffic, but is also connected to the wired side of the network. This permits the device to shuttle back and forth between the wired and wireless sides of the network. Figure 10–14 shows an access point device that is connected to a wireless portable personal computer.

Some wireless LANs use small individual transceivers, wall mounted or free standing, to attach individual computers or devices to a wired network. This permits some limited mobility with an unobstructed view of the transceiver for such devices.

Radio Waves

Wireless LAN technologies are implemented using infrared, spread spectrum, and narrowband microwave (radio). In an infrared LAN an individual cell is limited to a single room, because infrared light does not penetrate walls. Infrared can consist of either diffused or directed beam infrared. Its modulation technique is amplitude shift keying (ASK) and the access method is carrier sense multiple access (CSMA) or token passing. The data rate is from 1 Mbps to 10 Mbps and its range is from 50 feet to 200 feet, which is based on the physical layout.

A spread spectrum LAN makes use of the spread spectrum transmission technology. Spread spectrum is available in either frequency hop-

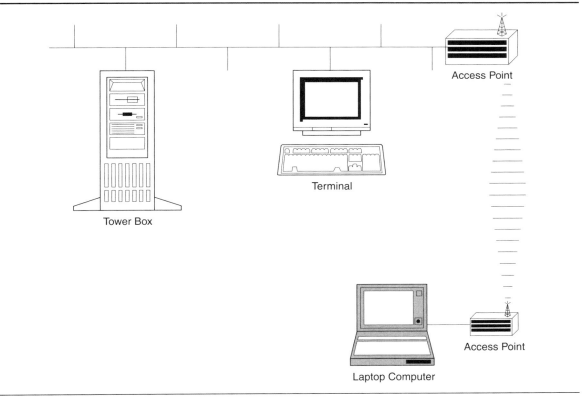

Figure 10–14
Wireless Portable Computer Connected to a Wired Network Access Point

ping or direct sequence. The modulation technique is frequency shift keying (FSK) and quadrature phase shift keying (QPSK), respectively, and the access method is carrier sense multiple access (CSMA). The data rate is 1 Mbps to 20 Mbps and its range is from 100 feet to 800 feet.

Narrowband microwave (radio) LANs do not use spread spectrum. Some of these products require FCC licensing. The modulation technique is FS/QPSK and the access method is ALOHA or CSMA. The data rate is 10 Mbps to 20 Mbps and the range is 40 feet to 130 feet.

10.7 EXTENDED WIRELESS LANS

In extended local area networks, an organization might use wireless components to extend the span of a LAN beyond normal distance limitations for wire- or fiber-based systems. Certain kinds of wireless networking equipment are available that extend LANs beyond their normal wire-based distance limitations or provide connectivity across areas where wire might not be allowed or be able to traverse. Wireless bridges are available that can connect networks up to 3 miles apart.

Such LAN bridges permit linking of locations, such as buildings or facilities, using line-of-sight broadcast transmissions. Such devices may make it unnecessary to route dedicated digital communications lines from one site to another through a communications carrier. Normally, up-front charges for this technology will be considerably higher, but the user will avoid the recurring monthly service charges from a carrier that can quickly make up this difference. Spread spectrum radio, infrared, and laser-based equipment of this kind is readily available from equipment suppliers.

Longer range, wireless bridges are also available, including spread spectrum solutions that work with either Ethernet or Token Ring over distances up to 25 miles. As with shorter range wireless bridges, the cost of a long-range wireless bridge may be justified because of the savings in communications costs it can realize over time. If appropriately connected, such equipment can transport both voice and data traffic.

personal communication network (PCN)

A low-power service, called **personal communication network (PCN)**, is available for wireless LAN access. The following applications are available on a PCN:

- Person-to-person calling
- Messaging
- Fax
- Wireless PBX
- Cordless telephone
- Mobile communications
- Laptop interfaces
- Closed user group communications
- Backbone access to existing services

PCN has the same characteristics as a PCS system, including portable handsets, handoff processes, rural area coverage in cells, and availability for networking purposes for inside or outside structures.

10.8 WIRELESS LAN TECHNOLOGIES

Wireless LANs make use of these four primary technologies for transmitting and receiving data:

- Infrared
- Laser
- Narrowband, single-frequency radio
- Spread spectrum radio

infrared

Infrared light can be used as a medium for network communications. It transmits in the light frequency rages of 100 gigahertz (GHz) to 1,000 ter-

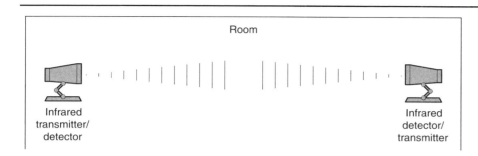

Figure 10–15
Line of Sight Infrared System

ahertz (THz). This technology is used in the remote control devices for television and stereo devices. Infrared can be broadcast in a single direction or in all directions, using a Light Emitting Diode (LED) to transmit and a photodiode to receive.

Infrared connections are limited to two types, **line of sight** and **diffuse**. Line-of-sight infrared systems (Figure 10–15) require that the transmitter and receiver of the infrared light be straight and in line with each other. Diffuse systems allow an infrared signal to bounce off a wall or other object.

line of sight
diffuse

Like radio waves, infrared can be an inexpensive solution in areas where transmission cable is difficult or impossible to install. It can also be used where there are mobile users, with the advantage that the signal is difficult to intercept without someone knowing. Transmission rates only reach up to 16 Mbps for directional communications, and are less then 1 Mpbs for omnidirectional communications. Infrared does not penetrate walls and can experience interference from strong light sources.

Computer networks can use infrared technology for data communications. For example, it is possible to equip a large room with a single infrared connection that provides network access to all computers in the room. Computers can remain in contact with the network while they are moved within the room. Infrared networks are especially convenient for small, portable computers, because infrared offers the advantages of wireless communication without the use of antennas.

Spread Spectrum

Radio communication is widely used as the medium for computer networks, including LANs, where users move around the office with portable computing devices. A transmitter/receiver, or transceiver, is installed in a central area that broadcasts signals to the computer devices. **Spread spectrum** is one method that uses this technique. It is also used in digital cellular communication systems and portable telephones. Spread spectrum distributes a signal over a wide range of transmission frequencies. These signals are hard to detect and if detected, difficult to modulate. This technique provides a high level of security. The main characteristic of

spread spectrum

spread spectrum is that the original signal is spread out to a very wide bandwidth, often over 200 times the bandwidth of the original signal.

Two different methods of implementing spread spectrum are in common use in wireless LANs. In the **direct sequence** method, the radio signal is broadcast over the entire bandwidth of the allocated spectrum. The transmitter and receiver both include a synchronized pseudonoise generator, which the receiver uses to detect the desired signal out of the resulting jumble. Most of the early wireless LAN products use direct sequence. The second method is **frequency hopping**. The transmitter and receiver are synchronized to hop between frequencies, stopping on each frequency for a few milliseconds. The amount of time per hop is called dwell time. Later-generation LANs use frequency hopping, which is an excellent method of handling interference. If the equipment finds an interfering signal, it marks that portion of the spectrum as busy and skips it. Both frequency hopping and direct sequence provide excellent security. Figure 10–16 highlights the key characteristics of a spread spectrum system. Input data enters the channel encoder, is modulated, and passed over the channel where it is demodulated and passed to the channel decoder (Stallings 2000).

Direct Sequence

In the direct sequence scheme, the data to transmit is altered by a bit stream that is generated by the sender. The bit stream represents every bit in the original data with multiple bits in the generated stream, which spreads the signal across a wider frequency band. If 16 bits are used to represent each bit of data, the signal is spread out 16 times its original bandwidth. The source generates a pseudorandom bit stream, called a chipping code, to modulate the original data and the destination generates the same bit stream to demodulate the transmission. Pseudorandom numbers are gen-

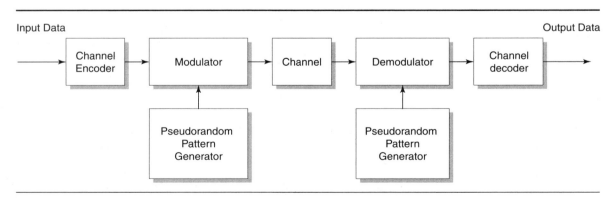

Figure 10–16
General Model of Spread Spectrum

erated by an algorithm, which uses some initial value called a seed. The algorithm is deterministic, which produces sequences of numbers that are not statistically random. The algorithm, however, is sufficient to pass the reasonableness test for randomness. The amount of spreading that is achieved is a direct result of the data rate of the pseudorandom stream.

Frequency Hopping
With the frequency hopping technique, the original data signal is not spread out but is transmitted over a wide range of frequencies that change at high-speed intervals. Both the transmitter and receiver jump frequencies in synchronization during the transmission, so a jammer would have difficulty targeting the exact frequency on which the devices are communicating, and would only hear unintelligible blips. Attempts to jam the signal would only impact a few bits of the transmission. The frequencies are derived from a table that both the sender and receiver follow. The process for frequency hopping is as follows:

- Binary data is fed into a modulator using some digital-to-analog encoding scheme.
- The resulting signal is centered around some base frequency.
- A pseudorandom number source serves as an index into a table of frequencies.
- At each successive interval, a new frequency is selected from the table.
- This frequency is then modulated by the signal produced by the initial modulator to produce a new signal with the same shape; however, it is now centered on the frequency chosen from the table.
- On reception, the spread spectrum signal is demodulated using the same sequence of table-driven frequencies and then demodulated to produce the output data.

Bluetooth Wireless Technology
A new wireless technology allows users to make wireless connections between various communications devices, such as mobile phones and desktop computers. This technology is called **Bluetooth**, and it uses radio transmission to transfer both voice and data in real time. The Bluetooth radio is built into a small microchip and operates in a globally available frequency band.

    The Bluetooth specification was formed by a special interest group (SIG) comprised of Ericsson, IBM, Intel, and Toshiba. It is designed to be an open standard for short-range systems. The sophisticated mode of transmission adopted ensures protection from interference and security of data. The Bluetooth specification has two power levels defined: a lower power level that covers a shorter personal area within a room and a higher power level that can cover a medium range, such as within a home.

margin note: Bluetooth

Figure 10–17
Bluetooth Piconets within a
Bluetooth Scatternet

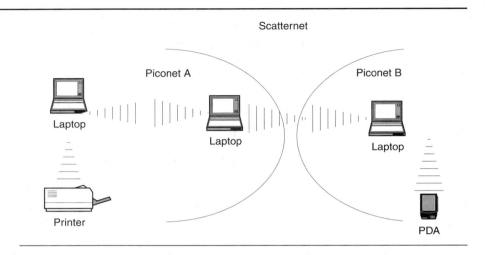

The Bluetooth technology supports both point-to-point and point-to-multipoint connections. Up to seven "slave" devices can communicate with a "master" radio. These so-called piconets can be established and linked together (Figure 10–17). All devices in the same piconet have priority synchronization; however, other devices can be set to enter at any time. The topology can be described as a flexible, multiple piconet structure.

Bluetooth includes support for both authentication and encryption, which is based on the use of a secret link key that is shared by a pair of devices. This secret link key is generated by a pairing procedure when two Bluetooth-compatible devices initially communicate. Bluetooth supports three security modes for a device, designated as modes 1,2, and 3. The security architecture used by Bluetooth is flexible—defining the methodology for authentication and encryption for the Bluetooth-specific protocol, while allowing other protocols transported via Bluetooth to use their own security features (Held 2001). The Bluetooth wireless networking specification is described in Chapter 7.

10.9 WIRELESS LAN SECURITY

Both wireless and wired networks are subject to the same security risks and issues. These include threats to the physical security of the resource, unauthorized access and eavesdropping, and attacks from within by authorized users. There are, however, two physical security concerns that must be addressed in a wireless implementation. These include the interception and reconstruction of radio transmissions and the theft or loss of system devices. These situations can result in eavesdropping episodes and unauthorized access and use of the wireless resources. While second and third generation (2G and 3G) wireless technologies continue to evolve and a security rating might be assigned to a specific network component, implementation specific parameters will remain key factors in defining the level of security actually achieved. To ensure the security of information stored or transmitted by a wireless device, compliance with an information security policy and associated standards is essential (Foose 2001).

When a WLAN is part of an Enterprise network, it provides an interface to a potential intruder that requires no physical intervention or mechanism. The basic security principles applicable to the wired networks also apply to the wireless environment, however there are unique issues that must be addressed. These issues include unauthorized access, integrity, denial of service, and vulnerabilities.

Unauthorized Access

Wireless systems provide opportunities for a number of vulnerabilities in confidentiality. These include browsing, eavesdropping, inference, leakage, masquerading, and traffic analysis. Each of these activities can have a negative impact on the organization's assets.

Eavesdropping is the ability of an intruder to intercept a message without detection. One of the most challenging security issues associated with wireless is the radio frequency (RF) emanation. Denying access to these signals is difficult, since the distance they travel often exceeds the physical control of the organization. This can allow passive interception by anyone with the proper equipment and a position that allows line-of-site of the emitted signals. Eavesdropping on these radio transmissions is relatively simple. When a transmission occurs over the frequency range, anyone with a receiver that can tune that frequency can eavesdrop. These devices are available on the open market and are reasonably priced. Neither sender nor receiver is aware that someone is listening to the transmission, since the interception of the signal is entirely passive.

eavesdropping

The wireless intruder can **masquerade** as a user, a serving network, or a home environment, and can misuse privileges of each. Intruders may impersonate a user to access services authorized for that user. The intruder may have received assistance from other entities such as the serving network, the home environment, or even the user. This could include part of the serving network's infrastructure, where the intruder uses an authorized user's access attempts to gain access to the resource. Users can abuse their privileges to gain unauthorized access to services or intentionally use their subscriptions without any intention of paying. This abuse of privileges can also extend to the serving network. The serving network could misuse authentication data for a user, which could allow an accomplice to masquerade as that user.

masquerade

Integrity

A number of activities can impact the integrity of the system's data and applications. These take the form of manipulations of user network traffic, control data, messages, stored data, and software. An intruder can modify, insert, replay, or delete user traffic on the radio interface. Note that this can also occur accidentally by a legitimate user. These same activities can also occur with control data. It is also possible that access to system entities can be obtained either locally or remotely and may involve breaching logical or physical controls.

Denial of Service

denial of service (DoS)

Wireless radio transmissions are vulnerable to **denial of service (DoS)** attacks. An intruder with a powerful transceiver can generate sufficient radio interference to interrupt WLAN communication. This attack can be conducted from beyond the perimeter or control of the organization. Equipment and components necessary to conduct this type of attack can be acquired from a number of sources. Protection against this threat is difficult and expensive, but it is easy to achieve with the proper equipment. DoS attacks may include physical intervention, protocol intervention, masquerading, and abuse of emergency services.

There are several methods of disrupting a network by physical means. Cables can be cut on wireless and wired networks. Signal jamming can interrupt the wireless interface. Power supplies to transmission equipment can be disconnected or damaged. Physical intervention can cause delayed transmissions on both wired and wireless networks. These situations are usually easy to fix, but can be difficult to locate.

Intruders can prevent the transmission of user and signaling traffic by introducing protocol failures. These protocol failures might be introduced by physical means. It is also possible for an intruder to masquerade as a network element that can intercept and block user traffic, signaling, or control data. Intruders may prevent access to services by other users and cause serious disruption to emergency services facilities by abusing the ability to make USIM-less calls to emergency services from 2G and 3G terminals.

Inference and Deception

In wired networks the wire can be traced to a particular device, however this is not possible in the wireless environment. This means that an effective authentication mechanism is crucial for WLAN security. Both parties must be able to authenticate identities and provide authorizations.

Wireless can be used as the initial point for a transitive trust attack if the intruder can fool the WLAN to trust the intruder's device. If successful, the intruder's mobile device becomes a network node, and may actually have access behind the firewall. This type of attack can be accomplished by using any device compatible with the WLAN from a location beyond the perimeter of the organization. The only real protection against this attack is to ensure there is a strong authentication mechanism for mobile devices accessing the network. Discovery of the unsuccessful attacks is dependent on the logging of unsuccessful logon attempts, however this discovery can be misleading because radio path problems can cause normal unsuccessful logon attempts.

Vulnerabilities

The 802.11b standard offers wired equivalent privacy (WEP) as a simple method to protect over-the-air transmissions between WLAN access points and network interface cards (NICs). WEP operates at the data link and physical layers (OSI Layers 1 and 2), and requires a secret key. This standard may provide sufficient security to discourage experimenters and casual eavesdroppers, but will not stop a determined attack.

There are several methods that can be used to protect the access point from theft and tampering. Because the access point is located in an open area, it is important to ensure its physical security. It can be mounted high on a wall or in the ceiling to discourage unauthorized tampering. It can be installed in a secure wiring closet, however it might be necessary to add an optional range extender antenna. Some access point devices can be optioned so that only authorized administrators can view or change settings (Ciampa 2001).

It is also possible to limit the ability of stations to associate with the WLAN. A WLAN must initiate an association process of communicating with the access point before a station can become accepted as part of the network. Prohibiting an active scanning response can solve this problem.

A wireless network can be compromised by a trivial method of wandering around. The attacker needs a laptop, a wireless network card, antenna, and a wireless network-sniffing program. The sniffer is looking for the service set identifier (SSID), which is the network name of the Wireless LAN. The attacker then uses this SSID to gain access through a DHCP-assigned IP address. This situation can be mitigated by an internal firewall or by using a security protocol like IPSec. More drastic measures require implementing building facilities that block radio waves from leaving the premises (McClure 2001).

Lastly, some access points allow for the creation of an **access control list (ACL)**, which is a method of authenticating legitimate WLAN users. A list of preapproved MAC addresses can be entered in the access point ACL table, which rejects all users who are not in the list.

access control list (ACL)

10.10 VIRTUAL LOCAL AREA NETWORK (VLAN)

A **Virtual Local Area Network (VLAN)** is a grouping of network devices not restricted to a physical segment or switch. VLANs can be used to restructure broadcast domains similarly to the way that bridges, switches,

Virtual Local Area Network (VLAN)

broadcast domain   and routers divide collision domains. [Note that a **broadcast domain** is a group of network devices that receives LAN broadcast traffic from each other (Hudson 2000)]. A VLAN can prevent hosts on virtual segments from reaching one another and can provide isolation from errant broadcasts as well as introducing additional security behind these virtual segments.

The benefits of using a VLAN centers on the concept that the network administrator can logically divide the LAN without changing the actual physical configuration. These benefits include better traffic control and increased security. Dividing the broadcast domains into logical groups increases security because it is more difficult for an intruder to tap a network port and determine the configuration of the LAN. Additionally, a VLAN allows the administrator to make servers behave as if they were distributed throughout the LAN, when in fact they can be physically locked up in a central location. Chapter 12 presents information concerning virtual network security, which is primarily focused on the Internet and VPN topologies.

The four VLAN assignment methods include Port-grouping, MAC-based, Layer-3-based, and Rule-based (Held 1997). The recent rule-based addition to VLAN creation methods is based on the ability of LAN switches to look inside packets and use predefined fields, and even individual bit settings as a mechanism for the creation of a virtual LAN. The ability to create virtual LANs via a rule-based methodology provides a virtually unlimited VLAN creation capability. Of these four methods, the port-grouping VLAN creation method provides the highest level of security since all stations must reside on the same VLAN.

Using a combination of switches and firewalls, a VLAN can be constructed that would provide connectivity and security between an organization's private network and those of business partners. Multiple VLAN can be configured to prevent connectivity among the multiple business partners (Elahi 2001).

VLAN Security

As VLANs are defined in the network, there are times when restrictions are needed on certain switch ports. For some organizations, the need to filter certain end nodes is required, which will allow special monitoring for configured ports. There may be instances in which VLAN security is justified by the integrity of the data that travels across the network. VLAN security allows administrators to filter certain confidential information transmissions, even at the station level.

There are special methods in which ports can be secured when configuring VLANs. Secure port filtering can be used to block input to an Ethernet port when the MAC address of the station attempting to access the port is different from the MAC address specified for that port (CCNA 2000).

Frame switches have VLAN capabilities that allow establishment of VLAN groups that correspond to an organization's security policy and also correspond to network management uses. VLANs can be established for diverse departments such as finance, personnel, engineering, and marketing. Controlled communication between these VLANs can make use of any additional appropriate access mechanisms presented in previous chapters.

10.11 METROPOLITAN AREA NETWORK (MAN)

A **Metropolitan Area Network (MAN)** is a group of data devices, or LANs, linked with each other within a city or large campus area. MANs operate in a manner similar to LANS, but over longer distances. MANs can be used to bridge the gap between WANs and LANs. A MAN could be implemented over a Fiber Distributed Data Interface (FDDI) ring or LEC facilities with LAN extenders and fiber bridges. Figure 10–18 shows two client-server systems that are using the FDDI technology to extend their access across a metropolitan area. The FDDI network is transparent to the users. Other devices can be used in place of the gateways.

Metropolitan Area Network (MAN)

Metropolitan Area Network Connectivity

Metropolitan area networks (MANs) are connections between LANs within a city. MANs may encompass several blocks or a 50-mile radius. For example, a bank in Charlotte keeps its account records in a nearby storage facility in another part of the city. Instead of transporting the records and files between the two sites, the bank leases high-capacity phone lines to transmit data images. The connections between these two sites are metropolitan area connections. MANs can be leased from a telephone company or constructed by the organization using copper, fiber-optic, or microwave-based services. They may also use the same services as WANs such as ISDN, T-1, and T-3. These local services consist of fiber multiplexers and fiber bridges. An example of such a service is BellSouth's Native Mode LAN Interconnection (NMLI). This service uses the embedded local FDDI architecture for its transport medium. Figure 10–19 depicts a simple MAN that uses LAN extenders as fiber bridges. Multimode fiber

Figure 10–18
Metropolitan Area Network Topology

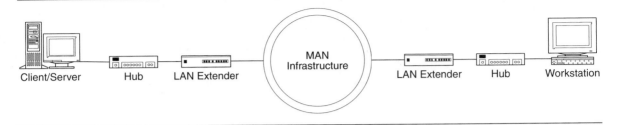

Figure 10–19
Transparent Connection of Two LANs using a MAN

connects the hub to the LAN extender and single-mode fiber is used to connect the LAN extender to the MAN cloud. The MAN is transparent to the user and the users may all be on the same LAN segment.

Metropolitan area networks are used to bridge the gap between WANs and LANs. Where MANs are deployed, they are usually used either as a lower-cost alternative to WANs or to extend the range of an existing LAN.

10.12 INTRANET

intranet

An emerging network paradigm arising out of the enormous growth of the Internet is the **intranet**. Many of the components and concepts that have been presented in the Internet section will apply for both the intranet and the extranet. The main difference that the user will encounter is the utilization of the various applications. Security threats are both internal as well as external, which requires the need to maintain security within an intranet as well as access from other networks. It is necessary to maintain this security without compromising network performance or user accessibility.

An intranet is a network of networks that is contained within an Enterprise. It may consist of many interlinked LANs and use leased lines in the WAN. Typically, an intranet includes connections through one or more gateway (firewall) computers to the outside Internet. The main purpose of an intranet is to share company information and computing resources among employees. An intranet can also be used to facilitate group- or teamwork and for interactive teleconferences. Figure 10–20 shows the intranet environment. Employees can access information such as directories, schedules, messages, and other corporate data that are useful, but need not be available to nonemployees.

An intranet uses open standards such as TCP/IP, HTTP, HTML, and other Internet protocols, and in general looks like a private version of the Internet. Today, an open design is much better than a closed proprietary design, because of the flexibility it gives customers.

Typically, larger enterprises allow users within their intranet to access the public Internet through firewall servers. These have the ability to

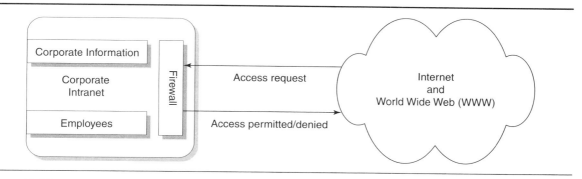

Figure 10-20
Intranet Environment

screen messages in both directions to maintain company security. When part of an intranet is made accessible to customers, partners, suppliers, or others outside the company, that part is called an extranet. The extranet will be discussed in detail later in this chapter.

Intranets use the TCP/IP standards architecture for internal corporate transmissions. They use routers to deliver messages within the company. They use Internet standards for e-mail, the Web, and other applications.

The popularity of intranets comes from a combination of the demands on businesses today and the ability of intranets to help companies meet those demands. The first step is to look at the demands on businesses today. Then it will be necessary to take a detailed look at the advantages of intranets and the leverage that vendors bring to the table.

There are several reasons why organizations are using intranets. One is the lure of the TCP/IP transmission technology, which is standardized, proven, highly scalable, and inexpensive. Having Internet access to the intranet will allow companies to move from proprietary applications, which lock them into a single vendor, to such open Internet applications as e-mail. With open application standards, companies can buy products from multiple vendors that will work together. Although the premise of the intranet is to support the sharing of corporate information with a limited or well-defined group, some information (e.g., products and services) is often shared with the general public over the Internet. Therefore, intranet sites have rapidly become commonplace over the Internet and the Web.

There are numerous justifications for an intranet. Most companies need to distribute vast amounts of internal documents to their employees, such as personnel directories, newsletters, memorandums, work schedules, and office policies. By making this information available to employees over an intranet, most hard-copy procedures and dissemination of information in the traditional manner becomes obsolete.

Intranet Security

A firewall is implemented in an intranet security system, which is designed to limit Internet access to a company's intranet. The firewall permits the flow of e-mail traffic, product and service information, job postings, and so forth, while restricting unauthorized access to private information residing on the intranet.

Firewalls are implemented with routers and proxy servers. These devices are discussed in Chapter 6. Remember that a proxy server is a firewall component that controls how internal users access the outside world and how Internet users access the internal network. In some cases, the proxy blocks all outside connections and only allows internal users to access the Internet, which means that the only packets allowed back through the proxy are those that return responses to requests from inside the firewall (Hioki 2001). Each organization must decide the level of Internet access that is acceptable to ensure that its Enterprise resources remain fully protected.

10.13 EXTRANET

extranet

An **extranet** is an Internet-like network that a company runs to conduct business with its employees, customers, vendors, and suppliers. Extranets include Web sites that provide corporate information to not only internal employees but also external partners, large customers, and particular suppliers through a secure access arrangement. An extranet is not available to the general public. It is called an extranet because it uses the technology of the public Internet (TCP/IP) and customers and suppliers often access it through the Internet via an ISP.

An extranet is a private network that uses the Internet protocols and the public telecommunications system to securely share part of an organization's information or operations with suppliers, vendors, partners, customers, or other businesses. An extranet has become the means for companies to engage in business-to-business (B2B) e-commerce on a global basis. It is a secure version of the Internet with VPN features. Chapters 12 and 13 provide in-depth discussions on VPN and e-commerce environments.

Collaborative Network

An extranet can be viewed as part of a company's intranet that is extended to users outside the company. It has also been described as a "state of mind" in which the Internet is perceived as a way to do business with other companies as well as to sell products to customers. The same benefits that HTML, HTTP, and SMTP/POP3 provide for Internet users can be applied to the extranet users.

Extranets are specialized virtual private networks that are similar to corporate IP-based intranets, however they are intended for controlled access by multiple organizations and other entities. This network allows business partners, suppliers, manufacturers, independent sales agents, and others to share information. An extranet can be extended to the general public for accessing certain types of information. Organizations build extranets to improve communication among key constituents, facilitate information distribution, and broaden access to each other's resources. It also provides a browser front-end to various corporate databases to expedite inventory tracking, invoicing, and supply-side management.

Collaborative extranet systems must access systems and databases behind the firewalls of multiple organizations, and all elements in this network must interoperate. Interoperability is facilitated by a number of technologies that operate under the following standards (Muller 2001):

- EDI INT
- X.509 v3 digital certificates
- Secure/Multipurpose Internet Mail (S/MIME)
- Lightweight Directory Access Protocol (LDAP)
- vCards
- Signed objects

The aim of these standards is to enable a comprehensive, interoperable infrastructure that permits secure transmission of information across a network of diverse users.

10.14 EXTRANET SECURITY ISSUES

Extranets are extensions of corporate intranets and usually allow access via the Internet, however sensitive data can be kept private via the use of a firewall. Private information and other resources can be kept off-limits by implementing a number of strategies, including packet filtering and intrusion detection. When implemented properly, extranets provide access to appropriate resources, while securing others on a selective basis.

The real security challenge involving intranets relates to partners whose business relationships are dynamically changing and complex. Today's partners may become tomorrow's competitors and could also be a current competitor. An extranet must permit dynamic changes in access control to guard against the loss of sensitive and private information. Two key security requirements for extranet access include user authentication and authorization.

The identity of a user wishing to access the extranet must be authenticated. This process is complicated when employees or business partners access information from multiple computers and from remote locations over the Internet. This means that remote sites require additional security protocols to limit server access. This gives rise to a number of issues, which include the following:

- There are often many Web servers in large organizations and users need access privileges for each server accessed.
- Users must remember passwords for many servers.
- Administrators must manage the access controls for each individual server.
- Many separate entries must be added or removed when a user's access privileges change or when there is user turnover.

A security process that lets the organization manage access controls for all centrally located server elements and present users with a single sign-on to the Web space would simplify security management.

After the user has been identified, the access privileges must be determined. Security policies must explicitly grant access rights to Web resources. This access control can be complicated if it is different for each Web server accessed. A centralized authorization framework would simplify this server administration.

A security management system must be easy to implement and administer, and must integrate easily into the organization's infrastructure. Complexities in security management increase the possibility of human errors, make the extranet difficult to navigate, and expose it to attack or misuse.

10.15 CONTENT MANAGEMENT

Security involves the protection of data and database resources, and also the contents of these assets. Content management is a critical issue with extranets since the owner loses control over the data after it is downloaded. Its ultimate destination and use is unknown.

Extranets lend themselves to the aggregation of tactical information, which can have a strategic value over time. This can become a security issue if content control and the time value of information are not considered as part of the security equation. Partners who are also competitors can realize an undue advantage if certain short-term data can be aggregated over the long term, which would provide a strategic advantage. This means that the same information should not be accessible in the same form to all partners.

An extranet is the only situation where an organization exposes its proprietary information to a semiopen audience. Sound content management means that the resource owner must never assume that the partner will be the final user of the information. Ease of information flow is inherent to achieving success, however it must be recognized that revealing too much information is a recipe for disaster. Content owners must, therefore, implement information security systems that permit content control and dynamic changes to minimize exposure to risk.

10.16 ELECTRONIC DATA INTERCHANGE (EDI)

Historically, buyers and sellers have communicated through **Electronic Data Interchange (EDI)**. EDI standards specify standard formats for a number of business documents such as invoices and purchase orders. Vendors have allowed their customers to log into a vendor host, check stock for availability and price, and create a purchase order. For example, pharmacists can place orders with local drug suppliers and the order will be delivered to the pharmacists the next morning. This process is typical of the extranet environment.

EDI is a series of standards that provide computer-to-computer exchange of business documents between different companies' computers over phone lines and the Internet. These standards allow for the transmission of items such as purchase orders, shipping documents, and invoices between an Enterprise and its suppliers. Figure 10–21 depicts an EDI Enterprise environment. In this example, EDI provides the customer with the ability to request product information from a vendor and allows the vendor to respond to the request with stock availability and cost.

In the past, organizations subscribed to the services of Value Added Network (VAN) providers to exchange business documents in electronic form. Software is now available that allows EDI usage over IP networks. This alternative takes the form of extranets that link to the information systems of business partners. Small operations, which could not afford the VAN EDI services, can now participate in this process at a low cost of entry. The common format for Web-based document exchanges is the HyperText Markup Language (HTML), and security is maintained during transmission over the Web through Secure HyperText Transfer Protocol (S-HTTP) or SSL protocol.

Electronic Data Interchange (EDI)

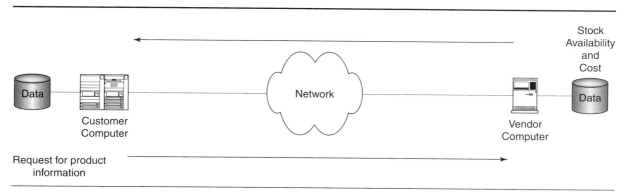

Figure 10–21
EDI Environment

10.17 COMPUTER TELEPHONE INTEGRATION (CTI)

Computer Telephone Integration (CTI)

Computer Telephone Integration (CTI) is a term for connecting a computer to a telephone switch. The computer issues the telephone switch commands for further processing of the call. CTI is currently implemented in three architectures—PBX-to-host interfaces, desktop CTI, and client-server CTI. Figure 10–22 is an example of a desktop implementation. There are different requirements to deliver CTI services for each of the three methods.

10.18 LAN, MAN, AND WAN COMPARISONS

Sometimes the boundaries between local area networks (LAN), metropolitan area networks (MAN), and wide area networks (WAN) are indistinct, which makes it difficult to determine where one network ends and another begins. A useful method is to determine the type of network by examining the four prominent network properties: medium, protocol, topology, and private versus public demarcation points.

- A LAN often ends where the medium changes at the demarcation point, such as wire-based networks to fiber-based networks. The designation can also change when there is a change in the layers that are impacted by the protocol being used. See Appendix A for the OSI model layers. A change in topology from a ring to a star often indicates a change in technologies. These are discussed later in this chapter.
- The boundary between private LANs and public WANs is the point at which LANs connect to the regional telephone network.

Figure 10–22
Desktop CTI Environment

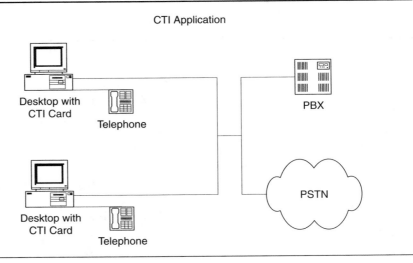

Table 10–1
LAN, MAN, and WAN Comparisons

Attribute	LAN	MAN	WAN
Network Coverage	Local—building or campus	Greater metropolitan area	Large geographic area
Complexity	Minimum	Moderate	Complex
Transport speeds	Relatively fast	100 Mbps	Slow
Cost	Moderate	Low	High
Ease of service	Easy	Easy	Difficult
Security	Easy to compromise	Compromise at interface	Good
Error quality	Good	Good	Poor (with copper)

The communications media could include coaxial cable, fiber-optic cable, twisted-pair wire, radio waves, or microwaves. Table 10–1 summarizes the relationship between LANs, MANs, and WANs. WANs are discussed in detail in Chapter 11.

A primary difference between LANs, MANs, and WANs is the distance over which devices communicate with other devices. A LAN is local in nature: It is owned by one entity and is located in a limited geographic area and most often in a single building. Local area networks are usually faster than WANs and have a lower error rate. In larger organizations, LANs can be linked between a complex of buildings in close proximity, which is referred to as the campus arrangement. It is easier to compromise a LAN because of the numerous cabling interfaces that are available.

Computers linked within a metropolitan area or city are part of a MAN. Metropolitan area networks have attributes of both LANs and WANs. The telecommunications facility for a MAN is usually transparent to the user. A MAN usually consists of mostly carrier facilities, therefore, the most vulnerable point is the physical interface at the user's location.

Devices that are connected between cities or different LATAs are part of a WAN. Wide area networks usually are slower than LANs and have a higher error rate. WAN facilities are the most secure because they are not usually available to the general public. A compromise in the network would probably occur in the wiring closet or demarcation point at the Enterprise location. Capital investments in WANs are high because the outside plant infrastructure is immense and facilities are duplicated to ensure network availability. This infrastructure is presented in Chapter 11.

10.19 STANDARDS

The standards process is important to all participants in the communications industry. Without standards, it would be impossible for multivendor, internetworking environments to exist today. Standards allow multiple

Table 10–2
Standards Organizations

Organization	Acronym	Mission/Contribution
International Standards Organization	ISO	OSI seven layer model
Comite Consultif International Telegraphique et Telephonique	CCITT	Telecommunications standards
International Telecommunications Union	ITU-T	CCITT successor
American National Standards Institute	ANSI	Information systems standards
Institute of Electrical and Electronics Engineers	IEEE	Local area network standards
Electronics Industries Association	EIA	Electrical signaling and wiring standards

vendors to manufacture competing products that will operate with a variety of networking protocols, media, and infrastructures. The end users, therefore, can be assured that hardware and software will operate as specified and will also interoperate successfully. Some of the significant standards organizations are listed in Table 10–2 along with their missions and contributions.

Of particular importance are the IEEE specifications. Table 10–3 provides common cable specifications that can be found in an Enterprise network. Baseband is a direct transmission method that uses a bandwidth whose lowest frequency is zero.

The Enterprise data communications system uses the various transmission media to provide connectivity to a LAN or remote data communication devices by passing information over a WAN or MAN. It will

Table 10–3
Media Comparisons

Specification	Medium	Speed	Segment length	Transmission technique
10Base5	50-ohm coax (10 mm)	10 Mbps	500 meters	Baseband
10Base2	50-ohm coax (5 mm)	10 Mbps	185 meters	Baseband
10BaseT	STP/UTP	10 Mbps	100 meters	Baseband
FOIRL	Fiber optic	10 Mbps	1000 meters	Baseband
10BROAD36	75-ohm coax	10 Mbps	1800 meters	Baseband

become evident that the type of medium is an important issue in the various broadband technologies and topologies.

10.20 INTERNET AND WEB RESOURCES

Bluetooth	www.bluetooth.com
Home RF Working Group	www.homerf.org
RSA Data Security	www.rsasecurity.com
Wireless LAN Association	www.wlana.com
WAP Forum	www.wapforum.org
Linktionary	www.linktionary.com
Dsniff	www.monkey.org/~dugsong/dsniff

■ SUMMARY

A LAN consists of a configuration of network and computer devices that share common resources, usually within an office building or campus. The LAN technology usually consists of an Ethernet, Token Ring, or LocalTalk network and uses TCP/IP, NetWare, AppleTalk, or DecNet software systems.

A wired LAN may consist of a Bus, Ring, or Star topology that uses OSI Layer 1 devices to provide connectivity to the organization's PCs and workstations. A client/server may be used to provide software and database support to the users.

Major interfaces consist of DTE and DCE specifications. Customer premise equipment (CPE) is used by the organization in conjunction with modem, DSU, and NT-1 communication devices to access the Internet and other corporate networks.

LAN devices that provide for local connectivity include bridges, firewalls, gateways, hubs, MAUs, repeaters, routers, switches, and transceivers. Devices that provide access to the network include modems, DSUs, and NT-1s.

A major component of a LAN is the media used to connect the components. Media options include copper wire, coaxial cable, fiber optics, and air. All these media are distance sensitive and copper wire is sensitive to EMI and RFI emissions. Wireless transmissions are subject to monitoring and eavesdropping.

Wired LANs and wireless LANS are susceptible to various types of attacks and intrusions from both internal and external sources. A number of solutions are available to the system's administrator for providing security to these resources.

A Metropolitan Area Network is a group of LANs and other local networks that are linked together within a large campus or city. FDDI is a common backbone technology for the topology. Fiber optics is the common media for connectivity.

Intranets and extranets are both used with LANs and MANs to provide connectivity between users, partners, suppliers, and customers. Numerous security issues must be addressed so a secure and safe environment is provided for all legitimate users on the Enterprise network. EDI and CTI technologies may be employed to provide additional support to the computer network environment.

Key Terms

Access control list (ACL)
Access point
Bluetooth
Bridge
Broadcast domain
Bus
Client/Server
Computer Telephone Integration (CTI)
Customer Premise Equipment (CPE)
Data Communications Equipment (DCE)
Data Service Unit (DSU)
Data Terminal Equipment (DTE)
Denial of Service (DoS)
Diffuse
Direct sequence
Eavesdropping
Electronic Data Interchange (EDI)
Ethernet
Extranet
Firewall
Frequency Hopping
Gateway
Hub
Infrared
Intranet
Local Area Network (LAN)
Line of sight
Masquerade
Metropolitan Area Network (MAN)
Modem
Multistation Access Unit (MAU)
Network Interface Card (NIC)
Network Termination 1 (NT-1)
Personal Communication Network (PCN)
Repeater
Ring
Router
Sniffer
Spread spectrum
Star
Switch
Token Ring
Transceiver
Virtual Local Area Network (VLAN)
Wireless Local Area Network (WLAN)

REVIEW EXERCISES

Questions and Analysis

Provide a short answer or in-depth analysis as required.

1. What are the most commonly used LAN data link technologies?

2. What are the most commonly used networking software systems?

3. What are the three common wired-LAN topologies?

4. What is the difference between a DCE and a DTE interface?

5. Why would an organization deploy a firewall?

6. Why is a gateway required in some topologies? Is there a security role for the gateway?

7. What security functions does a router provide?

8. What are the various types of LAN media?

9. Describe the various types of copper wire media.

10. Why is coaxial cable used instead of UTP?

11. What are the benefits of using fiber optic cable?

12. Provide a list of policies and procedures that should be enacted to ensure the proper level of security on a LAN.

13. What types of attacks can occur on a wired LAN?

14. What technologies are implemented to support the wireless LAN?

15. What applications are common on a personal communication network?

16. What are the four primary technologies for transmitting on a wireless LAN?

17. What are the two infrared connectionless methods used in wireless LANs?

18. What are the five components of a spread spectrum system?

19. Describe the Bluetooth technology.

20. What are the security issues with a wireless LAN? How are they different from a wired LAN?

21. Describe unauthorized access in a wireless LAN.

22. Describe eavesdropping in a wireless LAN.

23. How can an intruder masquerade as a legitimate user?

24. How is integrity compromised in the wireless LAN?

25. Why are wireless LANs vulnerable to denial of service attacks?

26. Describe inference and deception as it applies to a wireless LAN.

27. Describe the general vulnerabilities of a wireless LAN.

28. What are the benefits of creating a VLAN?

29. How are VLANs protected from attackers and intruders?

30. What is an intranet and why is it useful to the Enterprise?

31. What are the security issues with an intranet?

32. What are the security issues with an extranet?

33. Describe content management.

34. Provide a general comparison between LANs, MANs, and WANs.

35. Why are standards important the to LAN administrator?

Definitions and Terms

Provide the most correct answer from the list of key terms.

1. A standard for communicating business transactions between organizations.

2. An organization's internal network.

3. A network that provides connectivity to computers and other devices in close proximity.

4. A program used to capture traffic from the network.

5. A means in which LAN users on different physical LAN segments are afforded priority access privileges across the LAN backbone.

6. A computer that sits on the customer site and splits computing operations and applications between the desktop and one or more networked PCs or a mainframe computer.

7. Devices that sit between the DTE and the communications line. Includes modems, network interface cards (NICs), DSUs, etc.

8. The source or destination device that originates or receives data over communications networks.

9. A technique that uses a radio transmitter/receiver, or transceiver, installed in a central area, which broadcasts signals to the computer devices.

10. Infrared systems that require the transmitter and receiver of the infrared light be straight and in line with one another.

True/False Statements

Indicate whether the statement is true or false.

1. A local area network (LAN) consists of a number of central office devices that span a number of metropolitan areas.

2. A bus is an electrical connection that allows two or more wires to be connected.

3. A ring and a star network are basically the same topology, except for the speed of the network.

4. A client/server allows for file access, print functions, communication functions, and application programs.

5. CPE devices reside on the customer's premises, whereas DCE devices reside in the central office.

6. A firewall, a gateway, and a router can all be used to restrict access into an Enterprise network.

7. Hubs, repeaters, and multistation access units (MAUs) are all OSI Layer 1 devices and are points of possible security violations.

8. Bridges and switches can be used to segment a LAN, which is a technique for limiting network access to unauthorized personnel.

9. Copper wire (STP,UDP), coaxial cable (Coax), and optical fiber all exhibit the same level of risk for attacker intrusions or eavesdropping.

10. An active attack such as eavesdropping or monitoring can be focused toward e-mail or file transfers.

11. The access point in a wireless LAN is not a device susceptible to attackers.

12. Wireless LANs use infrared, LASER, radio, and microwave signals for transmitting and receiving.

13. Spread spectrum includes both direct sequence and frequency-hopping techniques for transmission.

14. The Bluetooth technology uses radio technology for both point-to-point and point-to-multipoint connectivity for long-range communications.

15. Wireless systems are subject to a number of unauthorized access techniques including eavesdropping and masquerading.

Activities and Research

1. A number of Web sites include applications supported by their LAN network products. Identify these sites and make recommendations for security precautions that might be important to the integrity and security of the systems.

2. Research and identify software products that provide for LAN security. Show the various features and benefits.

3. Research and identify hardware products that provide for LAN security. Identify the software products required to support the hardware.

4. Identify router, firewall, and gateway products that provide security to the LAN environment. Provide features and benefits of each.

5. Develop a research paper concerning wireless LAN security and integrity.

6. Identify the various types of attacks and intrusions that might occur on a LAN. Show how they would impact the Enterprise.

7. Prepare a matrix showing the limitations or LAN media. Indicate security issues for each.

8. Identify standards associated with LAN and MAN security. Describe why they are important to the security administrator.

9. Identify organizations that provide security support for LAN and MAN environments. Describe their roles.

10. Using all the concepts presented in Chapters 1 though 9 develop a secure LAN or MAN for a hypothetical organization.

CHAPTER

11

Wide Area Network Security

OBJECTIVES

Material presented in this chapter should enable the reader to

- Understand the magnitude and complexity of the wide area network infrastructure and operation.
- Learn how the various software and hardware components make up the wide area network infrastructure.
- Become familiar with the terms and definitions associated with the wide-area networking environment.
- Look at the issues relating to security and integrity of the wide area network.
- Identify the numerous standards and specifications that support the operation of the WAN and its many elements.
- Develop a strategy for security planning, development, implementation, and management oriented toward the wide area network environment.
- Understand how the WAN is used to provide connectivity between Enterprise networks.

INTRODUCTION

Data communications usually involve a computer with one or more terminals connected by communications lines or a number of computers interconnected with the Internet. The communications lines might be standard telephone lines, dedicated high-speed data communications lines for Frame Relay or ATM, or special arrangements that use telephone lines for services such as Integrated Services Digital Network (ISDN) or Asymmetrical Digital Subscriber Line (ADSL).

Wide area data links are typically used to provide point-to-point connections between pairs of systems that are typically located some distance from one another. Historically, most telecommunications facilities offered by common carriers provide an analog communications channel designed for the purpose of carrying telephone voice traffic. This is commonly referred to as **plain old telephone service (POTS)**.

plain old telephone service (POTS)

Telecommunications carriers now also provide specialized digital telecommunications circuits optimized for data transmission rather than voice transmission. There are various levels of digital service available, corresponding to different data rates. Common digital signal levels include DS0 (64 kbps), DS1(1.544 Mbps), and DS3 (44.736 Mbps).

Following a new trend, these private line and dedicated facilities are being replaced by the new broadband technologies. It is therefore essential that the Enterprise security professional know the components and operations of these new broadband technologies, which include Frame Relay, Digital Subscriber Line (DSL), and Asynchronous Transfer Mode (ATM). This chapter provides information on all facets of the wide-area networking environment and also provides insight into the issues relating to the protection of the Enterprise's computer and network assets.

Private networks continue to connect to the Internet, which raises security concerns with all network administrators. Hackers and intruders are a fact of life and constant vigilance is needed to protect the organization's assets and ensure that only legitimate users access the computer and network resources.

Previous chapters address the basics of all aspects of the Enterprise computer and networking environment and look at the many issues relating to security of those assets. This chapter brings all this together in describing security in the wide area network.

11.1 WIDE AREA NETWORK ENVIRONMENT

Wide Area Network (WAN)

A **Wide Area Network (WAN)** is a group of geographically dispersed data devices, or LANs, that can communicate with each other from different cities, different Local Access Transport Areas (LATAs), different states, and different countries. Its primary function is to tie together users who are widely separated geographically. Many of today's networks employ public telecommunications facilities to provide users with access to the resources of centrally located computer complexes and to permit fast interchange of information among users. Intelligent terminals, personal computers, workstations, minicomputers, and other forms of programmable devices are all part of these large networks. Communications between the various entities and organizations are completed over network media paths that consist of copper wires, coaxial cable, fiber optics, waveguide, and radio path. Each of these media offers different opportunities for security violations. An overview of the technologies that use these media, and the associated security issues, is presented in subsequent sections.

Many networks installed by data processing shops use a number of public telecommunications facilities, which allow Enterprise computer resources to communicate over long distances. These networks might be used to provide all users at remote locations access to the resources maintained in a centrally located computer complex. Wide area networks use telephone lines to connect businesses that are separated by long distances. For example, a warehouse in Chicago, connected to its regional sales office in Atlanta by a telephone, is a WAN connection. A WAN is not confined to a limited geographical area, as is a LAN. Various WAN connection configurations are available. Selection of an appropriate WAN connection depends on factors such as quality of service needed, speed, price, compatibility with the current computer systems at any given location, and the amount of traffic between locations. The types of WAN services and technologies available include

- T-1, T-3
- Frame Relay (FR)
- Asynchronous Transfer Mode (ATM)
- Integrated Services Digital Network (ISDN)
- Digital Subscriber Line (DSL) services
- Fiber Distributed Data Interface (FDDI)
- Wireless and Personal Computer Services (PCS)
- Fibre Channel (FC)
- Switched Multimegabit Data Service (SMDS)

Players in this environment also include the infrastructure services of the Synchronous Optical Network (SONET) and the Virtual Private Network (VPN). These backbone networks consist of many multiplexers and

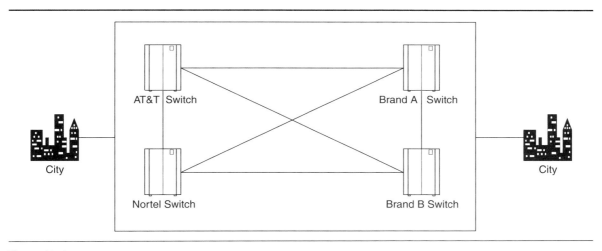

Figure 11–1
Wide Area Network Infrastructure

switching systems. Switching systems include configurations of AT&T, Sprint, MCI, and BellSouth systems. Figure 11–1 depicts a simple WAN in which numerous carriers provide seamless service across the entire network. A WAN is often used to connect similar and dissimilar LANs across multiple carriers and facility providers.

11.2 TELEPHONE NETWORK STRUCTURE

To understand the changing regulatory relationship between different telephone companies and their associated regulatory agencies, it is important to understand the physical layout of the basic telecommunications infrastructure. Several major classes of telecommunications equipment must fit together to form the communications network. The network is created by the various systems, sometimes autonomous, exchanging signals across the numerous interfaces. The major components that comprise the telecommunications infrastructure are as follows:

- Customer premises equipment (CPE)
- Subscriber loop plant
- Local switching systems
- Interoffice trunks
- Tandem switching offices
- Interexchange trunks
- Transmission equipment

Figure 11–2
Central Office Hierarchy

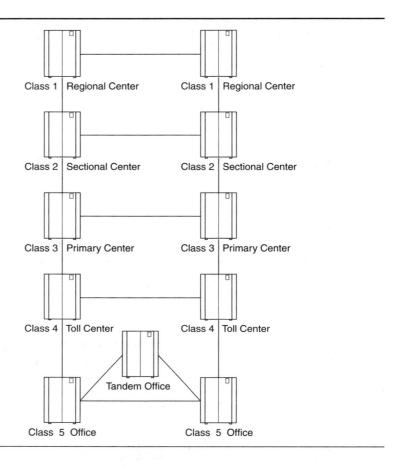

Public Switched Telephone Network (PSTN)
central offices (COs)

Local Access Transport Areas (LATAs)

Figure 11–2 illustrates the major components of the **Public Switched Telephone Network (PSTN)**, which are necessary to support long-distance dial-up service for data communications. The PSTN consists of **central offices (COs)**, tandem offices, end offices, and intermediate offices. Each of these offices performs specific functions in the network hierarchy. These offices are organized into 196 **Local Access Transport Areas (LATAs)** allocated during divestiture. A LATA is usually, but not always, equivalent to the area covered by a given area code.

Central Offices are categorized as Class 1 through Class 5. A residential or business call is first processed in the local central office, also known as an end office, wire center, or local office. Today's network, however, is becoming flatter with fewer central office classes. For our purposes, however, we will consider the five-level structure.

Central Office Classifications

Central offices are designated as Class 1 through Class 5 based on their functionality:

- Class 1 regional center is the highest capacity switching office in the PSTN hierarchy. This is the highest level toll office in AT&T's long distance switching hierarchy.
- Class 2 sectional center is the second highest capacity switching office in the PSTN hierarchy.
- Class 3 primary center is the third highest switching office in the PSTN hierarchy.
- Class 4 toll center is the fourth highest switching capacity office in the PSTN hierarchy.
- Class 5 office is the local switching office. It is sometimes called a wire center and is the office where your local access terminates.

Additionally, there is a Tandem Office the telephone company (Telco) uses to connect central offices when interoffice trunks are not available.

Note that local access is at the Class 5 office and long-distance calls go through a Class 4 office. Operator services such as information are processed in Class 4 offices. There are over 19,000 Class 5 offices and only 12 Class 1 offices.

The circuits between a residence or a business and the local CO are known as **local loops** or **access lines**. **Trunks** connect switches and are used to carry multiple voice-frequency circuits using a multiplexing arrangement between local telephone exchanges. The local exchange is a facility that belongs to the local Telco where calls are switched to their proper destination and local physical connections are terminated. Figure 11–3 depicts a simple local access configuration that consists of a telephone instrument connected to a public telephone switch via an access circuit.

local loops
access lines
trunks

SS7

Signaling System 7 (SS7) is an out-of-band signaling process used for supervision, alerting, and addressing within the telecommunications network. It separates call setup between switches from the actual voice paths. SS7 is an integral part of ISDN and provides for additional functions such as call forwarding, call screening, and call waiting.

Signaling System 7 (SS7)

Figure 11–3
Local Loop Connecting a Subscriber to a Central Office

SS7 is designed to be an open-ended common channel signaling (CCS) standard that can be used over a variety of digital circuit-switched networks. The overall purpose of SS7 is to provide an internationally standardized, general-purpose CCS system with the following primary characteristics:

- Suitable for operation over analog channels and at speeds below 64 kbps
- Suitable for use on point-to-point terrestrial and satellite links
- Designed to be a reliable means of transfer of information in the correct sequence without loss or duplication
- Optimized for use in digital telecommunications networks, using 64 kbps digital channels

The two major capabilities of SS7 include fast call setup via high-speed circuit-switched connections and transaction capabilities, which deal with remote database interactions.

11.3 TELECOMMUNICATIONS NETWORK FUNCTIONS

Numerous tasks and operations occur in the telecommunications network for it to perform its required functions. Some of these activities are simple; most are complex and require interactions with many other elements to successfully transport communications traffic. The following list includes the major categories that must be addressed:

- Network Management
- Security
- Recovery
- Addressing
- Synchronization
- Utilization
- Flow control
- Error detection/correction
- Interfacing
- Signal generation
- Handshaking
- Routing
- Formatting

Even though many of these activities and functions are automatic, managing the system is often time-intensive and requires a considerable amount of hands-on management from systems administrators and network technicians.

11.4 TELECOMMUNICATIONS NETWORK COMPONENTS

Switched Facilities

Telephone calls connected on the PSTN pass over switched or shared facilities. This type of network technology is referred to as circuit switching. Virtually all voice telephone calls are circuit switched. Dial-up modem access is also circuit switched. The facilities in the switched environment are only used for the duration of the call. A switched call has three phases: call setup, call transmission, and call termination. A three-phase call results in a lower cost because economies of scale are achieved in sharing facilities.

Dedicated Facilities

Dedicated lines, however, like private networks, are not shared. They are essentially available 24 hours a day for the customer agreeing to pay for them. Other names for private lines are point-to-point, direct, fixed, leased, and ringdown circuits.

Figure 11–4 depicts both circuit switched and dedicated circuit examples. The normal data communications devices such as modems are required to access the public network switches. The connectivity for dial-up access is temporary, whereas the dedicated access is permanent.

The terms for switched or dedicated networks, like many telecommunications terms, have taken on a life of their own and can be used in many different ways. Private lines, for example, may refer to tie lines, long-distance lines between cities, or even dedicated local telephone lines for one user behind a private branch exchange (PBX) or key system.

Many companies create or build their own networks. These private networks offer the advantages of guaranteed availability, greater reliability, and lower cost, if properly used. Private networks can be created

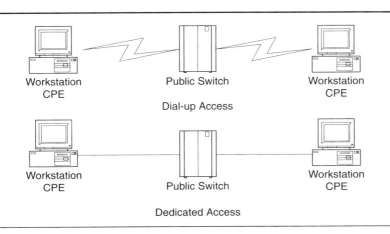

Figure 11–4
Circuit Switched and Dedicated Access

through a variety of services from a number of different vendors or actual network construction. They can be voice, data, or voice/data networks managed completely or partially by the customer. Often customers will migrate to a private network for the purpose of gaining increased control of their telecommunications services and realizing cost savings when they own right-of-ways and network infrastructures. Railroads and electric power companies often use their own right-of-ways for their communications facilities. Fiber optic cable may be buried in railroad beds or hung between power towers.

11.5 ELEMENTS OF A DATA COMMUNICATIONS SYSTEM

The four basic elements of a data communications system are a transmitter, a receiver, a transmission medium, and communications equipment. Standards relating to these elements are set forth in the **Open Systems Interconnection (OSI)** model developed by the **International Standards Organization (ISO)**. This seven-layer model is included in Appendix A.

Transmitters and receivers in this model are devices such as personal computers (PCs), workstations, printers, servers, mainframes, and minicomputers that can use and generate information. Such devices are categorized as **data terminal equipment (DTE)** or **customer premises equipment (CPE)**.

Data communications equipment (DCE) connects DTE to the network. A **Data Service Unit (DSU)** and a **modulator/demodulator (Modem)** are examples of DCE devices. Figure 11–5 depicts a common DTE/DCE interface arrangement by using modems to communicate across a PSTN. The access line is synonymous with the telephone company local loop. This access line (circuit) terminates at the customer's demarcation point (NI), usually in a wiring closet. This NI could be a point of eavesdropping by a sniffer device.

A frame or packet is the information unit most commonly used with these data networks. Both units contain end-user data or payload, which

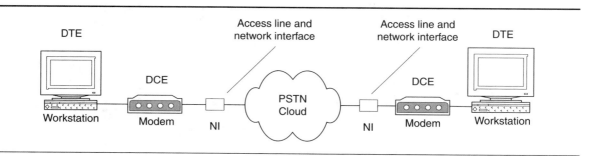

Figure 11–5
DTE/DCE Interface Arrangement

in turn is surrounded by additional control characters. Control characters are added by the communications logic in the transmitter and used by the communications logic in the receiver to implement the protocols and error checking systems. Packetizing is the process of adding overhead or management data to raw user data to ensure proper delivery. These frames and cells are transmitted over some medium between the various communications devices.

Transmission **media** may be classified as guided or unguided. With guided media, the electromagnetic waves are guided along a physical path on media such as twisted pair, coaxial cable, or optical fiber. The other option, unguided media, allows for the propagation of electromagnetic signals through air and a vacuum.

media

A requirements study must be conducted to determine the appropriate type of transmission medium. If there is noise in the environment, coaxial cable or fiber might be required. There are also distance and line-of-site requirements for the various transmission media, and certain standards associated with the transmission media that may be relevant for the system designs. These standards are presented later in this chapter.

Central Office Equipment

A central office (CO) is used as a centralized location for connecting subscriber's telephone lines. Several different hardware devices are used in the COs. These include the WAN switches, channel banks, digital access cross-connect switches (DACS), and multiplexers.

A WAN switch is located in a CO and is used to serve a large number of communication facilities. These switches can consist of equipment such as an AT&T 5ESS or a Nortel DMS10. Frame Relay and SMDS statistical time-division multiplexed (STDM) switches also fit in this category.

Channel Banks

The original channel bank combined twenty-four analog signal sources into a single DS-1 (1.544 Mbps) bit stream to transmit across either a public or private network. The newer D-4 (superframe) channel bank provides for both voice and digital data traffic. A data service unit (DSU) can be plugged into the D-4 channel bank to carry data at up to 56 kbps digitally.

Subscriber loop carrier (SLC96) is basically a channel bank that converts and multiplexes ninety-six voice signals into digital bit streams. It consists of a central office terminal (COT) and the remote terminal (RT). Five T-1 lines connect the COT and RT, with four primary T-1 lines carrying the bits for the ninety-six channels (4×24), and a spare T-1 line that is used for backup in the event of a failure. This method of redundant services is called protection switching. The SLC96 COT is colocated with the central office equipment and the RT is located near the customer's premises.

digital access cross-connect switch (DACS)

Digital Access Cross-connect Switch (DACS)

The **digital access cross-connect switch (DACS)** is a system that uses a time division multiplexing (TDM) scheme to switch and cross-connect digital bit streams. The number of input ports equals the number of output ports, which allows for a nonblocking connectivity. Therefore all circuits connected to the system can operate simultaneously without any overload. These devices are used by the Telcos for the following functions:

- Administration and testing
- Reconfiguration of ports on the system
- Rearranging of channels within a digital span
- Load balancing
- Network recovery and switching

The DACS also allows for linking traffic across multiple digital links and the insertion and deletion of bit streams into a channel (drop and insert). These devices were developed for use in a central office or point of presence (POP) to gain access to every digital bit stream at the DS-0 and DS-3 levels. A channel on a T-1, called a digroup, can be multiplexed onto a channel from a different T-1, or from one T-3 to another T-3. The DACS allows a considerable amount of flexibility in the manipulation of the central office facilities in the event of a major network failure. Figure 11–6 shows how multiple, partially filled T-1s can be multiplexed into full (24 DS0 channels) T-1s. Four partially filled T-1s can be concentrated into two T-1s.

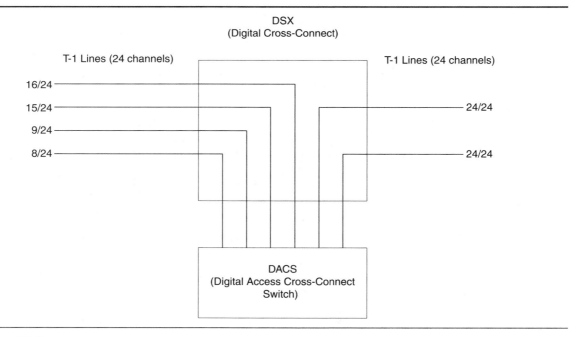

Figure 11–6
Concentration with a DACS

Multiplexers

The next step above the channel bank is the multiplexer (MUX). The MUX provides flexibility because services can be either channelized into twenty-four fixed time slots or nonchannelized. The MUX supports multiple interfaces via the use of a wide range of plug-in cards, which provide subrate speeds of less than the basic 64 kbps or super-rate channel capabilities above 64 kbps.

A T-1 MUX handles signals from different types of sources, including voice, data, video, and fax. A T-1 MUX can also support several modulation standards such as pulse code modulation (PCM) and adaptive differential PCM (ADPCM). The MUX may also allow for high-compression voice multiplexing at 8 kbps and 16 kbps, allowing a single 64 kbps channel to be used by up to eight voice devices (Stallings 2001).

11.6 INTERNETWORKING

Enterprise **internetworks** are typically constructed using three different elements: LAN data links, WAN data links, and network interconnection devices.

internetworks

LAN data links are normally confined to a single building or within a campus of buildings. They do not cross public thoroughfares and normally operate over private cabling. Local area network facilities, discussed in Chapter 10, are generally used to create many networks that allow any device on the data link to physically communicate with any other device on the data link. When LANS are located in widely separated locations, WAN data links are often used in conjunction with network interconnection devices to interconnect the LANs. Wide area network data links are most often used to implement point-to-point connections between pairs of network interconnection devices. There are a number of points in these network interface points from which Enterprise traffic can be intercepted. These candidates are discussed in subsequent sections.

Commonly used data transfer speeds over common carrier telecommunications facilities include 19.2 kbps and 56 kbps. Digital T-1 (1.544 Mbps) and T-3 (44.736 Mbps) facilities, also available from common carriers, are widely used in Enterprise internetworks. Wide-area networking facilities commonly used in computer networks include

- Common carrier telecommunications data circuits
- X.25 Packet-Switched Public Data Network circuits
- Frame Relay access
- Narrowband ISDN (N-ISDN) access
- Broadband ISDN (B-ISDN) access
- Distributed Queue Dual Bus (DQDB)/MAN access
- Switched Multimegabit Data Service (SMDS) access
- Synchronous Optical Network (SONET) facilities
- Fibre Channel connectivity
- DSL and Cable Modem access

To create flexible Enterprise internetworks, it is necessary to interconnect individual LAN data access links and WAN facilities. Several different types of devices can be used to accomplish this. The types of devices available for network interconnection can be divided into the following general categories:

- Bridges
- Routers
- WAN Switches
- Gateways

These devices contain firmware and software that allows them to make routing and transport decisions concerning communications traffic. Details concerning these devices are presented in Broadband Communications (Newman 2002). Information concerning these firewall devices is presented in Chapter 6.

11.7 NETWORK CONFIGURATIONS AND DESIGN

The design of most networks, including wide area networks, incorporates a variety of different topologies. Topology refers to the physical design and connectivity of a communications network. A topology is defined as the physical arrangement of nodes and links to form a network, including the connectivity pattern of the network elements.

The basic function of a communications network is to provide access paths by which an end user at one location can assess end users at other locations. The network designer must consider security when developing and implementing the Enterprise network.

centralized, decentralized distributed

There are three basic types of communications networks: **centralized**, **decentralized**, and **distributed**:

- A centralized network is a computer network with a central processing node through which all data and communications flow. This type of network requires dedicated hard-wired devices that operate using high-speed parallel or serial transmission techniques.
- Decentralized networking implies some processing distribution function using a host processor to control several remotely located processors. The host processor can off-load activities to other processors but it still maintains control. Data are often transported over serial links in a synchronous or asynchronous transmission mode.
- In distributed processing, a computer or node in the network performs its own processing and stores some of its data while the network manages communications between the nodes. This type of network could be considered a peer-to-peer network spread over

a large geographical area. It usually consists of many different types of hardware and software components. In distributed networks, all nodes in the network share responsibility for application processing, which requires that data format conversion and communications protocol-handshaking capabilities reside in each node.

11.8 ACCESS CIRCUIT TYPES

Whether the communications network topology is decentralized or distributed, there must be a method for providing connectivity between the processors or nodes. In today's environment, this connectivity takes the form of an access circuit or line. These access facilities are available from either Incumbent Local Exchange Carriers (ILECs), which are the Telcos, or Competitive Local Exchange Carriers (CLECs).

In telecommunications, a circuit is a discrete (specific) path between two or more points along which signals can be carried. Unless otherwise qualified, a circuit is a physical path consisting of one or more wire pairs (two wire/four wire) and possibly intermediate switching points. An access line or local loop is the local access connection or telephone line between a customer's premises and a carrier's **point of presence (POP)**, which is in the carrier's central office (CO). A POP generally takes the form of a switch or router. It can be a meeting point for Internet service providers (ISPs) where they exchange traffic and routes. Figure 11–7 describes the access line (local loop) and POP environment. The interexchange carrier (IXC) communicates with the LEC through the POP, and this POP device is often colocated with the carrier's equipment.

point of presence (POP)

Wiring Interface Standards

Several standards in the wide area network environment are associated with physical interfaces, transmission protocols, and routing protocols. The primary network device interface standards are listed in Table 11–1.

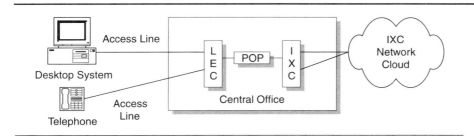

Figure 11–7
Access Line and POP Connectivity

Table 11-1
Interface Standards

Standard	Description
RS232-D	A set of standards specifying electrical, functional, and mechanical interface specifications for a 25-pin connector.
RS449	Originally designed to replace RS232-D to improve transmission capabilities. They are not compatible, since RS449 has a 37-pin connector with different functions. It uses twisted-pair cable.
X.21	The interface between DTE and DCE for synchronous operation on public switched networks.
V.35	An ITU-T standard for trunk interface between a network device and packet network that defines signaling for data rates greater than 19.2 kbps. The physical connector is a 34-pin rectangular M-block connector. V.36 is intended to replace V.35. RS422 and RS423 are the electrical specifications associated with V.35.

11.9 INTERFACES

Point-to-point networks involving the simple interconnection of two pieces of equipment are relatively simple to establish. They may use either a digital line with a Data Service Unit (DSU) or an analog line with a modem. Common interface standards for these devices include the RS232-D, RS449, X.21, and V.35 specifications.

Physical interface standards typically specify certain characteristics about the connection between data terminal equipment (DTE) and data communications equipment (DCE). The most commonly used serial DTE/DCE interface standard in the United States is the **RS232-D Electronic Industries Association (EIA)** standard.

RS232-D
Electronic Industries
Association (EIA)

If the RS232-D standard is implemented, the DTE and DCE is connected by a DB25 cable to each device, using a 25-pin or 9-pin connector (Figure 11–8). Each pin has a specific function in establishing communication between the devices. One drawback of the RS232-D standard is its limited bandwidth and distance.

X.21

The **X.21** interface standard is defined by ITU-T and uses a 15-pin connector. There are several significant differences between X.21 and the RS standards. First, X.21 was designed as a digital-signaling interface. The

Figure 11-8
RS232-D Interfaces

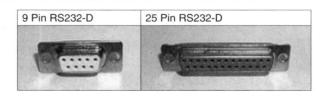

second involves control circuits in the interface. RS standards define specific circuits for control functions. However, X.21 puts more logic circuits (intelligence) in the DTE and DCE devices that can interpret control sequences and reduce the number of connecting pins in the interface.

The ITU-T **V.35** is a high-speed digital interface standard, which has been superceded by ITU-T standards V.36 and V.37. The standard describes synchronous DCE devices that operate at data signaling rates of 56 kbps, 64 kbps, and higher. This interface consists of a rectangular connector with thirty-four circuit pins. It is frequently used in Frame Relay and ATM installations.

V.35

An important interface concept is the **user-to-network interface (UNI).** This interface provides a physical point of separation between the responsibilities of the carrier and the customer. Any equipment and wiring on the user side of the UNI is the responsibility of the customer (unless a maintenance agreement makes equipment and wiring the responsibility of the carrier). Any equipment or lines on the network side of the UNI are the responsibility of the carrier. This is called the **network-to-network interface (NNI)**.

user-to-network interface (UNI)

network-to-network interface (NNI)

Another important interface concept is DTE/DCE. It provides a functional separation between different types of equipment, regardless of the side of the UNI on which they happen to be located. A DSU is an example of DCE, which may be owned by the customer and sits on the user side of the UNI. A computer is an example of DTE, but its modem is DCE. Both are located on the user side of the UNI. At the other end of a dedicated or dial-up line, or on the network side, there are DCE devices owned by the end users in the form of modems and DSUs. Figure 11–9 shows both the UNI/NNI relationship and examples of the DTE/DCE interface.

Physical network wiring configurations are point-to-point and multipoint. Point-to-point configurations have only one connection between the two terminals. Multipoint terminals share the connection with the

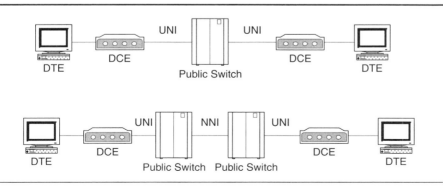

Figure 11–9
DTE/DCE and NNI/UNI

Figure 11–10
Point-to-Point and Multipoint Configurations

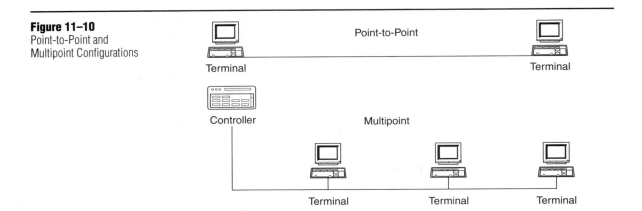

other terminals and require a controller to handle polling and prioritization. IBM synchronous/SDLC configurations can operate in a multipoint configuration. Figure 11–10 depicts both configurations.

Dedicated (Leased Line) and Dial-up Circuits
Dedicated lines often have the following characteristics:

- Fixed pricing
- Available 24 hours a day, 7 days a week
- Mileage sensitivity
- Managed by the user
- Secure service
- Good quality

Leased line service is available in numerous speeds. It is essential that the user be aware of these speeds, because a migration to a broadband product may be speed sensitive. Service is usually available in two categories: Digital Data Service (DDS) and high-speed services. Table 11–2 provides a list of generally available speeds.

A network is an arrangement of circuits. In a dial-up or switched connection, a circuit is reserved for use by one user, which lasts for the dura-

Table 11–2
Circuit Speeds

Digital Data Service (DDS)	2.4 kbps, 4.8 kbps, 9.6 kbps, 19.2 kbps, 28.8 kbps, 56 kbps
Digital Signal Hierarchy (DSx)	DS0 (64 kbps) DS1 (1.544 Mbps) DS3 (44.736 Mbps)

tion of the calling session. In a dedicated or leased line arrangement, a circuit is reserved in advance and can only be used by the owner or renter of the circuit. It is advantageous for a user to lease a dedicated line if security is an issue and if the circuit will be in use for most of the time. Occasional use would dictate a dial-up line; however, dial-up quality is usually less than dedicated quality.

The standard public telephone network provides switched or dial-up circuits, and using them for data is similar to making a telephone call. A temporary connection is built between data terminal equipment (DTE) as if there was a direct connection between the two devices. The circuit is set up on demand and discontinued when the transmission is complete. This technique is called circuit switching. An obvious advantage of circuit switching is flexibility. Usage charges are based on the duration of the call and distance, just as with a standard telephone call.

One factor that may impact the quality of a dial-up circuit is the availability of the actual network facilities used in routing the call when the connection occurs. Circuits may be variable in quality, and so can be good in one situation and marginal in another. Because of this variability, the data transmission speed that can be achieved dependably is less than with dedicated facilities. Using switched circuits exposes a subscriber to security risks. It is possible for unauthorized persons to dial in and access the computer facilities if security procedures are not adequate.

Figure 11–11 provides a general model of the difference between a private line, which is dedicated to the user, and switching, which is shared by numerous users.

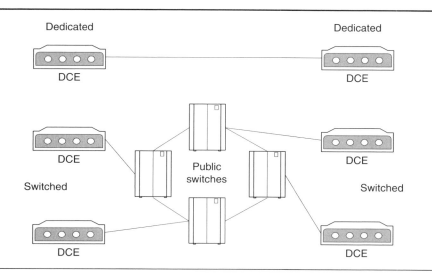

Figure 11–11
Dedicated/Switched Lines

Modem and Access Server Considerations

Users should not be allowed to install a modem, even temporarily, without proper authorization. Dial-up access through a modem is a good way to compromise the organization's network. All modem access should be logged and monitored on a regular basis. It is also a good idea to require one-time passwords when using dial-up access. It is also helpful if there is a single dial-in point into a modem pool, so that all users can be authenticated by the same method.

A different complement of modems should be used for any dial-out services, and both dial-in and dial-out services should be authenticated. If modems and access servers support callback, then it should be activated. With callback, after a user dials in and is authenticated, the system disconnects the call and calls back on the specified telephone number. This provides a level of security because the system calls back the actual user, not a hacker who might be masquerading as the user. As with a number of security mechanisms, callback can also be compromised.

Unsecure modems provide one of the easiest ways to illegally enter a network. To locate these modems, attackers can employ **war dialing**, which is a technique that dials different telephone numbers repeatedly looking for modem carrier tones. For war dialing, attackers will use telephone number ranges found on a number of sources, including Web sites and newsgroups. After discovering modems, the attackers look for systems without passwords.

Modems and access servers should be carefully configured and protected from intruders that might reconfigure them. Modems should be programmed to reset to the standard configuration at the beginning and end of each call, and modems and access servers should terminate calls cleanly. Servers should force a logout if the user hangs up unexpectedly (Oppenheimer 1999).

Wiring Closets

As discussed in Chapters 9 and 10, a wiring closet is a potential source of network vulnerability. Since WAN circuits terminate in this room and LAN cabling may also terminate in the same network hardware devices, there is ample opportunity for an intruder to gain access to both the LAN and WAN resources at this single point. The wiring closet often houses a patch panel and/or a punch-down block where physical cabling connections are terminated. If uncontrolled access is allowed, eavesdropping and monitoring can occur in a promiscuous mode.

The WAN demarcation point in the wiring closet is often located close to the communication system's equipment room. This area may house the various DCE devices that provide connectivity between the LAN and the WAN. Both the wiring closet and the equipment room must be secured against all unauthorized access.

11.10 WAN SWITCHES

WAN switches are essentially components of public or private communications networks. A switching system essentially contains a specialized computer that performs routing of signals over transmission routes. Examples of these WAN switches include AT&T #5ESS and Nortel DMS10. A communications network consists of numerous similar and dissimilar types of switches with various types of transmission links to connect them for seamless communication. Transport of signals through the switched telephone network is accomplished via **circuit switching** or **packet switching**.

circuit switching
packet switching

Circuit Switching

Circuit switching is like having your own highway (facility) on which your conversation can travel. As long as the connection stays open, no one else can use it. Most voice telephone calls are circuit switched. The process is as follows:

1. A request is made to the network from a user when a number is dialed.
2. A circuit path is established end-to-end through the telephone facility infrastructure.
3. The conversation or data transmission is completed.
4. The circuit path is released back to the infrastructure.

Note that the user has complete use of the facility as long as the connection is established. If the user hangs up and redials, the path will probably be a different one. This is what happens when you get a facility that is of poor quality, redial, and then get a good-quality facility.

Figure 11–12 provides an overview of the circuit switched environment. Each of the end devices connects to the network via some DCE device such as a modem. This connectivity is through a local CO, or an end office, wire center, or Class 5 office. The circuit switched network establishes a path through adjacent COs until connectivity is established to the other user. These COs, called nodes, can be classified as end nodes or intermediate nodes. The intermediate nodes connect other intermediate nodes and end nodes.

The connection path is established before transmission begins, therefore the channel capacity must be reserved in advance between the two end locations before any data can flow. All switches or nodes in the path must have available internal switching capacity to handle the requested connection. The switches have the intelligence to make these allocations and devise a route to provide this connectivity through the network. The Signaling System 7 (SS7) network provides this intelligence.

Circuit switching can be inefficient, because channel capacity is dedicated for the duration of the connection, whether data traffic is

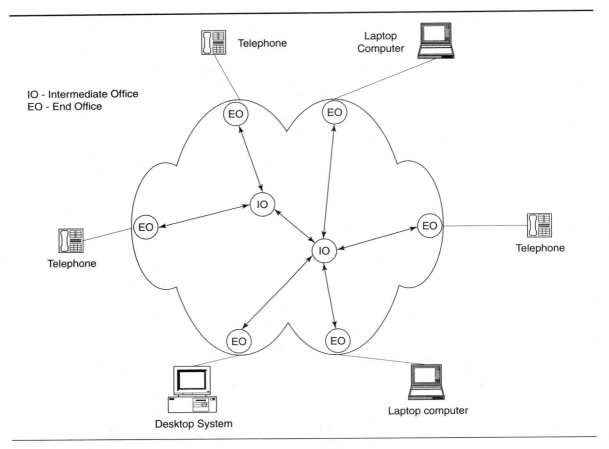

Figure 11–12
Circuit-Switched Network

present or not. The channel may be idle during most of the time when there is computer-to-computer communication. This efficiency improves for voice traffic, but it still does not approach 100 percent. Circuit switching was designed for voice traffic, but is now being used more for data transmissions.

Circuit-switching technology has been driven by voice traffic applications. Two primary requirements for voice traffic are no transmission delay and no variation in delay. A constant transmission rate must be maintained, because transmission and reception occur at the same signal rate for human conversations. The signal must also be of sufficiently high quality to provide intelligibility.

Circuit switching, with its widespread implementations, is well suited for the transmission of analog voice signals. Even though it is inefficient, circuit switching will remain an attractive choice for both wide and local area networking for some time. The main feature of circuit switching is transparency, which means that it appears as if there is a direct end-to-end connection with no special requirements of the end stations.

Packet Switching

Packet switching is like each user having individual railway cars, but sharing the same track (facility). The track (facility) takes the cars to their proper destination. Packet switching is popular because most data communications consist of short bursts of data with intervening spaces that usually last longer than the actual burst of data. Packet switching takes advantage of this characteristic by interleaving bursts of data from many users to maximize use of the shared communication network. This is illustrated in Figure 11–13, when three packets are transmitted and take different paths to the final destination.

Packet data networks (PDNs) are based on packet switching technology in which messages are broken down into fixed-length components called packets and sent through a network individually. A packet consists of the data to be transmitted and certain control information. Packet switching networks are designed to provide several alternative high-speed paths from one node to another.

packet data networks (PDNs)

This configuration provides a redundant, fail-safe capability and implies that the route that a specific message takes is in part a function of

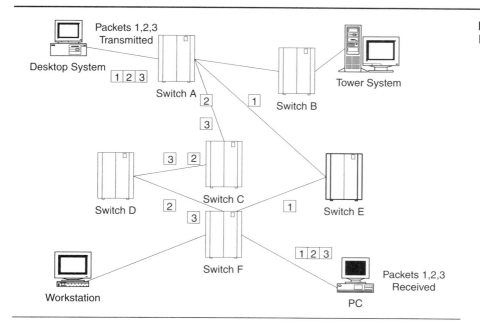

Figure 11–13
Packet Switched Network

the condition and traffic capacity of the various links on the network when the message is sent. The situation and the route taken by the message are dynamic.

Numerous packet switches are interconnected to form a WAN. A switch usually provides for multiple input/output (I/O) connections, making it possible to form many different topologies and to connect to several switches and computers. A WAN is usually not symmetric in design, because the interconnections among packet switches, and the capacity of each connection, are based on the expected traffic and the ability to provide redundancy in the event of failure.

A WAN allows many computers to send packets simultaneously, which is accomplished through a fundamental paradigm used with wide area packet switching called **store and forward**. This store and forward operation requires the switches to buffer packets in memory. The store operation occurs when a packet arrives, and is accomplished by placing a copy of the packet in the switch's memory and notifying the processor of the activity. The forward activity requires the processor to examine the packet and determine over which interface it should be sent on the path to its final destination.

A system that uses the store and forward paradigm can transport packets through the network as fast as the packet switch hardware and software allow. If multiple packets are sent to the same output device, the packet switch can hold the packets in memory until the output device is ready. Also, if the output device is busy, the processor places the outgoing packet in a queue that is associated with the device. To summarize, the store and forward technique allows a packet switch to buffer a short burst of packets that arrive simultaneously and transport them successfully to their destination.

Switching Techniques

The two approaches used to transport the packets are **datagram** and **virtual circuit**. Datagram is a connectionless service, whereas virtual circuit is a connection-oriented approach. In the datagram approach, each packet is treated independently, with no reference to packets that have already gone. Since each packet contains the destination address, each node looks at this address and makes a decision for the route to the next node. Thus, all packets of a particular message may not follow the same route and could arrive at the destination either out of order or not at all due to a number of circumstances. The destination node must determine if all packets have been received and if they are in the proper order. In this technique, each globally addressed message packet, which is treated independently, is referred to as a datagram.

In the virtual circuit approach, the packet switched network establishes what appears to be one end-to-end circuit between the sender and receiver. This appears very similar to the circuit switching approach where a path is chosen before any transmission takes place. Each packet contains a virtual circuit identification, instead of a destination address, and data. Each node of the pre-established route knows where to direct such packets; therefore, no routing decisions are required. As in circuit switching, all messages for that transmission take the same route over the virtual circuit, which has been set up for that particular transmission. The two DTE devices believe that they have a dedicated point-to-point circuit. At any time, each DTE can have more than one virtual circuit to any other DTE and each DTE can have virtual circuits to more than one DTE.

The main characteristic of the virtual circuit technique is that the route between stations is set up prior to data transfer. This, however, does not mean that there is a dedicated path, as in circuit switching. A packet is still buffered at each node and queued for output over a line, while other packets on other virtual circuits may share the use of the line. The difference from the datagram approach is that the node does not need to make a routing decision for each packet. It is made only once for all packets using that virtual circuit.

Note that the user has no input into the path that the packets take. The packet switching network makes all decisions based on several conditions. The packets may arrive at the destination out of order, but they will be reassembled into the correct order.

Access Requirements

Each vendor has a point of presence (POP) where connections are made to other vendors/telcos/Interexchange Carriers (IXCs), Local Exchange Carriers (LEC), Competitive Local Exchange Carriers (CLECs), and Incumbent Local Exchange Carriers (ILECs). The point of presence (POP) is a physical place where a carrier has a presence for network access and generally takes the form of a switch or router. A large Interexchange Carrier (IXC) will have a number of POPs, at which the local exchange carrier (LEC) networks interface to accept originating traffic and deliver terminating long-distance traffic. The interface is accomplished with switched and dedicated connections. Similarly, providers of X.25, Frame Relay, and ATM services have specialized POPs, which may be colocated with the circuit-switched POP for voice traffic. A POP is also a meeting point for Internet Service Providers (ISPs), where they exchange traffic and routes.

Connectivity between LECs is established with IXCs, such as AT&T, MCI, Sprint, and other carriers. Each carrier has a facility POP for interfacing with other carriers and the LECs. The term point of presence is changing to point of interface (POI). POP implies that a switching system is at that location; however, the POI can be in a closet in a basement or hotel, connected via a dedicated trunk to another part of the country. Regardless of whether the interface point is a POP or a POI, it is transparent to the subscriber.

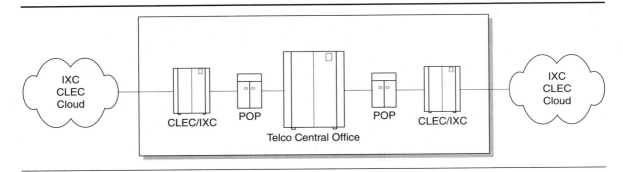

Figure 11–14
POP between IXC and Local Access

Figure 11–14 shows a POP environment where the various IXC, LEC, CLEC, and telco entities interface with each other.

11.11 THE INTERNET AND THE WEB

Internet

The **Internet**, sometimes called simply "the Net," is a worldwide system of computer networks. It is a network of networks in which users at any one computer can, if they have permission, get information from any other computer. Its development was funded by the Advanced Research Projects Agency (ARPA) of the U.S. government in 1969 and was first known as the ARPANET. The original aim was to create a network that would allow users of a research computer at one university to "talk to" research computers at other universities. A side benefit of ARPANET's design was that, because messages could be routed or rerouted in more than one direction, the network could continue to function even if parts of it were destroyed in the event of a military attack or other disaster.

Today, the Internet is a public, cooperative, and self-sustaining facility accessible to hundreds of millions of people worldwide. Physically, the Internet uses a portion of the total resources of the currently existing public telecommunications networks. Technically, what distinguishes the Internet is its use of a set of protocols called TCP/IP (Transmission Control Protocol/Internet Protocol). Two recent adaptations of Internet technology, the intranet and the extranet, also make use of the TCP/IP suite.

Both Digital Subscriber Line (DSL) and cable modem access offer broadband, "always on," connectivity to the Internet. By virtue of this permanent connection, networks tied to the Internet are vulnerable to attacks by hackers, risk virus infections, and expose internal data and systems to external sources. Numerous applications, including hardware and software firewalls, routers, and antivirus software, are addressing these security issues.

For many Internet users, electronic mail (e-mail) has practically replaced the Postal Service for short written transactions. Electronic mail is the most widely used application on the Net. You can also carry on live "conversations" with other computer users by using Internet Relay Chat (IRC). More recently, Internet hardware and software allows real-time voice conversations.

The most widely used part of the Internet is the **World Wide Web** (also **WWW** or the **Web**). Its outstanding feature is hypertext, a method of instant cross-referencing. Most Web sites contain certain words or phrases that appear in text in a different color than the rest of the text; often this text is underlined. When users select one of these words or phrases, they will be transferred to the Web site or page that is relevant to this word or phrase. Sometimes there are buttons, images, or portions of images that are "clickable." By moving the pointer over a spot on a Web site where the pointer changes into a hand, the user can click and be transferred to another Web site or page.

World Wide Web
WWW
Web

Internet and Web Basics

The Internet emerged from a U.S. government and military initiative to enable the interconnection of different, mainly UNIX-based, computer systems for intercommunication. UNIX was proclaimed to be the first portable operating system for computers, because it enabled software developed on a particular manufacturer's computer hardware to be easily ported to another manufacturer's hardware. Therefore it was natural to develop a means for easy transfer of data between systems. This led to **Transmission Control Protocol/Internet Protocol (TCP/IP)**.

Transmission Control Protocol/Internet Protocol (TCP/IP)

TCP/IP was quickly adopted by the academic community in the United States, and soon thereafter by academics worldwide, because it permitted rapid sharing of scientific information. It also permitted the electronic mail communication necessary for rapid discussion and analysis, both within and between university campuses.

In its original form, the Internet and the Internet Protocol (IP) provided a means for interconnecting computer servers together. The Internet addressing scheme allowed individual workstations, personal computers, or software applications running on either the server or any of the workstations to direct and send information to other applications on other servers or distant LANs. Being a unique address, the Internet address allowed an end user to be identified, no matter how many transit servers, routers, or networks would have to be traversed along the way. A suite of new protocols arrived with the Internet. These included simple mail transport protocol (SMTP), trivial file transfer protocol (TFTP), and file transfer protocol (FTP), which together form the basis of the Internet electronic mail service. These protocols are detailed in Chapter 7.

As the popularity of the Internet has grown, so have the numbers of servers and routers making up the network. **Internet Service Providers (ISPs)** have provided for dial-up access to the servers from private individuals using their PCs at home, and new information providers have furnished more Internet pages of information. **Browsers**, computer software that allows users to "surf" the Internet and the WWW, allow users to seek information from any of the connected servers by means of menu-driven screen software. These browsers, using the **Hypertext Markup Language (HTML)** tool, provide a hypertext search and browse capability.

Internet Service Providers (ISPs)

browsers

Hypertext Markup Language (HTML)

The WWW is a distributed hypermedia repository of information that is accessed with an interactive browser. A browser displays a page of information and allows the user to move to another page by making a mouse selection. Embedded into the Web page are hypertext or hypermedia documents, which are underlined text or highlighted graphical icons. By simply pointing and clicking on the hypertext or hypermedia, a link is made to that particular source document, which can also lead to the pathways of other hypertext links.

Web Access

Using the Web, users have access to millions of pages of information. Web "surfing" is done with a Web browser—the most popular of which are Netscape Navigator and Microsoft Internet Explorer. The appearance of a particular Web site may vary slightly depending on the browser used. Also, later versions of a particular browser are able to render more "bells and whistles" such as animation, virtual reality, sound, and music files, than earlier versions. **Java** and **JavaScript** are two common languages used in Web application design.

Java
JavaScript

A good place to start with understanding the Internet is to identify all of the components that make it up. The Internet is not a single network, but thousands of networks scattered around the globe. Universities, corporations, and other organizations own individual networks, which are linked together so that everyone can communicate with each other. Messages on the Internet are called **packets**. A device called a router makes it possible to send packets from one computer to another. Packets are transported from one network to another through a router at that network. The packet hops from one router to another across multiple networks until it eventually reaches the destination computer. The path it travels is called its route.

packets

Figure 11–15 shows the components and data flow involved in accessing the Internet from a user's PC. The personal computer user at 199.9.9.1 uses a dial-up modem to access the network that is usually represented by an ISP. The ISP routes the user's request to the appropriate Web server host (10.10.10.1) and then provides the response that is received from the Web server back to the user. The function of an ISP is explained later in more detail.

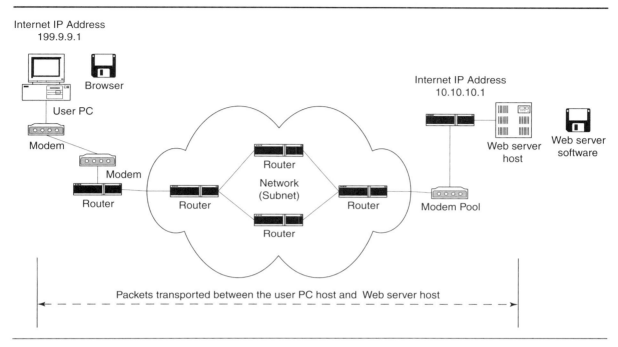

Figure 11-15
Internet Access

Push Technology

One of the most radical changes on the Web has been the creation of **push** technology. Traditionally, if several Web pages were of importance, there was no way of knowing when they changed. Users were required to download each page periodically to check for changes. Often, there would be no changes, so the downloading would be a waste of time.

In push technology, the effect is as if the Web servers periodically downloaded updated versions of subscribed Web pages to the user without any conscious effort. When reading a Web page, the user would know it was a very recent version. The Web servers would push the Web pages to the user, instead of the user having to pull them.

As an example, organizations use e-mail to push information to users without them asking for it (advertisements).

States and Cookies

HTTP was created with the concept that each request-response cycle would be separate. A Web server host would not remember the user from request to request: It would simply get the file that was requested and send it.

push

More complex actions, however, such as financial transactions, consist of a sequence of HTTP request-response cycles that build on one another. Each action in the sequence creates a state (condition). The next action builds on that state to create a new state. RFC 2109, HTTP State Management Mechanism, provides information on this process.

To help Web servers track a user's history, browsers allow Web servers to store brief text files on the user's PC hard disk drive. These short text messages are called **cookies**. When the user sends a message to the Web server, the Web server can also get the cookie to see what has been done in the past.

Many Web sites will attempt to place a cookie on a client's system. A client can have hundreds of cookies resident at any time, which can become a nuisance because they may constantly cause pop-ups, which interrupt the current online activity. Web browsers have a feature for controlling cookies, however this feature may also cause the user grief. Cookies are somewhat controversial because they allow a Web server to store files on the hard disk drive. They are text files, however, not programs that could contain viruses.

Uniform Resource Locator

All Web pages and Internet resources have a unique address called a **Uniform Resource Locator (URL)**. A URL is analogous to a card catalog number that references a library book. It permits its user, through the use of a client program, to retrieve documents from the Internet server. A browser extracts from the URL the protocol used to access the item, the name of the computer on which the item resides, and the name of the item.

The first part of the URL specifies the **Hypertext Transfer Protocol (HTTP)**. It is the Web's main protocol for transporting HTML documents between clients and servers. A colon and two slashes follow the protocol. The "www" and the remainder of the URL indicate you are accessing a Web server in search of a particular file and also its path, which is specified after any single slashes. An example of a URL is http://www.weather.com.

Search Engines

The most common tool used for navigating the Web is called a search engine. **Search engines** are programs used to search for virtually any form of intelligence imaginable. These are explained in Chapter 7. Access to search engines is provided by ISPs and commercial online services. When a connection to the Web is first established, the home page typically has the ISP's proprietary search engine and may list links to other popular search engines such as Yahoo, Lycos, and Alta Vista.

Domain Name System

The **Domain Name System (DNS)** is a method of mapping domain names and IP addresses for computers linked to the Internet. This means that DNS resolves symbolic names to their corresponding IP addresses. The

Table 11–3
U.S. Internet Domains

Domain	Description
com	Commercial organization
edu	Educational institute
gov	Government organization
int	International organization
mil	Military organization
net	Network or service provider
org	General organization

system was developed in 1986 to replace a central registry of IP addresses used to identify host computers connected to the ARPANET. There are currently nine "root" DNS servers on the Internet. The DNS structure is a hierarchical tree consisting of domains. A **domain** is a node and its descendent nodes on a network. A domain name is the unique name of a particular node within the domain, and domain names are made up of subdomains, which are nodes leading to a particular node. Subdomains are separated by a period. On the Internet, URLs refer to the domain names for a computer or a group of computers. These domains are separated into categories, which are denoted by the last element in the domain name. For example, in the domain name "bob.faculty.atl.devry.edu," the user *bob* is a faculty member at the Atlanta DeVry campus. The root node *edu* is at the top of the hierarchy. Table 11–3 lists various types of Internet domains used in the United States.

domain

A complete listing of country top-level domain names can be viewed at http://www.uninett.no/navn/domreg.html.

11.12 INTERNET NETWORK COMPONENTS

To connect to an Internet site, the user normally first connects to an ISP. When dialing into an ISP, access is usually gained via a router owned by that ISP. The ISP also has a router connected to the Internet. This second router is the interface into the entire Internet. The backbone of the Internet in the United States consists of companies called **network service providers (NSPs)**. The NSPs are interconnected via a number of networks.

network service providers (NSPs)

The levels of Internet access can be separated into five categories:

- Level 1: Interconnection—Network Access Points (NAPs).
- Level 2: National Backbone
- Level 3: Regional Networks
- Level 4: Internet Service Providers (ISPs)
- Level 5: Consumer and Business Market

On the top level are the NAPs where major backbone operators or NSPs interconnect to establish the core concept of an Internet. Level 2 is the national backbone operator. The network of networks spreads out from this point.

The third level of the Internet consists of regional networks and the companies that operate regional backbones. Typically, these companies operate backbones within a state or among several adjoining states much like the NSPs. They typically connect to a national backbone operator, or increasingly to several NSPs, to be on the Internet. Some have a presence at an NAP, which is used to extend this network to smaller cities and towns in their areas. Businesses are connected to those points with direct access connections, where they usually maintain dial-up terminal banks to offer 28.8 to 56 kbps dial-up SLIP/PPP connections to consumers. In many cases, regional networks are much more extensive than national backbones, but on a smaller geographic scale.

At the fourth level of the Internet are the individual ISPs. These vary from small operations to large organizations, some with up to 100,000 dial-up customers. They generally do not operate a backbone or even a regional network. Connections are leased from a national backbone provider or a regional network operator. They may offer service nationally, but use the POPs and backbone structure of their larger backbone operator associate. Several large providers, such as EarthLink and MindSpring, operate at this level. Generally, these companies operate an equipment room in a single area code, lease connections to a national backbone provider, and offer dial-up connections and leased connections to consumers and businesses in the local area. The focus is on customer service, configuration, and training, and often at lower rates.

The fifth level of the Internet is the consumer and business market. Many companies set up dial-up ports at their offices for employees to connect from home or on the road.

11.13 ACCESS AND TRANSPORT

A carrier is an organization that transmits telecommunications traffic for a price. Figure 11–16 shows the two types of carriers involved in the United States—ISPs and NSPs.

Accessing the Internet

To access the resources available on the Internet, the user must connect the computer to the network. For most users today, this means subscribing to an ISP. ISPs provide access using modems, ISDN service, ADSL, or other dial-up connection services. Other types of connections are possible but they may be beyond the means of most individuals. These include the large-bandwidth connections of DS-3 (44.736 Mbps) or OC-3 (155.52 Mbps) that are provided to large corporations and governmental organizations.

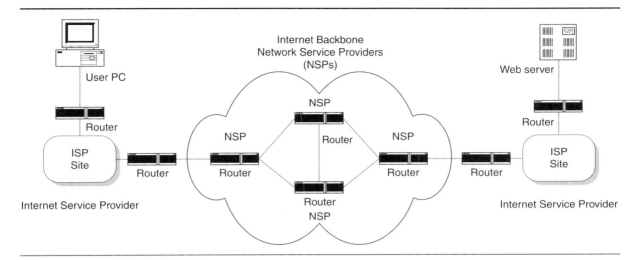

Figure 11–16
ISPs and NSPs

Figure 11–17 shows various access technologies that can be used to connect to the Internet. The ISP can provide one-stop shopping for all network access, including client-server hardware and software and security requirements. Access from the full-service client to the ISP is 56 kbps Frame Relay service. The Internet Presence Provider (IPP) provides design, management, and maintenance. The IPP connects to the ISP via 56 kbps Frame Relay service and provides dial-up access to client PC-5. Client PC-2 uses ISDN service. Client PC-3 has router access via T-1 service to the ISP. Client PC-4 has router access to the ADSL service.

Internet Service Providers

When users connect to the Internet, they first connect to an ISP. Figure 11–17 shows that when dialing into an ISP, the user dials into a router owned by that ISP. The ISP also has a router connected to the Internet backbone. This second router is the gateway to the entire Internet. Several sites on the Web provide listings of ISPs. At least one site indicates more than 8,000 ISPs worldwide.

Be careful when shopping for an ISP because they too are potential victims of attackers. Their level of security can quickly decide a customer's level of security. If the upstream feed is compromised, the attacker can inspect all data bound for the Internet. In other words, private e-mail can be read, Web form submissions viewed, and downloaded files intercepted. There is also the potential of information being stolen and connections

Figure 11-17
Clients-Server Internet Connectivity

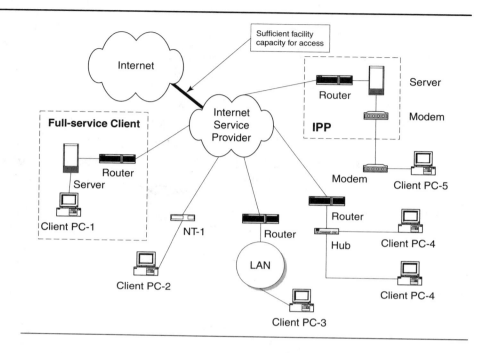

being hijacked. The ISP is a possible haven for a hijacker to lurk, so be careful (Goncalves 2000).

Network Service Providers

As described earlier, the backbone of the Internet consists of several competing NSPs that are interconnected. Although NSPs compete for ISP business, they must work together to provide total interconnection. To connect to an NSP, the ISP must pay the NSP a monthly fee.

Figure 11-18 provides a view of the overall client-server architecture for Internet connectivity. The Internet gateway configuration has no special software loaded on the client PCs. The server provides local LAN transport protocol to TCP/IP translation and client software for supported Internet services. In the internet-enabled client configuration, every client has TCP/IP and local LAN transport protocol loaded. Internet client software is also loaded on each client.

11.14 IP ADDRESSING

As with any other network layer protocol, the IP addressing scheme is integral to the process of routing IP datagrams through an internetwork. Each IP address has specific components and follows a basic format. These IP addresses can be subdivided and used to create addresses for subnetworks.

Wide Area Network Security

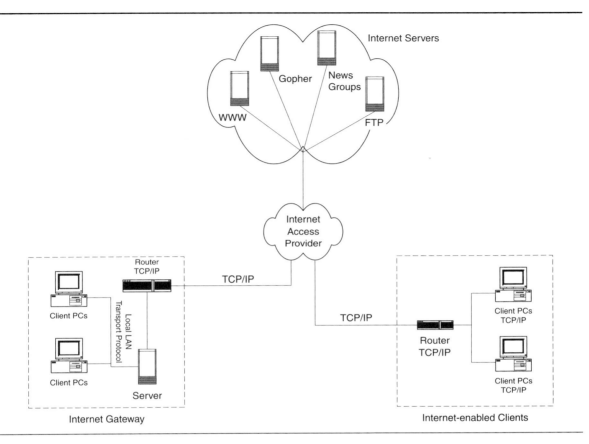

Figure 11–18
Internet Gateways, Services, and Clients

The 32-bit IPv4 address is grouped 8 bits at a time, separated by dots, and represented in decimal format (otherwise referred to as dotted decimal notation). The minimum value of an octet is zero and the maximum value is 255. Figure 11–19 provides the basic format and an example of an IP address (171.10.10.1).

Internet Addressing

Any computer attached to the Internet is called a host. When users are connected to the Internet with a home PC, they become a temporary WWW host. To contact someone via telephone requires a telephone number. On the Internet the equivalent is the Internet address. Currently, the Internet address (IPv4) is a string of 32 bits (0,1s). This address is the IP address. Each host has its own unique IP address just like each telephone has its unique number.

Figure 11–19
IPv4 Address Format

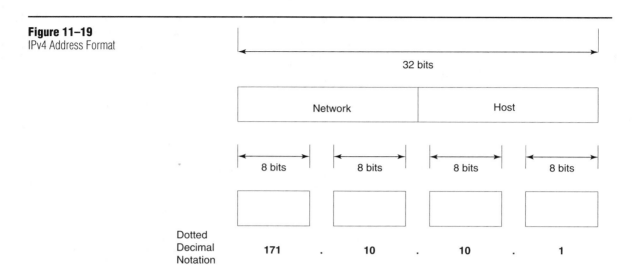

The IP address is converted into four 8-bit strings (segments). An example of an address is 188.23.44.5. The four decimal segments are separated by decimal points, called dots, and thus the name dotted-decimal notation. Unique IP addresses are assigned by the **Internet Assigned Numbers Authority (IANA)**. A host name is defined by the Internet, which consists of several labels of text separated by dots. This substitutes for the Internet address. The user can use either the Internet address or the host name when accessing the Internet site.

Internet Assigned Numbers Authority (IANA)

A new version of Internet address (IPv6) will allow for 128 bits to accommodate the enormous growth expected. IPv6 was adopted by the IETF in 1994 and a strategy for migration to it was recommended. Figure 11–20 provides the general layout of the IPv6 format (Coulouris 2001).

Users want to ensure that their network access and transactions are secure. IPv6 provides for security measures in authentication and privacy.

Figure 11–20
IPv6 Format

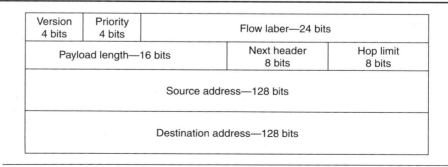

A login process is required, and if the user is validated the receiver can be assured that the traffic is legitimate and was not modified in transit. Packets that leave a site can be encrypted and authenticated at the receiving site. IPv6 is in the testing phase and should be available in the near future.

Internet Security

Computer and network managers develop documentation called a baseline that inventories everything concerning the Enterprise resources and assets. This baseline development is especially important in the event of an attack from the Internet. This data and information can be used to reconstruct the systems after a disaster occurs that is caused by some intruder or attacker. While the organization's personnel are developing this information, potential attackers may be attempting to develop a profile of the organization. The Enterprise staff must ensure that they do not contribute to such a potentially damaging endeavor.

These attackers are looking for specific information that can aid their cause. Items of relevance can be related to the Internet, intranets, extranets, and remote access elements. Details that can assist an attacker includes the following:

- IP addresses reachable by the Internet
- TCP and UDP services running on which systems?
- System architecture
- Access control mechanisms
- Intrusion detection systems
- Routing tables and access control lists
- Network management information
- Demographics
- Organization structure
- Personnel lists
- Contact lists with telephone numbers
- Network equipment type, model, and configurations

This looks like a formidable list, however an attacker has the capability of gathering this information. Maybe the organization should hire this person! Don't help the attacker by providing any of this information on the organization's Web page.

11.15 WORLD WIDE WEB (WWW)

Today, the WWW is the premier application for most Internet users. The WWW is a collection of servers accessed via the Internet that offer graphical or multimedia presentations about a company's products, personnel, or services. Web browsers integrate e-mail and newsreaders and support an increasingly interactive, visual, and even animated interface to the Internet. Companies wishing to use the WWW as a marketing tool establish a Web site on the Internet to publicize the address of that location. The Web site and Web server presentation design, implementation, and management can be done in-house or be contracted out to a professional Web site development and management service.

Almost any of the current data processing and data communications applications are candidates to ride the Internet. Some major application areas are as follows:

- Customer service
- Distance learning
- Electronic commerce
- Internet access
- Telemarketing
- Web design
- Web hosting

Reasons why people use the Web include the following:

- Browsing
- Business
- Chatting socially with other Web users
- Entertainment

Virtually all classes of businesses can use the Internet in the following ways:

- Advertising via Web pages and e-mail
- Communicating via e-mail
- Selling goods and services
- Selling stocks or other financial instruments
- Transferring documents via electronic data interchange (EDI)
- Providing travel services
- Providing reference materials

11.16 SECURITY IN THE WAN

Obviously, the WAN is massive and there are numerous opportunities for an intruder or attacker to cause disruption and damage to the assets. The responsibility of the security administrator is to block or mitigate these efforts. This task is enormous and ongoing. Chapters 1 though 9 present information concerning the many facets of Enterprise security and techniques and methods that can be deployed to assist the administrator in providing security in the WAN.

As with LAN security, discussed in Chapter 10, attention must be directed toward protecting physical components, facilities, and buildings that house the various hardware and software components of the wide area network. This includes the massive databases and terminal inventories that go to support the WAN infrastructure.

Of particular interest are the processes necessary to restore network functions during a power outage or disaster. Disaster recovery and contingency planning is a major part of network security and network integrity assurance activities. Without the WAN infrastructure, the Enterprise network will not function.

Security Issues

WAN security means employing security operations to achieve three major goals: authentication, integrity, and confidentiality. The types of Internet security services available are as follows (Black 2000):

- Access control—Prevention of unauthorized use of a resource
- Authentication—Assurance that traffic is sent by legitimate users
- Confidentiality—Assurance that a user's traffic is not examined by unauthorized users
- Integrity—Assurance that received traffic has not been modified after the initial transmission
- Nonrepudiation—Inability to disavow a transaction

These services and related topics are presented in considerable detail in Chapters 1 through 8.

Two primary methods for encrypting network traffic on the Web are **S-HTTP** and **SSL**. Details concerning these protocols are presented in Chapter 4. Secure HTTP is a secure version of HTTP that requires that both client and server S-HTTP versions are installed for secure end-to-end encrypted transmission. It is based on public-key encryption and provides security at the document or application level. It uses digital signature encryption to assure the documents possess both authenticity and message integrity. SSL is described as wrapping an encrypted envelope around HTTP transmissions. Whereas S-HTTP only encrypts Web documents, SSL can be wrapped around other Internet service transmissions, such as FTP, Telnet, and Gopher, and including HTTP. SSL is a connection-level encryption method providing security to the network link.

S-HTTP
SSL

Another Internet security protocol, named **Secure Courier**, is directly aimed toward securing and authenticating commercial financial transactions. It is based on SSL and allows users to create a secure digital envelope for transmission of financial transactions over the Internet. Secure Courier also provides consumer authentication for the e-merchants inhabiting the commercial Internet (e-commerce).

Secure Courier

An Internet e-mail specific encryption standard that also uses digital signature encryption to guarantee authenticity, security, and message integrity of received e-mail is called PGP (Chapter 4). PGP overcomes inherent security loopholes with public/private-key security schemes by implementing a web of trust, in which e-mail users electronically sign one

Table 11–4
Summary of Web Threats

	Threats	Consequences	Countermeasures
Authentication	User impersonation Data forgery	Misrepresentation of user Belief that false information is valid	Cryptographic techniques
Confidentiality	Eavesdropping Server information theft Client data theft Client-server connection information Network config information	Information loss Privacy loss	Encryption Web proxies
Denial of Service	Killing user threads Flooding with bogus requests Filling up disk or memory DNS attacks	Annoying Disruptive Prevent user productivity	Difficult to prevent
Integrity	User data modification Trojan horse browser Memory modification Message modification in transit	Loss of information Compromise of machine Vulnerability to all other threats	Cryptographic checksums

another's public keys to create and interconnected group of public-key users (Goldman 1999).

Web Threats

Web threats can be categorized as either active or passive. Active attacks include impersonating another user, altering messages in transit between client and server, and altering information on a Web site. Passive attacks include eavesdropping on network traffic between browser and server and gaining information on a Web site that is normally restricted.

Web threats can also be grouped according to threat location; that is, whether it involves a Web server, Web browser, or network traffic between the browser and server. Issues of server and browser security are the purview of computer system security, whereas issues of traffic security are part of network security. Table 11–4 provides a summary of security threats pervasive in the Web environment (Stallings 2000).

11.17 INTERNET AND WEB RESOURCES

SecuritySearch.net   www.securitySearch.net
ESecurityOnLine.com   www.esecurityonline.com
Electronic Privacy Info   www.epic.org
   Center

Rivest Cryptography and Security	theory.lcs.mit.edu/
IETF	web.mit.edu/network/
Yahoo Security and Encryption	dir.yahoo.com/Computers_and_Internet/Security_and_Encryption
Counterpane Internet Security	www.counterpane.com/hotlist.html
RSA Laboratories	www.rsasecurity.com/rsalabs/faq
Security Resource Net	www.nsi.org/compsec.html
ITAA	www.itaa.org/infosec
Distributed DoS Attacks	staff.washington.edu/dittrich/misc/
AntiOnline	www.antionline.com
Microsoft Security	www.microsoft.com/security/default.asp

■ SUMMARY

The WAN is the national network that connects all of the geographically dispersed LANs and customers across the United States. Service categories include private lines, dial-up access, high-speed communication facilities, and wireless systems. The PSTN infrastructure includes five classes of offices that provide trunking between offices and local loops to customer locations.

The telecommunications network provides a multitude of functions, including network management, security, routing, traffic control, and network recovery. Central office equipment includes the WAN switch, channel banks, DACS, common equipment, and multiplexers. Common transfer speeds offered by the Telcos include low speed dial-up, DS-0, DS-1, and DS-3. WAN facilities provide the following communication services:

- Common carrier data circuits
- X.25 packet-switched circuits
- Frame Relay access
- ISDN access
- SONET facilities for FDDI
- DSL capabilities

WAN switches provide for circuit switching and packet switching environments. Circuit switching allows to the user to establish a communications path for the entire transmission, whereas packet switching traffic can take many paths to the desired destination. There are tradeoffs for both methods.

The Internet is a major component of the WAN. The Internet is made up of NSPs and ISPs that provide connectivity between users and between users and providers. The Internet is composed of the following levels:

- Network access points
- National backbone
- Regional networks
- ISPs
- Users and the business market

TCP/IP in conjunction with IP addresses is used to facilitate communication between all participants on the Internet. The present IPv4 addressing scheme is being enlarged to IPv6 to accommodate the additional Internet users.

Security is a major issue with the Internet. Opportunities exist for serious damage to computer and network resources if they are not protected from malicious users that ply the Internet. Security services for the Internet offer the following benefits:

- Access control
- Authentication
- Confidentiality
- Integrity
- Nonrepudiation

Key Terms

Access line
Browser
Central Office (CO)
Centralized
Circuit switching
Cookies
Customer Premises Equipment (CPE)
Data Communications Equipment (DCE)
Datagram
Data Service Unit (DSU)
Data Terminal Equipment (DTE)
Decentralized
Digital Access Cross-connect Switch (DACS)
Distributed
Domain
Domain Name System (DNS)
Electronic Industries Association (EIA)
Hypertext Markup Language (HTML)

Hypertext Transfer Protocol (HTTP)
International Standards Organization (ISO)
Internet
Internet Assigned Numbers Authority (IANA)
Internet Service Provider (ISP)
Internetworks
Java
JavaScript
Local Access Transport Area (LATA)
Local loop
Media
Modulator/Demodulator (Modem)
Network-to-Network Interface (NNI)
Network Service Provider (NSP)
Open Systems Interconnection (OSI)
Packet Data Network (PDN)
Packets
Packet Switching

Plain Old Telephone Service (POTS)
Point of Presence (POP)
Public Switched Telephone Network (PSTN)
Push
RS232-D
Search engine
Secure Courier
S-HTTP
Signaling System 7(SS7)
SSL
Store and Forward

Transmission Control Protocol/Internet Protocol (TCP/IP)
Trunk
Uniform Resource Locator (URL)
User-to-Network Interface (UNI)
V.35
Virtual circuit
War dialing
Web
Wide Area Network (WAN)
World Wide Web (WWW)
X.21

REVIEW EXERCISES

Questions and Analysis

Provide a short answer or in-depth analysis as required.

1. Describe the WAN in general terms.
2. What types of services are available in the WAN?
3. What are the major components of the telecommunications infrastructure?
4. What are the five central office designations?
5. What is the function of SS7?
6. What are the telecommunications network functions?
7. What are the elements of a data communications system?
8. Identify the WAN facilities commonly used in computer networks.
9. What is the difference between centralized and decentralized communication networks?
10. What is the function of a POP?
11. Comment on wiring interface standards. Why are they needed?
12. What is a UNI? How is different from a NNI?
13. What are the traffic speeds allowed by DDS and DSO transmissions?
14. Why should the administrator be concerned with modem implementations?
15. Why are wiring closets and demarcations a point of interest for the security manager?
16. Compare and contrast circuit switching and packet switching.
17. What is a datagram? How does it relate to a virtual circuit?
18. Discuss why the virtual circuit approach may be superior to the datagram approach.
19. Provide a general description of the Internet.
20. What is PUSH technology?
21. What are a state and a cookie? Are they dangerous to the user?
22. Identify the major Internet domains.
23. What are the five categories (levels) of the Internet?
24. What is the function of an ISP?
25. What is the function of network service providers?
26. What is the function of the IANA?
27. What Internet security services are available for users?

28. What are the types of information that an attacker would be gathering?
29. Discuss and identify Web threats.
30. Discuss the issues of WAN security.

Definitions and Terms

Provide the most correct answer from the list of key terms.

1. An Internet mechanism that lets site developers place information on the client's computer for later use.
2. A name used to locate a computer, as in *www.weather.com*.
3. A network utility used to resolve system names and IP addresses.
4. The protocol used to transmit data across the WWW.
5. The protocol for identifying documents on the Web.
6. A tool used to locate modems and repeat dial tones within a range of telephone numbers.
7. A wide-area information retrieval initiative to provide universal access to Internet documents.
8. A group of geographically dispersed data devices, or LANs that can communicate with each other from different cities, different Local Access Transport Areas (LATAs), different states, and different countries.
9. An out-of-band signaling process used for supervision, alerting, and addressing within the telecommunications network.
10. This interface provides a physical point of separation between the responsibilities of the carrier and the customer.

True/False Statements

Indicate whether the statement is true or false.

1. A WAN is a group of geographically dispersed networks that can communicate across a wide area.
2. The PSTN consists of a number of different communication facilities, each with a wide range of capabilities in the network hierarchy.
3. Three basic types of communication networks include distributed, centralized, and decentralized.
4. The POP and demarcation point are synonymous access points.
5. The UNI and NNI interfaces are important considerations for security issues because the NNI is subject to hacker access.
6. A dial-up access is more secure from attackers than is a private line.
7. A technique employed by network attackers is called war dialing, which requires repeatedly dialing different telephone numbers, looking for a dial tone.
8. Packet switching and circuit switching are both used in the public networks, however circuit switching is generally the method selected for data transport.
9. The Internet uses the unreliable datagram approach to transport network traffic.
10. Push technology allows for unsolicited advertisements on the Web.
11. A Web server can store a cookie on a user's PC only for the duration of the on-line connection.
12. Internet domains include org, edu, net, com, and tip.
13. An ISP connects to an NSP for Internet connectivity.
14. Two primary methods for encrypting network traffic on the Web are HTTP and SSL.
15. Eavesdropping on network traffic is an active Web threat.

Activities and Research

1. Sketch the components of a communications system, including DCE and DTE devices.
2. Arrange for a tour of a local central office. Identify the components presented in this chapter.
3. Identify the major standards and specifications associated with the WAN. Prepare a brief statement on each.
4. Identify the major standards and specifications associated with the Internet. Show why they are important.
5. Produce a drawing that depicts UNI, NNI, DCE, and DTE interfaces.
6. Identify the services offered by the local communications carrier, and include the various transmission speeds offered.

7. Show the difference between the transmission of the same packets over packet switched and circuit switched networks.

8. Identify ISPs that operate in the local area. Determine the NSPs that support these ISPs.

9. Write a synopsis for TCP/IP.

10. Compare and contrast IPv4 and IPv6.

11. Identify browsers and compare their features and benefits.

12. Identify the various cookies that are resident on your PCs.

13. Identify a PUSH document on your PC.

14. Identify the various search engines available on your PC. Look at the operation of each.

15. Look up the IANA Web site and identify the number of organizations represented.

16. Research and identify security offerings that can be used to protect the Enterprise assets and resources from threats originating on the WAN. Prepare a matrix of cost and benefits of each.

CHAPTER

12

Virtual Networks Security

OBJECTIVES

Material included in this chapter should enable the reader to

- Understand VPN technology and its relation to the current broadband market. Identify the various types of virtual private networks.
- Become familiar with the terms and definitions that are relevant in the Enterprise network environment. Look at the similarities between VPNs and other network technologies.
- Understand the function and interaction of a VPN in the Enterprise network environment. Look at the issue of tunneling and the security techniques that must be deployed in a VPN.
- Identify components that comprise the VPN networking environments. Look at router and firewall components that are part of the VPN.
- Identify the issues and considerations that must be part of the VPN deployment process.
- Become familiar with the various protocols that are part of the VPN infrastructure.
- Look at the various security issues that must be considered when using the Internet as a backbone network.

■ INTRODUCTION

Employees, business partners, and customers remotely access many diverse Enterprise networks across private networks. Organizations are looking at the Internet as a vehicle for transporting data reliably and securely. Because the Internet is public and not secure, organizations are reluctant to use it to transport sensitive information. IP-based **virtual private network (VPNs)**, however, offer the security of a private corporate network and the economy and simplicity of the Internet. Measures designed to prevent unauthorized remote access include the use of VPN connections between the network and the remote users. The encryption techniques offered by VPNs are designed to prevent unauthorized users from discovering the content of remote messages.

virtual private network (VPNs)

A VPN is a private data network that makes use of the public telecommunications infrastructure, which maintains privacy through the use of tunneling and security procedures. The VPN is so named because an individual user shares communications channels with other users. Switches are placed on these channels to allow an end user access to multiple end sites. A VPN can be contrasted with a system of owned or leased lines that can only be used by one company. The connection between sender and receiver acts as if it were completely private, even though it uses a link across a public network to carry information.

The idea of the VPN is to give the organization the same capabilities at much lower cost by using the shared public infrastructure rather than a private one. Telephone companies have provided secure shared resources for voice messages. A VPN makes it possible to have the same secure sharing of public resources for data. Figure 12–1 depicts a VPN between Charlotte and New Orleans that uses the Internet for transport. Access to the Internet cloud occurs both at Charlotte and then again at New Orleans, which saves on transport costs.

Virtual networks are economical for large organizations that have a considerable amount of on-net calling. Most of the features of a dedicated private network can be provided. Organizations are interested in using private virtual networks for both extranets and wide area intranets.

Using a VPN involves encrypting data before sending it through the public network and decrypting the data at the receiving end. An additional level of security involves encrypting not only the data but also the

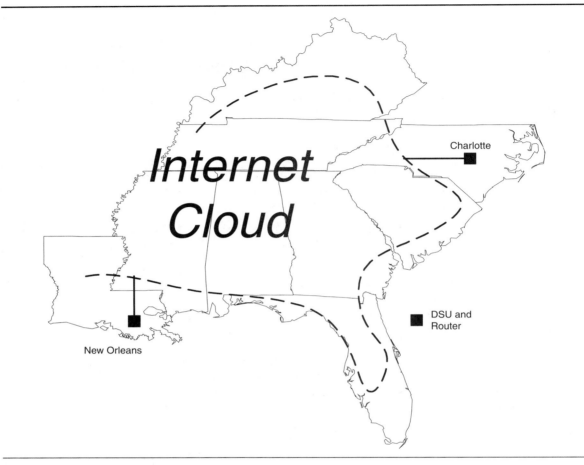

Figure 12–1
VPN Over the Internet

originating and receiving network addresses. A VPN combines the advantages of private public networks by allowing a company with multiple sites to have the illusion of a completely private network while using the public network to carry traffic between sites.

The voice private network question is greatly affected by virtual networks being offered by the major Interexchange Carriers (IXCs). A VPN operates as if it was composed of voice private circuits; however, it is actually part of the IXC's switched network. The VPN architectures of the IXCs are all similar, although the terminology may differ somewhat. As VPNs spread and grow, many organizations are implementing VPN services provided by the communications carriers.

The use of private and wide area networks will continue to be an option for carrying Enterprise network traffic. Attackers and intruders will also continue to require the attention of the network administrator. Security must be at the forefront of virtual network design and implementation. This chapter addresses these important issues.

12.1 THE NEW WORKFORCE

Many employees are trading in their desktop PCs in the office for a laptop to use on the road or at home. Organizations want remote employees to easily gain access to their networks and computer resources, however a heightened awareness of security issues has raised some red flags. There is the need for security tools that deter hackers and intruders and also prevent employees, partners, and others from obtaining access to sensitive corporate information. Telecommuting employees tend to make fewer backups of their files than do those in traditional operations, which can result in the loss of valuable, expensive, and irreplaceable data. Organizations are responding to these new situations by implementing virtual networks that use a variety of technologies.

Organizations with multiple geographic locations and telecommuting employees can use one of two approaches to building a computer network:

- Private network connectivity where the circuits are leased.
- Public Internet connectivity with access through an ISP.

The primary advantage of using leased circuits to interconnect locations is the privacy afforded. No other organization has access to a leased circuit, so the transmissions are secure and private. The primary advantage of using the Internet for connectivity is the low cost. Instead of paying for leased circuits, the organization only pays for local ISP access. It can become very expensive to lease private circuits for many telecommuters, so the alternative of a virtual network is attractive (Comer 2001).

12.2 VIRTUAL NETWORKS

The virtual private network allows an individual user to share communications channels with other users. Switches are placed on these channels to allow an end user access to multiple end sites. Ideally, users do not perceive that they are sharing a network with each other, thus the term *virtual private network*. Users think that they are the only ones on the network, although they are not.

Virtual networks allow multiple secure data streams to be sent simultaneously over networks while maintaining sufficient information to provide service differentiation, billing, and switching. They can operate over Internet Protocol (IP), Frame Relay, or Asynchronous Transfer Mode (ATM) networks, or over the Internet. Increasingly, carriers provide virtual private networks (VPNs) as managed products that run on high-speed backbones.

Two fundamental components of the public network make VPNs possible. The process of tunneling enables the virtual part of a VPN and security services keep the VPN data private.

VPN is a relatively new term in the computer/communications industry, but the ideas behind the concept are not new. Public X.25 networks have offered VPN services for years, and switched T-1 services also offer VPN-like features. However, it will become evident that several of the emerging technologies offer more powerful VPNs than these older technologies.

VPNs represent temporary or permanent connections across a public network, such as the Internet, that make use of special **encryption** technology to transmit and receive data meant to be impenetrable to anyone who attempts to monitor and decode the packets transmitted across the public network. The connection between the sender and receiver acts as if it were completely private, even though it uses a link across a public network to carry information. This makes something virtually public act as though it were private, thus a *Virtual Private Network*.

The User Environment

Even if virtual networks are not being considered as complete replacements for existing WANs, they may be worth investigating for telecommuter remote access or for access between smaller branch offices and headquarters locations. VPN services, whether they use a proxy server, router, gateway, or firewall, reside outside of the Enterprise system.

The client side of the network may consist of a Windows PC with an IP stack and a piece of client software provided by the VPN vendor. Remote users run their normal Internet Service Provider access software, but when users attempt to connect to a host, the VPN client software begins encrypting all of the data sent between the host and client.

Windows NT 4.0 and Windows 95 support a special TCP/IP protocol called **Point-to-Point Tunneling Protocol (PPTP)**. PPTP permits a user running Windows 98, for example, to dial into a Windows NT server running the **Remote Access Server (RAS)**, and it supports the equivalent of a private, encrypted, dial-up session across the Internet. Similarly, a VPN could be established permanently across the Internet by leasing dedicated lines to an Internet Services Provider (ISP) at each end of a two-way link and maintaining ongoing PPTP-based communications across that dedicated link. This means that organizations can use the Internet as a private dial-up service for users with machines running Windows 95 and Windows NT 4.0 or as a way to interconnect multiple LANs across the Internet, one pair of networks at a time.

Dial-up use produces two advantages:

- It is not necessary to install several modems on a RAS so users can dial directly into the server machine; instead they can dial into any ISP. This saves money on hardware and systems management.
- Remote users can usually access the RAS by making only a local phone call, no matter where in the world they might be located, as long as they can access an ISP locally. Distance from the RAS no longer matters, which saves money on long-distance telephone charges.

The VPN uses PPTP or other equivalent protocols to extend the reach of private networks across public one—easily, economically, and transparently—by using on-demand, dial-up connections across the Internet.

12.3 VIRTUAL NETWORKS OVERVIEW

Some InterExchange Carriers (IECs) offer VPNs to their subscribers, a service that requires support from a CO tandem switch. The typical tandem switch has a nonblocking switching fabric, which provides access primarily to CO trunks. Digital trunks terminate in digital interface frames that couple incoming T-1 or T-3 bit streams directly to the switching network. Peripheral equipment detects signaling, and the central processor sets up a path through the switching network from the incoming time slot to an outgoing time slot, which it assigns to an outgoing digital channel. The digital switch acts as a large time-slot interchange device that is transparent to the bit stream in the terminating circuits.

A virtual network is one that operates as if it is composed of switched private lines, but in reality is derived by shared use of a carrier's switched facilities. The database for a VPN is contained in a **service control point (SCP)**; that is, a computer connected to tandem switches by 64 kbps data links. The switches in a virtual network are known as **service switching points (SSPs)**.

service control point (SCP)

service switching points (SSPs)

A virtual private network handles calls in the following three manners:

- Dedicated access line to dedicated access line (DAL)
- Dedicated access line to switched access line (DAS)
- Switched access line to switched access line (SAL)

A DAL-to-DAL call bypasses the LEC's access charges in both the originating and terminating direction, reducing the cost significantly. The DAL-to-SAL is handled on a VPN the same way it is handled with a customer using conventional DAL service. The access charge is eliminated in the originating, but not the terminating, direction. For SAL-to-SAL calls, the VPN handles calls like regular long-distance calls except for features and restrictions.

Virtual private networks also emulate many features of electronic tandem networks. They offer a full restriction range such as blocking calls to overseas locations, selected area codes, central office codes, or even selected station numbers. If the virtual network is used in conjunction with account codes, calls can be restricted for certain station numbers.

12.4 VPN TYPES

A VPN is a dedicated network link that is provided over a shared network, using encryption and authentication techniques to provide privacy and security. It provides customer connectivity deployed on this shared infrastructure with the same policies as a private network. VPNs allow traffic from many sources to traverse the same network in private data streams. The data streams can be differentiated, permitting routing to different destinations and provisions for different levels of service. The shared infrastructure can provide a service provider with IP, Frame Relay, ATM backbone, or Internet access, however many VPNs use the IP protocol over Frame Relay or ATM backbones.

There are three types of VPNs, which align with how businesses and organizations use them, namely access VPN, intranet VPN, and extranet VPN. Further, three types of IP LANs include Frame Relay, CPE-based IP, and network-based IP (Dooley 2001).

Access VPN

access VPN

An **access VPN** provides remote access to a corporate intranet or extranet over a shared infrastructure with the same policies as a private network. Access VPNs enable users to access corporate resources when, where, and however they require. Access VPNs encompass analog, dial, ISDN, DSL, mobile IP, and cable technologies to securely connect mobile users, telecommuters, or branch offices.

Intranet VPN

intranet VPN

An **intranet VPN** links corporate headquarters, remote offices, and branch offices over a shared infrastructure using dedicated connections. Businesses enjoy the same policies as a private network, including security, quality of service (QoS), manageability, and reliability.

Extranet VPN

extranet VPN

An **extranet VPN** links customers, suppliers, partners, or communities of interest to a corporate intranet over a shared infrastructure using dedicated connections. Businesses enjoy the same policies as a private network, including security, QoS, manageability, and reliability.

Frame Relay VPN

A Frame Relay VPN uses Permanent Virtual Circuits (PVCs) linked to Internet gateways. Quality of Service (QoS) is good, however the need to manually configure PVCs between sites makes many-to-many connectivity difficult.

CPE-based VPN

The CPE-based VPN provider deploys VPN access routers onsite using several protocols, which can include PPTP, L2TP, or IPSec. A variety of equipment can be used, including IP VPN routers, standard routers with added VPN software, IP VPN gateways that work with routers, IP VPN appliances, or firewalls acting as IP VPN gateways.

Network-based VPN

The newest method is the network-based IP VPN, which is delivered from the provider's point of presence (POP), over a standard communications access line. The only CPE required is a standard router. This allows the carriers to house all the VPN functionality in their network without establishing and managing expensive onsite CPE.

12.5 THE INTERNET VPN

There are compelling economic and reliability reasons for considering using the Internet as a virtual network. However, there are also compelling reasons for organizations to avoid this alternative for replacing their current communication's configurations. These reasons include security and performance issues.

Connecting the organization's network to the Internet exposes it to attacks from crackers, hackers, and other intruders who are beyond the control of the private network. There are several security mechanisms that can be deployed to create a significant barrier between the public Internet and the organization's private network. These mechanisms include the use of a firewall, a proxy server, an authentication server, and encryption techniques. There are some firewalls currently available that combine all of these features in a single product (Held 1997). Firewalls are discussed later in this chapter.

Rather than depend on the traditional telcos for dedicated leased lines or on Frame Relay's PVCs, another option is to use an Internet-based VPN. This would use the open, distributed infrastructure of the Internet to transmit data between corporate sites. Companies using an Internet VPN would set up connections to the local connection points, or points of presence (POP) of their Internet Service Provider (ISP) and let the ISP ensure that the data are transmitted to the appropriate destinations via the Internet. The rest of the connectivity details for the network and Internet infrastructure would be the ISP's responsibility. The link created to support a given communications session between sites is dynamically established, reducing the load on the network. Permanent links are not part of the VPNs structure. The bandwidth required for a session is not allocated until required and is released for other uses when a session is completed.

Because the Internet is a public network with open transmission of data, Internet VPNs include the provision of encrypting data passed between VPN sites, which protects data against eavesdropping and tampering by unauthorized parties. Issues still outstanding for Internet VPNs are guaranteed performance and security. Additional information concerning the Internet VPN can be found in RFC 2764—A Framework for IP-Based Virtual Private Networks.

An Internet VPN should only be part of an organization's security plan. Securing tunnels for private communications between sites will be useless if employee passwords are openly available or if other holes are in the security of the overall network. When implementing a VPN, it becomes necessary to integrate VPN-related security management of keys and user rights into the organization's security policies.

If the organization is going to operate in the international scene, additional issues must be addressed. The U.S. government and many foreign governments restrict the export of advanced encryption software. Mobile users using 128-bit encryption to secure transmissions in the U.S. are still largely restricted from using more than 56-bit encryption when traveling abroad (Kosiur 1998).

12.6 VIRTUAL NETWORKS CONNECTIVITY

One of the most important advantages offered by carrier VPN providers is they can connect VPNs over their own backbone networks, completely avoiding connections through the Internet. Such connections allow organizations using VPNs to realize similar security and reliability, which is available on a private network.

In VPNs, *virtual* implies that the network is dynamic, with connections set up according to the organizational needs. Unlike the private lines used in traditional virtual private networks, Internet VPNs do not maintain permanent links between the endpoints that comprise the corporate network. Instead, a connection is created between the two sites when needed and is torn down when no longer required. This makes the bandwidth and other network resources available for other uses.

The term virtual also means that the logical structure of the network is formed only from the company's devices, regardless of the physical structure of the underlying network. Devices such as routers and switches that are part of the ISP's network are hidden from the devices and users of the company's private network.

The connections comprising the company's VPN do not have the same physical characteristics as the hard-wired connections used on the Local

Area Network. Hiding of the ISP and other infrastructures from the VPN applications is made possible by tunneling, one of the two fundamental components that comprise the VPN architecture. The other fundamental component is security services.

Tunneling

A key component of VPN is **tunneling**, which is a vehicle for encapsulating packets inside a protocol that is understood at the entry and exit points of a given network. These entry and exit points are defined as tunnel interfaces. The tunnel itself is similar to a hardware interface, but is configured in software. Figure 12–2 depicts a VPN that uses a tunnel across the Internet.

End-user software capabilities and ISP support have resulted in the division of tunnels into two classes: voluntary and compulsory. **Voluntary tunnels** are created at the request of the user for a specific use. **Compulsory tunnels** are created automatically without any action from the user, without allowing the user any choice in the matter. Within the compulsory category are two subclasses: static and dynamic. The static tunnels can be subdivided again, into realm-based and automatic classes.

When using a voluntary tunnel, the end user can simultaneously open a secure tunnel through the Internet and access other Internet hosts via basic TCP/IP without tunneling. The client-side endpoint of a voluntary tunnel resides on the user's computer. Voluntary tunnels are often used to provide privacy and data integrity for intranet traffic being sent over the Internet.

Because compulsory tunnels are created without the user's consent, they may be transparent to the end user. The client-side endpoint of a compulsory tunnel typically resides on a RAS. All traffic originating from the end user's computer is forwarded over the PPTP tunnel by the RAS. Access to other services outside the intranet would be controlled by the network administrators. PPTP enables multiple connections to be carried over a single tunnel.

Because a compulsory tunnel has predetermined endpoints and the user cannot access other parts of the Internet, these tunnels offer better

tunneling

Voluntary tunnels
Compulsory tunnels

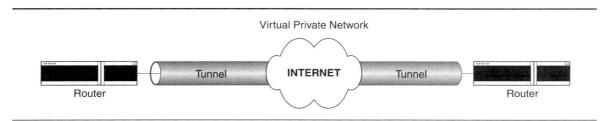

Figure 12–2
VPN Tunnel Across the Internet

access control than voluntary tunnels. If corporate policy states that employees cannot access the public Internet, then a compulsory tunnel would keep them out of it while still allowing them to use the Internet to access the VPN.

Static compulsory tunnels typically require dedicated equipment or manual configuration. These dedicated, automatic tunnels might require the user to call a special telephone number to make the connection. On the other hand, in realm-based, or manual, tunneling schemes, the RAS examines a portion of the user's name, called a realm, to decide where to tunnel the traffic associated with that user.

A more flexible approach would be to dynamically choose the tunnel destination on a per-user basis when the user connects to the RAS. These dynamic tunnels can be set up in PPTP.

Tunneling has two definitions, one is a LAN term and the other an Internet term. As a LAN term, tunneling means to temporarily change the destination of a packet to traverse one or more routers that are incapable of routing to the real destination. This book uses the Internet term, which says that tunneling is used to provide a secure, temporary path over the Internet. As an example, a telecommuter might dial into an ISP, which would recognize the request for a high-priority, point-to-point tunnel across the Internet to a corporate gateway. The tunnel would be established, effectively making its way through other, lower-priority Internet traffic. Figure 12–3 illustrates a tunneling network example with tunneling between security gateways and between a security gateway and a client workstation.

Tunneling that involves three types of protocols include (1) the passenger protocol, (2) the encapsulating protocol, and (3) the carrier protocol. The passenger protocol is that which is being encapsulated. In a dial-up scenario, this protocol could be **Serial Line Internet Protocol (SLIP)**, **Point-to-Point Protocol (PPP),** or text dialog. The encapsulating protocol is used to create, maintain, and tear down the tunnel. The carrier protocol is used to carry the encapsulated protocol. IP is the first carrier protocol used

Serial Line Internet Protocol (SLIP)
Point-to-Point Protocol (PPP)

Figure 12–3
LAN and Client VPN Tunnels

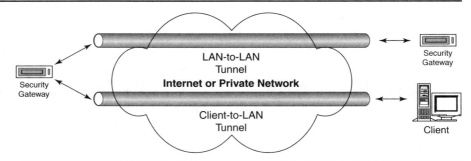

by the L2F Protocol because of its robust routing capabilities, ubiquitous support across different media, and deployment within the Internet.

No dependency exists between the L2F Protocol and IP. In subsequent releases of the L2F functionality, Frame Relay, X.25, and ATM virtual circuits could be used as a direct Layer 2, carrier protocol for the tunnel.

To create a tunnel, the source end encapsulates its packets in IP packets for transit across the public network. For VPNs, the encapsulation may include encrypting the original packet and adding a new IP header to the packet. At the receiving end, the gateway removes the IP header and decrypts the packet if necessary, forwarding the original packet to its destination. The new IP header includes an **authentication header (AH)** and an **encapsulating security payload (ESP)**. As presented in Chapter 7, the AH provides support for data integrity and authentication of IP packets. Security features are as follows (Stalling 2000):

- The data integrity feature ensures that undetected modification to a packet's content is not possible.
- The authentication feature enables an end system or network to authenticate the user or application and filter traffic accordingly.
- Address spoofing attacks are prevented.
- Replay attacks are defeated.

Both AH and ESP support transport and tunnel mode communications.

authentication header (AH)
encapsulating security payload (ESP)

Security Services

Equally important to a VPN's use is the issue of privacy or security. In its most basic use, the *private* in VPN means that a tunnel between two users on a VPN appears as a private link, even when running over shared media. For business use, especially for LAN-LAN links, private has to mean security that is free from prying eyes and tampering. VPNs need to provide four critical functions to ensure the security of data. These functions are as follows:

- **Authentication**—Ensuring that the data are coming from the claimed source.
- **Access control**—Restricting unauthorized use from gaining admission to the network.
- **Confidentiality**—Preventing anyone from reading or copying as the data travels across the network.
- **Security**—Ensuring that no one tampers with the data as the data travel the network.

authentication

access control

confidentiality

security

Although tunnels ease the transmission of the data across the network, authenticating users and maintaining the integrity of the data depend on cryptographic procedures such as digital signatures and encryption. These procedures use shared secrets called keys, which have to be managed and distributed with care.

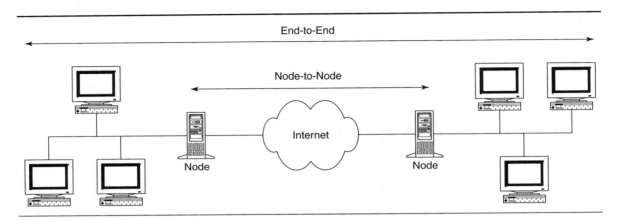

Figure 12–4
End-to-End versus Node-to-Node Security

Security is implemented at the lower levels of the OSI model: the Data Link and Network layers. Deploying security services at these OSI layers makes much of the security services transparent to the user. Implementation of security at these levels can take two forms, which affect the user's responsibility for securing personal data. Security can be implemented either for end-to-end communications, such as communication between two computers, or between other network components, such as firewalls and routers. The latter case refers to node-to-node security and is depicted in Figure 12–4.

Using security on a node-to-node basis can make the security services more transparent to the end users and relieve them of some of the heavy-duty computational requirements such as encryption. Node-to-node security, however, requires that the networks behind the node be trusted networks, which means that they are secure against other attacks of unauthorized use. End-to-end security, because it involves each host, is inherently more sound than node-to-node security. End-to-end security, however, increases the complexity for the end user and can be more challenging to manage.

The traditional security policy for a centrally located computer system identified all the assets in the organization's information infrastructure that required protection. With the advent of distributed information systems, security policies had to embrace departmental LANs as well. This meant adding policies for accessing resources belonging to different departments who had diverse requirements. Since security must be woven into any distributed system design, a number of questions need to be asked before proceeding with any distributed system. Some key questions are as follows (Kosiur 1998):

- Which Internet services does the organization plan to use?
- Which departments within the organization will be part of this network?
- Will access to network services be accessible remotely or locally?
- What security methods will be required and supported?
- What are the risks for providing distributed access?
- What is the cost to provide secure distributed access?
- How will security impact usability?
- What will be the availability of the network resources?
- What are the backup capabilities of the network?
- Will users and employees require any training to use the system?

12.7 VPN CONNECTIVITY AND DESIGN

Figure 12–5 depicts a simple VPN network. The VPN server for each network is connected through the Internet or public network via tunneling. The remote workstation is also connected to the VPN server through the Internet or public network via the tunneling technique.

Basically, a virtual circuit connection is set up on a network between a sender and a receiver in which both the route for the session and the bandwidth are allocated dynamically. VPNs can be established between two or more LANs or between remote users and a LAN.

Access is dedicated (on-net) or switched (off-net). Calls placed over the network are rated in three categories: (1) on-net to on-net, (2) on-net to off-net, and (3) off-net to off-net. The on-net portion of the calls does not incur LEC access charges, and thus reduces the cost of the call.

A VPN consists of both hardware and software components. Major hardware devices include routers and firewalls that contain some software component.

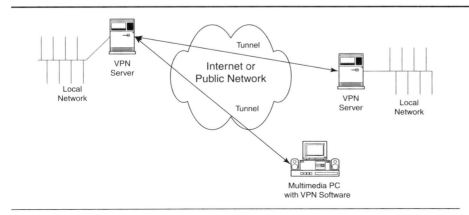

Figure 12–5
VPN Network with Tunneling

12.8 VPN HARDWARE AND SOFTWARE

After the user has a connection to the Internet or the public access network, the important network devices for the VPN are the ones that control access to the protected LAN from remote and external sources. The external source might be another of the corporate LAN, a mobile worker with a laptop, or a corporate partner. VPN access devices should be able to handle all of these situations; however, not all are equally adept at handling the different connectivity situations.

VPN hardware and software can be placed at various locations in the network. These include security gateways, policy servers, and certificate authority holders, all of which prevent unauthorized intrusions. This section looks at firewalls and routers, which are implemented in the VPN to provide access control.

Firewalls and Routers

firewalls

Firewalls have long been used to protect the LAN from other parts of an IP internetwork by controlling access to resources on the basis of packet type, application type, and IP address. Figure 12–6 shows the locations of routers and firewalls in the VPN. These devices can be placed anywhere between the DCE device and the LAN.

A firewall is a device acting as a network filter that restricts access to a private network from the outside. It implements access controls based on the contents of the packages of data that are transmitted between two parties or devices on the network. A firewall setup on a network Internet server can deny the use of unsafe Internet services and can use filtering to

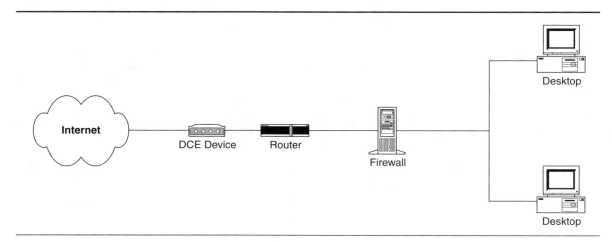

Figure 12–6
Firewall and Router Locations

allow Internet traffic to enter a network only from trusted sites. Internet server firewalls can also filter out unsigned Java and ActiveX applets and known viruses embedded in e-mail attachments.

Internet server firewall protection is valuable, but it does not guarantee complete network security against external attacks. Network Internet server firewalls do not prevent attacks that come through authorized remote access facilities. To block such attacks, it is necessary to place additional firewalls on the computers that communicate with network remote access facilities. Without this protection, attackers may gain control over these remote computers and use them to breach the security of the Enterprise network.

There are three main classes of firewalls: packet filters, application and circuit gateways (proxies), and stateful multilayer inspection firewalls. Each of these will be discussed in detail.

Packet Filtering Firewalls

Packet filtering firewalls were the first generation of firewalls. **Packet filters** track the source and destination address of IP packets permitting packets to pass through the firewall based on rules set by the network manager. Two advantages of packet filter firewalls are they are fairly easy to implement and they are transparent to the end users. They can, however, be difficult to configure properly, particularly if a large number of rules have to be generated to handle a wide variety of applications and users.

packet filters

Packet filtering often requires no separate firewall, because it is often included in most TCP/IP routers at no extra charge; but it is not the best firewall security that can be implemented. One of its deficiencies is that filters are based on IP addresses, which can be forged, and not authenticated user identification.

Packet filters can be used as part of the organization's VPN because they can limit the traffic that passes through a tunnel to another network, based on the protocol and direction of traffic. For example, it is possible to configure a packet filter firewall to disallow File Transfer Protocol (FTP) traffic between two networks, while allowing Hypertext Transfer Protocol (HTTP) and Simple Mail Transfer Protocol (SMTP) traffic between the two, further refining the granularity of the control on protected traffic between sites.

Application and Circuit Gateways

Application and circuit gateways enable users to utilize a **proxy server** to communicate with secure systems, hiding valuable data and servers from potential attackers. The proxy accepts a connection from the other side and, if the connection is permitted, makes a second connection to the destination host on the other side. The client attempting the connection is never directly connected to the destination. Because proxies can act on different types of traffic or packets from different applications, a proxy firewall (server) is usually designed to use proxy agents. In this case, an

proxy server

agent is programmed to handle one specific type of transfer, such as FTP or TCP traffic. The more types of traffic that must pass through the proxy, the more proxy agents need to be loaded and running on the machine.

Circuit proxies focus on the TCP/IP layers, using the network IP connection as a proxy. Circuit proxies are more secure than packet filters because computers on the external network never gain information about internal network IP addresses or ports. A circuit proxy is typically installed between the company's network router and the public network (that is, the Internet), communicating with the public network on behalf of the company's network. Real network addresses can be hidden because only the addresses of the proxy are transmitted to the public network.

Circuit proxies are slower than packet filters because they must reconstruct the IP header to each packet to its correct destination. Also, circuit proxies are not transparent to the end user because they require modified client software.

Stateful Multilayer Inspection

Stateful Multilayer Inspection (SMLI)

A firewall technique called **Stateful Multilayer Inspection (SMLI)** was invented to make security tighter while making it easier and less expensive to use, without slowing performance. SMLI is the foundation of a new generation of firewall products that can be applied across different kinds of protocol boundaries, with an abundance of easy-to-use features and advanced functions.

SMLI is similar to an application proxy in the sense that all levels of the OSI model are examined. Instead of using a proxy, SMLI uses traffic-screen algorithms optimized for high-throughput data parsing. With SMLI, each packet is examined and compared against known states of friendly packets.

One advantage of SMLI is that the firewall closes all TCP ports and then dynamically opens ports when connections require them. Stateful inspection firewalls also provide features such as TCP sequence-number randomization and User Datagram Protocol (UDP) filtering. These firewalls, however, must be supplemented with proxies to support other important functions such as authentication.

Many firewall vendors include a tunnel capability in their products. Like routers, firewalls have to process all IP traffic. Because of all the processing performed by firewalls, they are ill-suited for tunneling on large networks with large amounts of traffic. Combining tunneling and encryption with firewalls is probably best used on small networks with low volumes of traffic. Also, like routers, firewalls can be a single point of failure for the VPN.

Routers

router

A **router** is an intelligent device that connects like and unlike LANs and WANs. Routers are protocol sensitive and typically support multiple protocols. Routers usually operate at Layer 3 of the OSI model and are responsible for deciding which of several paths network traffic will follow.

To do this, a router uses a routing protocol to gain information about the network and algorithms to choose the best route based on several criteria known as routing metrics.

Because routers examine and process every packet that leaves the LAN, it is natural to include packet encryption on routers. Vendors of router-based VPN services usually offer two types of products: either add-on software or an additional feature of a coprocessor-based encryption engine. However, adding the encryption tasks to the same box as the router increases the risks of losing access to the VPN if the router has a failure.

Many of the requirements for an encrypting router are the same as those for firewalls. Encrypting routers are appropriate for VPNs if they have the following features:

- Supports both transport mode and tunnel mode IP Security (IPSec).
- Restricts access by operations personnel to keys.
- Supports a cryptographic key length that best matches the security needs.
- Includes separate network connections for encrypted and unencrypted traffic.
- Supports the default IPSec cryptographic algorithms.
- Supports automatic rekeying at regular periods or for each new connection.
- Logs failures when processing headers and issues alerts for repeated disallowed activities.

Because routers are generally designed to investigate packets at Layer 3 of the OSI model and not to authenticate users, it will probably be necessary to add an authentication server in addition to the encrypting router to create a secure VPN. This server will provide the authentication functions for the VPN.

Another possible VPN solution is to use special hardware that is designed for the tasks of tunneling and encryption. These devices usually operate as encrypting bridges that are typically placed between the network routers and the WAN links. Although most of these hardware devices are designed for LAN-to-LAN configurations, some products also support client-to-LAN tunneling.

VPN Software

Most of the essential functions of VPNs are performed by software on server or client computers. A representative sample of VPN software vendors includes Microsoft, Novell, and Symantec. Beyond basic VPN operations, software products are also available for monitoring and management functions. IBM's VPN Manager provides features such as monitoring and logging, troubleshooting, and policy simulation.

To provide virtual private networking capabilities using the Internet as an Enterprise network backbone, specialized tunneling protocols were developed to establish private, secure channels between connected systems. Four protocols were originally suggested as VPN solutions. Three are designed to work at the OSI Layer 2:

- Microsoft's point-to-point Tunneling Protocol (PPTP)
- Cisco's Layer 2 Forwarding (L2F) protocol
- A combination of the two called Layer 2 Tunneling Protocol (L2TP)

The only VPN protocol for OSI Layer 3 is IPSec, which has been developed by the Internet Engineering Task Force (IETF) over the past few years. An effort is underway to have the IETF combine the two rival standards of PPTP and L2F to form Layer 2 Tunneling Protocol (L2TP). One shortcoming of this proposal is that it does not deal with security issues such as authentication. Figure 12–7 depicts the use of tunneling protocols to build a virtual private network. Various tunneling protocols are used to pass traffic across the VPN, a process that is transparent to the end users. The PPTP, L2F, L2TP, and IPSec protocols are explained in Chapter 7.

12.9 VPN IMPLEMENTATION

Difficulties with in-house VPNs have been slowing their adoption by businesses. VPN software does not always work smoothly with existing hardware and software. Public Key Infrastructures are complex and suffer from interoperability between equipment from different vendors. These difficulties are exacerbated when organizations extend VPNs to form extranets, connecting them with suppliers and other partners.

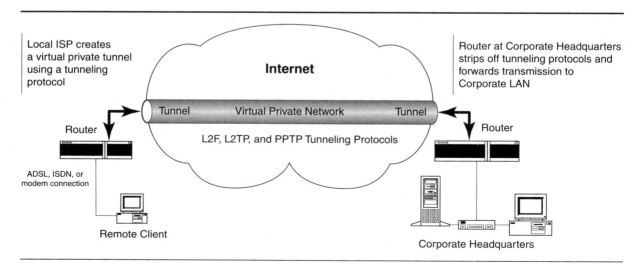

Figure 12–7
Using Tunneling Protocols

Compared to VPNs set up and operated by corporate network personnel, carrier managed VPNs offer the advantages of lower workloads for in-house Information Technology (IT) staff. Since VPN carriers can provide an entire VPN solution for customers—including design, installation, and support—the reduction in requirements for in-house expertise can be substantial.

Carrier Implementation

Carriers today use two basic strategies for building VPNs. The first is to build the VPN at the customer location, using customer equipment to provide VPN capabilities. The network company installs and manages the equipment, ensuring the VPN operates properly. This is called a CPE-based VPN. The alternative strategy is to provide all VPN connections through the carrier's POPs with a gateway device at the customer's location. This is called a network-based VPN and it is seen as providing the greatest possibility for future growth. A network-based VPN provides the same services as a CPE-based VPN, including application proxy firewall, encryption, intrusion detection, and tunnel termination. It, however, integrates these functions into a single carrier-based platform, permitting support of many subscribers simultaneously with customized VPN services.

Carriers implement many VPNs within a single physical network. A corporate-wide ATM WAN provided over a public ATM service network is one example of a VPN. Shared network resources are assigned in fair proportion to the bandwidth required by subscribers and users receive specific QoS guarantees.

In a VPN, a single access circuit from the site to the network is usually sufficient, because multiple virtual circuits and paths can be provided via multiple users from a site to their destinations. For example, each virtual circuit or path can be allocated a peak rate equal to the access circuit but the sum of average rates must be less than the access circuit.

Figure 12–8 illustrates this concept by showing how users A, B, and C at location one have a single physical circuit into their premises ATM device. This ATM device converts these inputs to three ATM virtual circuits and then transmits them over a single physical link between network ATM switches to the destination-based ATM switch (ATM End System). These individual user virtual circuits are logically switched across the ATM network to the destination premises device, where they are delivered to the physical access circuit of the end user. Note that while this single circuit provides good circuit aggregation, it can also be a single point of failure for all users accessing the network, so having backup alternative facilities is a good idea.

The availability of a public switched ATM service, with its bandwidth-on-demand sharing capabilities, will likely lure users away from private point-to-point networks and onto public networks. Users attain both the

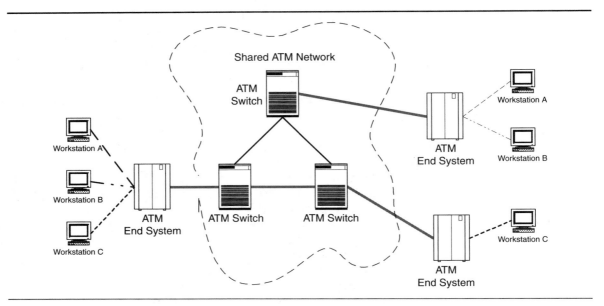

Figure 12–8
Virtual Circuits over a Shared ATM Network

increased reliability of very large public network platforms and cost savings over the private line operations. The cost savings result from the economies of scale of the larger backbone infrastructure with its high availability and resiliency to failure through built-in alternate routing.

Virtual Dial-up Service

The major elements of a virtual dial-up service include authentication and security, authorization, address allocation, and accounting. For the virtual dial-up service, the ISP pursues authentication to the extent required to discover the user's apparent identity and the desired corporate gateway. No password interaction is performed at this point. As soon as the corporate gateway is determined, a connection is initiated with the authentication information gathered by the ISP. The corporate gateway completes the authentication by either accepting or rejecting the connection.

In a virtual dial-up service, the burden of providing detailed authorization based on policy statements is given directly to the remote user's corporation. By allowing end-to-end connectivity between remote users and their corporate gateway, all authorization can be performed as if the remote users are dialed into the corporate location directly.

For the virtual dial-up service, the corporate gateway can exist behind the corporate firewall and allocate addresses that are internal.

Because virtual dialup is an access service, accounting for connection attempts is of significant interest. The corporate gateway can reject new connections based on the authentication information gathered by the ISP, with corresponding logging. Because the corporate gateway can decline a connection based on the ISP information, accounting can easily draw a distinction between a series of failed connection attempts and of brief successful connections.

Dial-up Components

A dial-up service can be implemented to access the VPN infrastructure. The four components of this scenario are the remote user, the **Network Access Server (NAS)**, the Internet Service Provider (ISP), and the corporate gateway.

Network Access Server (NAS)

Remote users access the corporate LAN as if they dialed directly into the corporate gateway, although their physical dial-up is through the ISP NAS. Figure 12–9 provides a step-by-step illustration of this process.

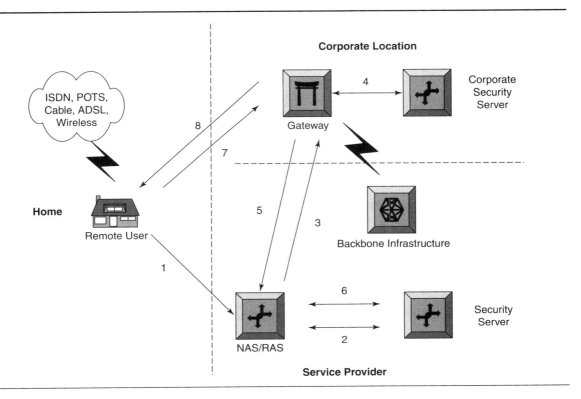

Figure 12–9
Dial-Up VPN Communications

To illustrate how the virtual dial-up works, the following eight steps describe what might happen when a remote user initiates access:

1. Remote user initiates a PPP connection and the NAS accepts the call.
2. The NAS identifies the remote user.
3. The NAS initiates an L2F tunnel to the desired corporate gateway.
4. The corporate gateway authenticates the remote user and accepts or declines the tunnel.
5. The corporate gateway confirms acceptance of the call and the L2F tunnel.
6. The NAS logs the acceptance and traffic (optional).
7. The corporate gateway exchanges PPP negotiations with the remote user. The IP address can be assigned by the corporate gateway at this point.
8. End-to-end data is tunneled from the remote user to the corporate gateway.

12.10 VPN STANDARDS

Currently there are no overall standards outlining the software and hardware components on a VPN. Every vendor that provides a VPN service uses methods best supported by its own hardware platforms and software applications.

Microsoft, 3Com, and several other companies have proposed the standard protocol, PPTP, and Microsoft has even built the protocol into its Windows NT server. VPN software such as Microsoft's PPTP support, as well as any security software, would usually be installed on a company's firewall server. Cisco uses network access servers (NASs) and routers with its internetwork operating system (IOS) software to provide a virtual dialup functionality. This functionality is based on the L2F Protocol IETF draft RFC. The PPTP is the most recent of the remote access protocols defined by the IETF in RFC 1171. PPTP provides on-demand, multiprotocol support for VPNs. PPTP is an extension of PPP that requires two connections to be made. The Layer 2 Tunneling Protocol (L2TP) is a related IETF standard that addresses VPN packet encryption technologies.

A number of IETF working groups are currently addressing issues relating to VPNs. These efforts can be viewed on the IETF Web site at www.ietf.org/html.charters.

IPSec Standards

Internet Protocol Security (IPSec)

Internet Protocol Security (IPSec) standards are designed to ensure secure VPN communications on the Internet. They ensure confidentiality, integrity, and authenticity of data communication. IPSec procedures encrypt each data packet with a choice of algorithms and it sends the packets to an IPSec-compliant device at the receiving end. Both packet and message encryption methods are included in the IPSec standard, as are various VPN authentication and management methods.

IPSec relies on cryptography to protect communications in a variety of environments, including communication links between computers on private networks, links between corporate sites, and links between dial-up users and corporate LANs. It is also used between trading partners for e-commerce applications. IPSec is a tunneling protocol designed for both IPv4 and IPv6. One tunnel can be created to carry all traffic, or multiple tunnels can be created between the same endpoints to support a variety of TCP services.

Implementation of IPSec has been slow, partly because it was originally designed for IPv6 and because there are interoperability problems between vendor products. Encryption is processor intensive and may not be supportable in some networking environments.

12.11 VPN APPLICATIONS

VPNs have become popular because they give both mobile users and telecommuting employees remote access to corporate networks at a lower cost than traditional WANs. They reduce the need for remote access servers and modem banks at the corporate location, and dial-in traffic can be handled via the same method as other Internet traffic.

Virtual networks are economical for large companies that have a considerable amount of on-network calling. Most of the features of a dedicated private network can be provided. Locations can call each other with an abbreviated dialing plan, calls can be restricted from selected areas or country codes, and other dialing privileges can be applied based on trunk group or location. Special billing arrangements are provided. Details of calling activity are furnished online or on magnetic media, which allows the company to analyze long-distance costs.

VPN applications can be considered for the following situations:

- Long-distance usage in excess of 200,000 minutes a month
- Multiple locations, at least one of which is large enough to justify one or more T-1 access lines
- The needs for tie-line services, such as extension number dialing, without the volume to justify a dedicated tie-line network
- Remote users such as clients who are dialing ISDN/PSTN from either home or a remote location
- Internet Service Provider (ISP) access
- Corporate Gateway that provides services to remote users
- Mobile workers that use dial-in connections to ISPs

12.12 VPN ISSUES AND CONSIDERATIONS

Although there is proven demand for Internet-based VPNs, the market is still in its relative infancy, as protocols and devices continue toward standardization. Internet VPNs offer more flexibility and scalability than other alternatives. There is also a reduction in the requirements for both technical support and communications equipment.

Internet VPNs do not use permanent virtual circuits or dedicated leased lines like traditional WANs; instead, the VPN uses the open architecture of the Internet to share data between different locations. Since information is moved along on an untrusted public network, Internet VPNs use encryption technologies to keep data secure and private from unintended recipients. Organizations that employ Internet-based VPNs establish POPs with ISPs so that data can be sent to appropriate destinations.

IP-based VPNs can provide secure communications over packet networks, such as WANs, intranets, and the Internet. Without the use of VPN technology, communications over these networks could be subject to unauthorized interception. Standards that provide compatibility between IP-based VPNs have been developed. These provide a technical basis for wider IP-based VPN availability and use.

VPN hardware and software now make it possible for providers to separately support traffic from different customers, guaranteeing the quality of VPN for corporate customers, capturing information for billing VPN services, and sharing revenues between carriers when VPNs traverse more than one carrier network. The continuing development of Multiprotocol Label Switching (MPLS) is providing further impetus to VPN growth.

There are four technology protocols that can be used in securing VPNs (these protocols are presented in Chapter 7):

- Point-to-Point Tunneling Protocol (PPTP)
- Layer 2 Forwarding Protocol (L2F)
- Layer 2 Tunneling Protocol (L2TP)
- IP Security Protocol (IPSec)

PPTP, L2F, and L2TP are designed primarily for dial-up VPNs, whereas IPSec is used for LAN-to-LAN networks.

Corporations have concerns when using the Internet to conduct business. The design and deployment of an Internet VPN can be impacted by the following issues:

- Reliability
- Congestion
- Throughput
- Security
- Interoperability
- Bottlenecks
- IP address issues

- Performance
- Multiprotocol support
- Integrated solutions

Using VPNs involves some important disadvantages. VPNs have not solved the issues of reliability, congestion, and throughput, which are all problems for corporate networks. Internet VPNs, specifically, still have problems with guaranteed performance and security. There is currently a restriction on the amount of media supported on VPN networks. Encryption is fundamental to maintaining privacy and integrity on the VPN. The management of secret keys can be problematic as the number of corresponding parties grows. Someone in the organization must also manage the infrastructure for distributing digital certificates.

Specific problems and difficulties include the following:

- Different organizations may use the same IP addresses in their IP networks. When network address translation (NAT) is used, IPSec may not be able to verify originator addresses.
- Dynamic routing changes are frequent in many networks. Many VPN devices use statically configured addressing and routing that would require manual updating.
- There are problems with VPN software working with existing networking hardware and software. Differences in interface cards and software versions can require modifications specific to each end user's configuration.
- Public key interfaces (PKI) are complex and there is an interoperability problem between different vendor's hardware. The IPSec protocol may require the same vendor's equipment at every node. PKI difficulties may increase when implementing extranets that connect to suppliers and other partners.

12.13 SECURITY ISSUES

Security has been a concern for virtual network users; however years of research has resulted in solutions that allow transmission of sensitive data over these networks (Ramteke 2001). It is now possible for employees to access data remotely from almost anywhere there is Internet connectivity. It is also possible for customers to access Web sites and transfer credit card information without being concerned about security.

Ensuring security, however, is a major issue when implementing virtual networks, and should not be treated lightly. Security problems can be solved with various techniques, such as tunneling, when implementing a VPN. Securing tunnels for private communications between corporate sites will do little if employee passwords are openly available or if other holes in the security of the network exist. VPN-related security management—of keys and user rights—must be integrated into the rest of the organization's security policies. As noted earlier, several areas of security that are implemented in VPNs are authentication, encryption, integrity, nonrepudiation, and content filtering.

Authentication
Authentication is a vital part of a VPN's security structure. Before users can connect to a site, they must first authenticate themselves; that is, they must be able to prove they are who they say they are. Passwords offer only weak forms of authentication, however VPNs use stronger forms. Strong authentication implements at least two authentication methods, which could include a key or token card, a password, or some biometric method.

The most secure VPN communications use authentication methods to identify the recipient before the message is sent. The purpose of authentication is to ensure messages are sent to legitimate recipients. Chapter 2 provides an in-depth discussion of the identification and authentication process. The simple types of authentication of passwords and PINs are too vulnerable to attackers. Certificates can be used to enhance the protection of the VPN.

Digital certificates based on public/private key pairs are used in message encryption. Each certificate contains a private key that identifies the recipient. A trusted certificate authority (CA), such as an Enterprise network or a commercial certificate organization, produces the private key. The sender can verify the recipient certificate by using a matching public key to decrypt it.

Encryption
In normal packet network use, packets containing information are sent over the network and collected at the receiving address. The message content is removed from each packet and reassembled to form the original message. One method of protecting VPN messages from interception is to use packet encryption. The contents of the packets sent over the network are encrypted, and when received at the far end, are decrypted to the original message. This prevents anyone who intercepts the message along the transmission path from understanding the content.

A more secure method is to encrypt an entire VPN message at the source and send the encrypted message in the form of packets. These packets are collected at the receiving end and the encrypted message is reassembled and then decrypted. Message encryption allows the use of powerful encryption methods without slowing VPN communications. It is also possible for a VPN to use both packet and message encryption because they need not interfere with each other.

There are a number of encryption techniques that can be employed to protect the security of VPN transmissions. These include hashing, secret-key cryptography, public-key cryptography, certificates, and IPSec and SSL protocols. Chapter 3 provides an in-depth analysis of these techniques.

Integrity

The danger exists that someone will change IP packets in transit. **Integrity** checks ensure that packets have not been destroyed, changed, or reordered in their passage from the sender to the receiver. Powerful encryption methods must be used since simple encryption is subject to penetration by attackers. The Data Encryption Standard (DES), presented in Chapter 3, is the most popular encryption method for VPNs, however other encryption methods are also acceptable. VPN standards use public-key encryption. All messages sent to a given user are encrypted using a public key designed specifically for that recipient. Public keys can be made widely available and there is no need to take special measures to protect them since they cannot be used to decrypt messages. Message recipients have a unique and secret private key that matches their public key and only that key can be used to decrypt received messages. The public key method avoids the security exposure caused by exchanging encryption keys between sending and receiving parties.

integrity

Nonrepudiation

What is **nonrepudiation** and why is this concept important to the success of virtual networks? Nonrepudiation is a mechanism that prevents a user from denying a legitimate, billable financial transaction. This mechanism provides for monitoring of all network endpoints during the course of the exchange so any applicable charges might be supported in the event they are challenged. It provides a means of proving that a given message came from the party who sent it.

nonrepudiation

Content Filtering

A danger exists where a user is authenticated who may send unauthorized requests to the network resource. To prevent inappropriate behavior after authentication, it is necessary to pass each subsequent message from the user through a **content filter** (Panko 1999). Some companies check transmission for improper content such as viruses. If the computers used at the sending and receiving ends of the VPN are vulnerable to outside attacks, the security offered by the VPN can be undermined. One widely used method of securing these computers is to employ firewalls. Many products now bundle firewall and VPN protection into one solution.

content filter

Hacker Vulnerabilities

Tests on VPN security have revealed a number of vulnerabilities and holes that must be corrected to ensure a secure environment. These are oriented primarily toward PPTP client/server interactions and include the following (McClure 2001):

- Clients connecting to networks via PPTP servers could act as a backdoor onto these networks.

- The control channel for negotiating and managing connections is unauthenticated and is vulnerable to DoS and spoofing attacks.
- Eavesdroppers can obtain useful information from the control channel traffic because only the data payload is encrypted.
- Reuse of send and receive session keys can make the network vulnerable to a common cryptographic attack.
- Use of session keys to encrypt network data, which are generated from user-supplied passwords, decreases the practical key strength.
- MS-CHAP relies on legacy cryptographic functions that have previously been defeated.

12.14 VOICE OVER INTERNET PROTOCOL (VoIP)

In the past, Information Technology departments did not wish to be involved in the voice service requirements for an organization. This led to splitting the telecommunications functions between a department for voice services and a department for data processing and data communications services. In today's high-technology world, it will not be long before voice and data are both delivered over the organization's data network using digital signaling technology. This service is called **Voice over Internet Protocol (VoIP)**. As the quality of VoIP improves, as standards are developed, and as the technology matures, VoIP will be used to replace the analog telephone set. Figure 12–10 provides a general network overview of a VoIP environment.

VoIP Overview

Traditionally, voice traffic was transmitted over a circuit switched network and data traffic was transmitted over a packet-switched network. VoIP is the transmission of voice traffic over an IP network but it is not the same as **Internet Telephony**. VoIP is voice transmitted as packets over a data network, whereas Internet Telephony is voice sent as packets over the public Internet. The Internet Protocol (IP) is a standard for data transmission that is based on the packet switching technology. The three main classes of IP networks in operation today include the following:

- IP networks owned and managed by carriers (operators)
- Open and public Internet
- IP networks for closed user groups such as WANs and intranets

Currently, voice communication is provided primarily over circuit-switched networks, which is a technology that lends itself to services requiring high quality and minimum delays. IP networks, however, are designed for the transmission of data, where delays and occasional data loss is less critical. When deploying VoIP, vendors must therefore overcome a number of issues to ensure acceptable performance. Despite the complexities, the current trend is toward using corporate IP networks for

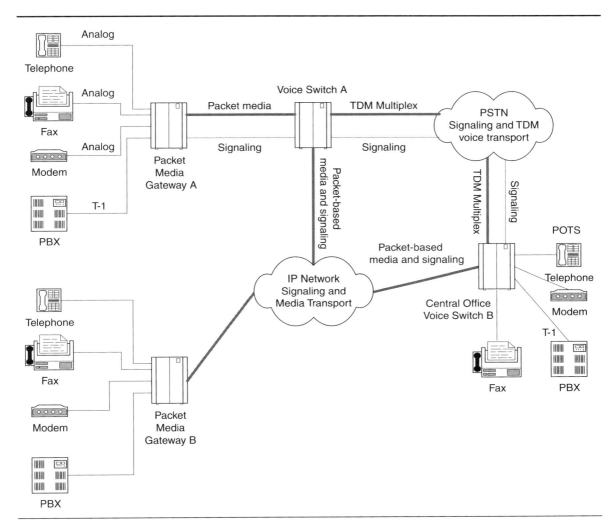

Figure 12–10
VoIP Network Environment

voice and data transmission. This increased interest in VoIP on the corporate network is driven by the opportunity to combine today's data and voice networks into one single network. This combination could result in significant cost reductions, an improved ability to manage the network, and easier integration with support and maintenance systems.

IP Telephony Technology Overview

Today's networks consist of circuit-switched and packet-switched facilities. Most voice traffic is circuit switched and is transmitted over a public-switched telephone network (PSTN). The speed that voice traffic is transmitted consists of an aggregate rate of 64 kbps. A direct connection between two connection points provides a permanent 64 kbps link for the duration of the call. This link is not available for any other purpose during this connection. PSTN provides low delay or latency and is bidirectional, thus allowing for two-way or full-duplex conversations to take place.

In a packet-switched network, data are divided into packets, each one with a destination address. When the packets are transmitted through the network, the addresses are read at each network node for routing information. At the destination, the packets are reassembled and resequenced. Depending on the traffic levels and congestion in the network, packets may take different routes to the end destination. Packet switching provides a virtual circuit connection and is usually half duplex. There is no dedicated connection as is required for circuit switching, therefore it is called a connectionless network.

The common standard for interconnection between technologically diverse networks is the TCP/IP suite of protocols. In the TCP/IP stack, IP provides for the transportation of information; and TCP is concerned with fragmentation of the message, retransmission if required, acknowledgment of delivery, flow control, and reassembly. The layers above TCP are the application specific protocols such as Simple Mail Transfer Protocol (SMTP) for e-mail. Table 12–1 summarizes the functionality of the TCP/IP protocol suite.

The process for transmission of voice traffic over an IP network is as follows:

- The caller uses the PSTN and dials the access number of the IP voice gateway.
- On authentication of the calling party, the caller dials the number of the desired destination.
- In the gateway, the voice signal is digitized, compressed, and converted into IP packets.
- The IP packets are transmitted over an IP network that is also shared with other IP traffic.

Table 12–1 Functionality of TCP/IP Suite Layers

TCP/IP Layer	Function
Application Layer	Manages the details related to specific applications. The application program elects the kind of transport needed and passes it to the transport layer.
Transport Layer	The software segments the data to be transmitted into small packets and adds addresses. There are two transport protocols—UDP and TCP.
Network Layer (Internet)	Affects the movement of packets through the network and manages the routing of packets from node to node. Examples include SMTP and FTP.

IP Telephony Gateway

The gateway is located between the circuit-switched and the packet-switched networks and performs the functionality for enabling voice traffic to be transmitted across different networks and technologies. Voice calls are digitized, encoded, compressed, and packetized in the originating gateway. At the destination gateway, the process is reversed. These gateways typically provide a specified number of analog or digital port interface connections on one side and a 10 Mbps or 100 Mbps Ethernet interface connection on the other side. The gateway provides the following functions:

- The interface and signaling between the networks
- Voice processing functions such as call setup and tear-down
- Translation between telephone numbers and IP addresses
- Compression and decompression of voice signals
- Packetizing
- Echo control
- Silence suppression
- Forward error correction
- Jitter-buffer techniques
- QoS

Via the H323v2 protocol, the VoIP gateways are connected to a VoIP gatekeeper, which serves as a system controller. The VoIP gatekeeper provides for caller authentication, call accounting information, billing plans, and routing tables.

VoIP Quality

Phone calls over the Internet rarely achieve the quality of circuit-switched calls, however the quality is improving. The quality of VoIP calls appears to be as good as or better than that of cellular calls and many users will accept the reduction in quality for a reduction in cost.

Quality of IP telephony is impacted by latency between the two parties. There is also a variation in this delay, which is called jitter. When the amount of delay changes as the call progresses, the effect is annoying. Real-time Transport Protocol (RTP) is supposed to reduce this jitter problem.

12.15 INTERNET AND WEB RESOURCES

IETF	www.ietf.org/html.charters
IBM	www.ibm.com
Microsoft	www.microsoft.com
Novell	www.novell.com
RSA Security	www.rsasecurity.com
Symantec	www.symantec.com
VPN Consortium (VPNC)	www.vpnc.org

■ SUMMARY

Virtual network technology allows a company with multiple sites to have a private network but use a public network as a carrier. Although the company can use the public network as a link between its sites, this technology restricts traffic to packets that can travel only between the company's sites. Even if an outsider accidentally receives a copy of a packet, virtual network security technology prevents the understanding of its contents.

To build a Virtual Private Network (VPN), a company buys a special hardware and software system for each of its sites. The system is placed between the company's private, internal networks and the public network. Each of the systems must be configured with the addresses of the company's other VPN systems. The software will then exchange packets only with the VPN systems at the company's other sites. To guarantee privacy, VPN encrypts each packet before transmission.

The systems administrator uses various tools to ensure the security and integrity of the VPN. Three types of firewalls are used in a VPN: packet filters, proxies, and stateful inspection systems. Each of these systems differs in the security they provide, their difficulty in configuration, and their performance.

In addition to configuring the VPN system at each site, a network manager must also configure routing at the site. When a computer at one site sends a packet to a computer at another site, the packet is routed to the local VPN system. The VPN system examines the destination, encrypts the packet, and sends the result across the public network to the VPN system at the destination site. When the packet arrives, the receiving VPN system verifies that it came from a valid peer, decrypts the contents, and forwards the packet to its destination.

Three types of IP VPNs include Frame Relay, CPE-based IP, and network-based IP. An important advantage of carrier VPN providers in they can connect VPNs over their own backbone networks, which avoids interfacing to the Internet. Carrier-based VPNs provide the same security and reliability as a private VPN. Users of carrier VPNs must be careful to maintain sufficient expert staff to evaluate and monitor the effectiveness and security of the VPN operations they entrust to outside providers.

Basically, a VPN combines the advantages of private and public networks by allowing a company with multiple sites to have the illusion of a completely private network and by using a public network to carry traffic between locations. To take full advantage of VPN technology and realize the potential savings, it is necessary that all organizations become familiar with the technology.

Voice traffic can be transmitted over an IP-based network—called Voice over IP (VoIP). The current trend is toward using corporate IP networks for voice and data transmission. This increased interest in VoIP on the corporate network is driven by the opportunity to combine today's data and voice networks into one single network. This combination could result in significant cost reductions, however security issues must be addressed.

Key Terms

Access control
Access VPN
Authentication
Authentication Header (AH)
Compulsory tunnel
Confidentiality
Content filter
Encryption
Encapsulating Security Payload (ESP)
Extranet VPN
Firewall
Integrity
Internet Protocol Security (IPSec)
Internet Telephony
Intranet VPN
Network Access Server (NAS)

Nonrepudiation
Packet filters
Point-to-Point Tunneling Protocol (PPTP)
Point-to-Point Protocol (PPP)
Proxy server
Remote Access Server (RAS)
Router
Security
Serial Line Internet Protocol (SLIP)
Service control point (SCP)
Service switching point (SSP)
Stateful Multilayer Inspection (SMLI)
Tunneling
Virtual Private Network (VPN)
Voice over Internet Protocol (VoIP)
Voluntary tunnel

REVIEW EXERCISES

Questions and Analysis

Provide a short answer or in-depth analysis as required.

1. Describe the new workforce.
2. What are the two approaches for a telecommuting network?
3. Provide a general overview of a VPN.
4. Why would a company want to implement a VPN?
5. What concerns do organizations have about using the Internet to conduct business?
6. What are two advantages of dial-up VPNs accessing Remote Access Servers that are used to connect to ISPs.
7. What are the three basic types of VPNs? How are they used?
8. Two fundamental concepts comprise the VPN architecture. Provide a brief description of each.
9. Define tunneling. How is it used in the VPN environment?
10. What is the difference between voluntary and compulsory tunnels?
11. What are the three types of protocols used in tunneling? How are they used?
12. Draw a simple sketch of tunnels in the Internet environment.

13. What is the encapsulation process? What is the format of an IP packet that has been encapsulated?

14. What are the four security functions that are necessary for securing data across the VPN?

15. Give examples of questions that must be asked when implementing a distributed network.

16. How would an Internet VPN be implemented? What part would an ISP and POP play in this connectivity?

17. Provide several reasons why anyone would want to implement an Internet VPN.

18. What are the hardware and software components that might be found in a VPN?

19. Where would a proxy server be deployed? What is its purpose?

20. Where is a network access server used in a VPN?

21. Where does an ISP fit in the VPN environment?

22. What is a firewall? What are the three main classes of firewalls? How are they different?

23. What is a router? Give a general description of the part that a router plays in the VPN environment.

24. How does the functionality of a router differ from a firewall? What is the difference between a firewall and a gateway server?

25. Why would an encrypting router be beneficial in the VPN environment?

26. Four protocols were originally suggested for use in VPNs. List them and give a brief description of each.

27. How are VPNs implemented in the carrier environment?

28. Installing a VPN in the organization involves a number of issues. Discuss several in the context of a particular industry.

29. What is meant by authentication? What is the difference between authentication and authorization?

30. Discuss the encryption process as it applies to system integrity.

Definitions and Terms

Provide the most correct answer from the list of key terms.

1. Limiting the flow of information from the resources of a system only to authorized persons, programs, or processes in a network.

2. The process of validating the claimed identity of an end user or a device.

3. A header included in IPSec, which provides authentication and integrity of the IP packet.

4. Ensuring that information is not disclosed to people who are not explicitly intended to receive it.

5. A method of scrambling information in such a way that it is not readable by anyone except the intended recipient.

6. A system, based on either hardware or software, used to regulate traffic between two networks.

7. A property of a cryptographic system that prevents a sender from denying later that a message was sent or a certain action performed.

8. The limiting of TCP/IP traffic, based on the type of traffic, source and destination addresses, and ports.

9. Like a firewall, it is designed to protect internal resources on networks that are connected to other networks such as the Internet.

10. A piece of computer hardware that sits on a corporate LAN. Employees use the PSTN to get access to their e-mail, etc.

11. An OSI Layer 3 intelligent device that connects like and unlike LANs and WANs.

12. A way of ensuring data on a network is protected from unauthorized use.

13. Provides a secure, temporary path over the Internet or another network.

14. The foundation of a new generation of firewall products that can be applied across different kinds of protocol boundaries, with an abundance of easy-to-use features and advanced functions.

15. A software-defined network offering the appearance, functionality, and usefulness of a dedicated private network.

True/False Statements

Indicate whether the statement is true or false.

1. The database for a VPN resides in an SSP.
2. Three types of IP LANs include an access VPN, an intranet VPN, and an extranet VPN.
3. Tunneling allows for the encapsulating of packets that flow across a network.
4. An authentication header and an encapsulating security payload is part of an IP payload frame.
5. Four security functions provided by VPNs include authorization, access control, confidentiality, and security.
6. A firewall is a filtering device that can be employed to restrict network traffic.
7. Three main classifications of firewalls include packet filters, proxies, and SMLI devices.
8. Circuit proxies are more secure than packet filters; however, circuit filters are also faster than packet filters.
9. An encrypting router can be used in the place of a firewall if the device incorporates support for VPNs.
10. Four protocols used in VPNs include PPP, PPTP, IPSec, and L2TP.
11. Four components for VPN dial-up service include the NAS, the ISP, the corporate gateway, and the remote user.
12. IPSec standards are designed to ensure secure VPN communications over the Internet.
13. IPSec is designed primarily for dial-up VPNs, whereas PPTP is designed for LAN-to-LAN networks.
14. Security methods implemented in VPNs include repudiation, decryption, authorization, and content filtering.
15. VoIP is implemented over the Internet by using analog technology.

Activities and Research

1. Identify standards and specifications that are associated with VPNs. Create a matrix that identifies common categories of features and functions.
2. Research and prepare a two-page report concerning the pros and cons for implementing a VPN.
3. Develop a list of ten security issues that must be considered when using a VPN. Describe why they are important.
4. Identify hardware and software components that are utilized in a VPN. Show how they are used.
5. Research and provide a comparison matrix of firewall products used in a VPN. Show vendor, product description, features, and cost.
6. Compare the protocols suggested for VPN solutions. Show the advantages and disadvantages of each.
7. Contact the various carriers and determine if VPNs are offered in your locality. Identify costs and implementation issues.
8. Design and price a VPN solution for some hypothetical organization.
9. Identify applications that could be used over a VPN for a particular business sector (e.g., transportation). Show the benefits of each.

CHAPTER

13

Distributed Systems Security

OBJECTIVES

Material included in this chapter should enable the reader to

- Become familiar with the distributed networking environment and the security issues that must be addressed.
- Recognize the various technologies and security techniques that are used in the e-commerce environment.
- Learn the terms and definitions associated with the electronic commerce industry.
- Become familiar with the various financial and online transaction systems that are part of the e-commerce environment.
- Identify the hardware and software components that comprise the e-commerce computing network.
- Become familiar with the threats and other malicious activities that can impact e-commerce applications.
- Evaluate the various countermeasures that can be deployed to provide for secure electronic commerce.

INTRODUCTION

With the advent of the Internet connectivity to local area network topologies such as intranets and extranets, it should not be a surprise that commercial interests would want to exploit communications by network computers by way of e-commerce. The implications and ramifications to the organization and its resources are far reaching.

Personal information such as credit card and social security numbers must be protected. Major issues include protection of data and information storage and the methods used to ensure confidentiality and integrity. The Enterprise organization must apply security principles and methods, which include authentication and authorization techniques. Integrity of the database and the software that processes the transactions must be assured.

One aspect of e-commerce is using the Internet for credit card purchases of many items including airline tickets, computer hardware and software, and books. E-commerce can also secure transactions, authorize payments, and move money between accounts. A number of processes are used to accomplish these transactions.

Because millions of dollars are at stake with e-commerce transactions, their security is a major issue in today's business environment. When considering e-commerce security, an organization must thoroughly evaluate all aspects of its network access, since users expect a secure and private exchange of information.

The boom in the e-commerce arena has spawned a plethora of general security and secure-payment schemes for credit card users and financial institutions. The main players are SET, SSL, and TLS protocols. In addition, a number of proprietary schemes also exist that serve special purposes, such as EDI, or provide security features within a specific product line.

Security management becomes both more difficult and more critical when users have multiple accounts on different computer systems. With systems that provide an open connectivity to the Internet, it is essential that security systems be in place that can provide for a central administrative capability. A broad range of authentication and authorization services and management tools are needed for an effective security program. This is especially true when transporting data from one platform in an Enterprise network to a different platform in another.

This chapter explores the subject of electronic commerce and its implication to the computer and networking operations that would process e-commerce transactions. Considerable emphasis is placed on the methods and procedures that can be used to ensure integrity, privacy, and security of the organization's assets and resources.

13.1 THE DISTRIBUTED COMPUTING ENVIRONMENT

How does one describe the distributed computer and networking environment? There are many entities and components that makeup this complex environment. Indeed, information presented in all the previous chapters is relevant and part of this discussion. There are a number of terms and definitions that must be understood before proceeding. They include the following (Held 1996; Newton 1999):

- **Distributed application**—An application in which some of the processing, storage, and control functions are situated in different places and connected by transmission facilities.
- **Distributed architecture**—A bus or ring LAN that uses shared communication media and shared access methods.
- **Distributed computing environment**—A comprehensive integrated set of services that supports the development, use, and maintenance of distributed applications.
- **Distributed data processing**—A data processing arrangement where the computers are decentralized, or located apart from each other.
- **Distributed environment**—A network environment, or topology, where decision making, file storage, and other network functions are not centralized but instead are found throughout the network.
- **Distributed file system**—A type of file system where the file system itself manages and transparently locates pieces of information from remote files and distributes files across a network.
- **Distributed management environment**—A compilation of technologies now being selected by the Open Software Foundation (OSF) to create a unified network and systems management framework, as well as applications.
- **Distributed name service**—A technique for storing network node name information throughout the network; the information can be requested from, and supplied by, any node.
- **Distributed network**—A communications network comprised of two or more processing centers. The processing may be distributed between each center and each processor must be in communications with all others.
- **Distributed processing**—A network of computers where the processing of information is initiated in local computers, and the resultant data is sent to a central computer for further processing with the data of other local systems.

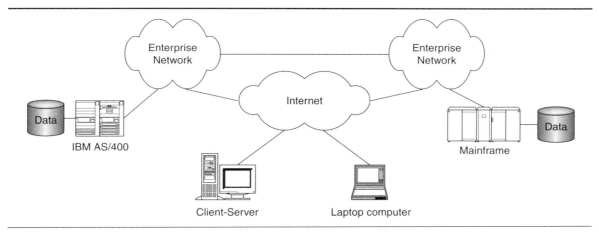

Figure 13–1
Distributed Enterprise Computer Network

As depicted in Figure 13–1 a distributed Enterprise computer network can include a number of different platforms located across a wide geographic area, accessible over a corporate network or the Internet. These include, but are not limited to, laptop computers, servers, and computer systems.

The distributed environment contains a variety of complex components and, therefore, has numerous security issues. A concept that fits in all these environments is electronic commerce (e-commerce). Until recently, e-commerce depended mostly on expensive, private, value-added networks. They were highly secure, however they were difficult to implement and were proprietary. Using the Internet is more open, however it is also more vulnerable.

13.2　E-COMMERCE OVERVIEW

E-commerce over the Internet allows for credit card purchases of such items as airline tickets, computer hardware and software, books, and miscellaneous products. E-commerce also provides methods for securing transactions, authorizing payments, and moving money between accounts. One process used to accomplish e-commerce transactions is called **electronic data interchange (EDI)**. EDI is a process whereby standardized e-commerce documents are transferred between diverse and remotely located computer systems (Figure 13–2). These documents include purchase orders and invoices. The form and format of EDI documents may be defined by vendor specifications or standards organizations, such as ITU-T and ANSI. Electronic mail (e-mail) also plays a major role in e-commerce.

EDI and other forms of business transactions have been taking place over public and private networks for some time. The financial system,

e-commerce

electronic data interchange (EDI)

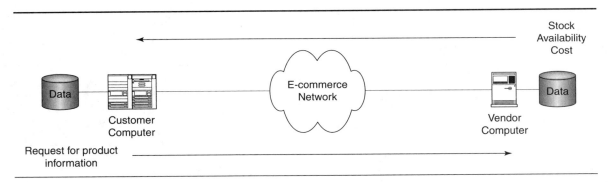

Figure 13–2
EDI Environment

which consists of banks and bank branches, executes transactions and moves money over the telecommunications network on a regular basis. Credit- and debit-card transactions are approved over dial-up lines or dedicated telephone circuits. Millions of stock market transactions take place every day using electronic means. All of these transactions must possess a high degree of integrity.

Because these transactions are occurring over a public network that is prone to abuse, methods and technologies must be in place to ensure security. Most buyers and sellers using the Internet have had no previous relationship; they do not know who to trust. The element must be established in some way.

13.3 E-COMMERCE INFRASTRUCTURE

Chapter 10 presents background information on intranet and extranet LANs that form the infrastructure for Internet and Web-based commerce activities. This infrastructure must be designed to support commerce activities between organizations, within organizations, and between organizations to clientele (Stallings 2001).

Intranet

intranet

As presented in Chapter 10, **intranet** activity takes place between members of the same organization. The intranet must extend not just to workers at the central location, but to all remote sites and to telecommuters. This becomes complex because WAN access, Internet access, and even dial-up access may be required. Chapter 11 describes the components of the wide area network. It is necessary to provide offsite workers with the same level of service as local users, while maintaining the security and integrity levels of a private network. The intranet requirements are easy to meet because all locations are known and the application usage is predictable.

Extranet

Chapter 10 also provides details concerning the development and utilization of the extranet topology. An **extranet** typically includes Web sites that provide information to internal employees and have secure areas to provide information to customers and external partners like suppliers, manufacturers, and distributors. Most business-to-business (B2B) applications are built on a secure extranet, where a logon and password is required for access. Security is essential because the organization is opening up its network to a variety of outsiders. The infrastructure must allow certain data to flow freely while blocking confidential data and unauthorized access. Security and integrity of the intranet can be assured by the use of a **firewall** component.

Figure 13–3 depicts intranet configurations at both the corporate location and the vendor location. These two diverse networks communicate with each other over an extranet using the EDI application.

Firewalls are intended to provide protection for an internal network against unauthorized access from an untrusted network such as the **Internet**. This is accomplished by blocking or preventing some traffic and passing or permitting other traffic. Because a firewall primarily filters traffic, a common implementation is to employ packet filtering on the interfaces by using routers. Detailed descriptions of firewall functions and solutions are described in Chapter 6.

World Wide Web

A business-to-consumer network is the most visible type of e-commerce. It is also the most familiar to the average user since it is accessible by way of the **World Wide Web (WWW)**. The business-to-consumer network is based on the use of Web sites from which consumers can make purchases of goods and services. It is difficult to ensure a high level of security and integrity on this network. The consumers and the vendors may not be known. The originating location of the transaction and its transmission path may be unknown. There are many challenges for the computer and network administrator.

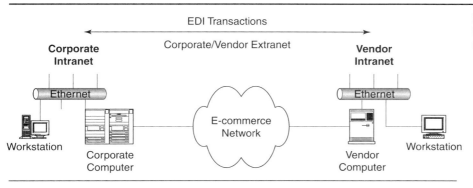

Figure 13–3
Intranet and Extranet Connectivity

computer telephone integration (CTI)

Computer Telephone Integration

Computer telephone integration (CTI), presented in Chapter 6, provides a link between the voice systems and databases on the computer systems. E-commerce transactions can be completed over the Internet or over the ubiquitous telephone system. Agents working in an ACD call center process many of these voice-oriented transactions. The database records are accessed and modified from both sources, therefore security issues relating to computer systems must also apply to voice systems. An overview follows that provides an understanding of the relationship of CTI to e-commerce.

Figure 13–4 depicts the components that are part of a CTI system. Note that the primary devices are the PBX and/or an automatic call distributor, call center, computer system, and database. Access to the CTI system is via a LAN, MAN, or WAN topology. These topologies are discussed in Chapters 10 and 11. Novell created Telephony Services API (TSAPI) and Microsoft created Telephony API (TAPI) for CTI applications. TSAPI was enhanced in 1996 by a vendor consortium called Versit. CTI protocols allow a computer to manipulate telephone devices that are connected to telephone circuits. In some cases, telephony devices might be connected over a network, but the signals that cross the network are for control.

interactive voice response (IVR)

automatic voice recognition (AVR)

Interactive voice response (IVR) is a CTI application. It is the front-end computerized operator that guides the caller through keypad options. It is possible to have prompts and messages changed based on programmed times and dates. An enhancement to IVR is **automatic voice recognition (AVR)**.

Figure 13–4
CTI Environment

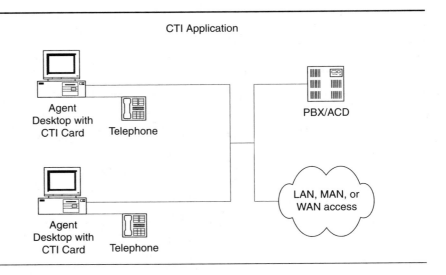

Universal in-boxes are another aspect of CTI. Also, CTI can provide services such as voice mail, faxing, and e-mail on a company-wide basis. Information is stored on disk and is accessible to users who dial into the server by telephone or workstation access.

Screen synchronization (**screen pops**) is a major function of the CTI system. It relieves the call-center agent from asking for a customer identifier such as an account number or social security number and then entering it via the keyboard. Additionally, if automatic number identifier (ANI) is used to identify the caller device, cost savings are realized by a reduction in time required to acquire the customer's database record (Green 1996).

screen pops

An important part of CTI is the ability to collect information from a telephone call and store it in a database for future use. When a call requires more than one session, the agent can pick up the call where the previous conversation stopped. CTI makes it possible to provide personalized customer services based on knowledge of a caller's preferences, which are stored in the computer's database. This database is an important asset to the organization and must be protected from malicious activities. This information, in the wrong hands, could prove costly to the organization.

13.4 E-COMMERCE SECURITY

There is considerable apprehension concerning the lack of security on the Internet. There are a number of initiatives to set security standards for electronic commerce. To transact business, organizations must have a means to conduct secure, confidential, and reliable transactions. To achieve this goal, four cornerstones of secure electronic transactions must be present. These include the following attributes:

- Possess the ability to prevent unauthorized monitoring of traffic.
- Prevent the content of messages from being altered after transmission and provide a method to prove they have or have not been altered.
- Determine whether a transmission is from an authentic source or a masquerade.
- Prevent a sender from denying the receipt or transmission of a message (nonrepudiation).

Cryptographic systems provide a method to transmit information across an untrusted communication system without disclosing the content of the information to a snoop. Cryptographic systems provide confidentiality and can also provide proof that a transmission's integrity has not been compromised. Chapter 3 provides details concerning specific cryptographic systems.

cryptographic systems

Securing Electronic Transactions

Individuals, organizations, industries, and commerce use the Internet to transmit e-mail, purchase goods and services, transact banking, and perform other unique services that depend on secure transmissions. There are methods that can provide secure transmissions for each (Coulouris 2001).

E-mail often contains confidential information such as credit card numbers or social security numbers. It is essential that the contents and the sender of a message be authenticated. An e-mail might be considered a legal document when conducting a transaction such as responding to an official competitive bid, therefore it is essential that both the sender and the recipient be assured of the integrity of the transmission.

Many network users purchase goods and services over the Internet. Buyers select the items and authorize payment using the Web. These items are then delivered through the normal shipment channel. Typical items include books, CDs, and auction items. Software and recordings can be delivered immediately by downloading through the Internet. All of these transactions must be validated and tracked.

It is now possible to conduct banking transactions through an electronic bank. These banks offer users virtually all the services and functions provided by conventional banks. Functions include checking of account balances and statements, transferring funds between accounts, and electronic payments. Authentication and authorization are two security functions that must occur with these transactions. The user must be identifiable and have authority to access a particular account with a particular transaction.

There are a number of Internet transactions that fall in the general category. The Internet lends itself to the supply of small quantities of information and other services to many users. The development of the Web as a high-quality publishing medium depends on the extent to which information suppliers of books and documents can obtain payments from users for that information. The use of the Internet for voice and video-conferencing is another service that is likely to be supplied when it is paid for by end-users.

Transactions such as these can be safely executed only when protected by appropriate security policies and mechanisms. A client must be protected against the disclosure of credit and debit card numbers during transmission and against a fraudulent vendor who obtains payment without delivering the goods or service. Conversely, vendors must obtain payment before releasing the items selected for purchase, whether they are supplied online or through a normal delivery method. In addition, the cost to achieve the required protection must be reasonable in comparison to the value of the transaction.

Security Services

security services

Security services provide a range of services and protocols for e-commerce, including private communications, access controls, security in business transactions, and electronic cash management. Users who engage in business transactions over the Internet need secure connections. This level of security is provided by a secure channel service called **Secure Sockets Layer (SSL)** protocol. With SSL, the client and server use a handshaking technique to agree on the level of security required during a session. Authentication takes place over a secure channel and all information transmitted during the sessions is encrypted. The protocol authenticates the client, then exchanges a master key, which is used to encrypt subsequent data exchanges.

Secure Sockets Layer (SSL)

With the growth of e-commerce, security over the Internet has become increasingly important. **Secure Electronic Transaction (SET)**, presented in Chapter 4, was designed to replace the SSL standard and provide electronic customers and merchants with the assurance that e-commerce can be conducted in a secure manner. Note that SET is an encryption standard for secure credit card authorization over the Internet. SET offers the prospect of lower business costs, because nonsecured credit card transactions on the Internet carry high fees and are generally considered high risk by credit card companies.

Secure Electronic Transaction (SET)

The SET standard addressed the following major business requirements:

- Guarantee confidentiality of payment information
- Guarantee integrity of all transmitted data
- Provide cardholder authentication
- Provide vendor authentication
- Provide security practices to protect e-commerce transactions
- Develop an open protocol
- Provide for interoperability between software and network components

Many problems have occurred with SSL, therefore a new standard such as SET is needed. With SSL, online merchants encrypt confidential information during transmission, so there is not a method of preventing fraudulent use of this information. SET provides increased security to prevent such occurrences, since it authenticates both merchants and cardholders. The specification guarantees that transaction content is not altered during the transmission between the originator and the recipient. Also unlike SSL, SET allows impromptu shopping expeditions by consumers equipped with digital certificates.

To meet the business requirements of e-commerce, SET incorporates the following features (Stallings 2000):

- Cardholder account authentication
- Confidentiality of information
- Integrity of data
- Merchant authentication

Private Communication Technology (PCT) is an alternative to SSL. It provides many of the same functions as SSL, but uses different keys for

Private Communication Technology (PCT)

Secure Hypertext Transfer Protocol (S-HTTP)

authentication and encryption. **Secure Hypertext Transfer Protocol (S-HTTP)** goes a step further by providing a method to attach digital signatures to Web pages. This verifies to users that the pages are indeed from the intended source. S-HTTP is useful in workflow and document routing applications where documents must be signed and verified using digital signatures. S-HTTP will promote the growth of e-commerce as its transaction security features will promote spontaneous commercial transactions. Additional protocol details are presented in the secure sessions and transaction system section of Chapter 4.

New technologies and tools have been developed to make Internet e-commerce secure, however they lack acceptance. The industry has been working to develop the methodologies that will make public acceptance more widespread. Overall security has increased on the Internet. Vastly improved firewall technology provides a method of keeping intruders out while letting real customers in.

Electronic Frontier Foundation (EFF) CommerceNet eTRUST

The **Electronic Frontier Foundation (EFF)** and **CommerceNet** are working to overcome some of the issues users have with trust on the Internet. They have implemented **eTRUST**, an initiative meant to establish more consumer trust and confidence in electronic transactions. The purpose of this initiative is to build a logging system that consumers will associate with trust and to make the eTRUST brand known worldwide.

CommerceNet brings together a number of organizations to focus on the advancement of e-commerce worldwide. It accelerates the development and implementation of important new technologies and practices that will impact the way companies conduct business. CommerceNet serves as a focal point for the collaborative development of solutions that will affect all businesses in the future.

13.5 FINANCIAL TRANSACTIONS

Financial transactions are the lifeblood of e-commerce, which is why it is vitally important to keep these transactions secure. Unfortunately, unsecured and insufficiently secured financial transactions are more commonplace than most consumers realize. While industry statistics indicate improvements in security technologies have been implemented, many organizations are still not using them to their fullest advantage. Studies have indicated financial losses can be attributed to computer-related breaches, stolen credit card numbers, and stolen logon passwords.

To mitigate these security breaches, many organizations have implemented security programs that are supposed to protect them and their customers from data theft, fraud, and misuse. All is not well, however, because internal policies, monitoring of security measures, and follow-up of security breaches are often disregarded.

Technology alone will not protect either an entire organization or a lone financial transaction. If there is a vulnerable spot within an Enterprise's security framework, an intruder will find a way to access and exploit it. For this reason, multiple security technologies, combined with a well thought out and monitored security plan, must be implemented to ensure financial transactions are secured and protected.

It is expensive to maintain a comprehensive security system for financial transactions, however it is much more expensive to maintain one that is inadequate. If an organization is going to participate in the e-commerce game, it must provide whatever security systems are necessary to offer an adequate level of support. The components most often used to provide this support includes the following technologies (Ledford 2001):

- Digital signatures
- Public Key Infrastructure (PKI)
- Secure Socket Layer (SSL)
- Authentication certificates
- Firewalls
- Border managers
- Biometrics
- Passwords

If an organization does not have the available capital to improve or develop the infrastructure for electronic transaction security, outsourcing is an alternative. Ideally outsourcing should be considered as an interim solution, with a plan to provide these functions in house in the future.

Payment Protocols

The Secure Electronic Transaction (SET) protocol was developed so merchants could automatically and safely collect and process payments from Internet clients. The SET protocol is presented in Chapter 4. Basically, SET secures credit card transactions by authenticating cardholders, merchants, and banks. It also preserves the confidentiality of payment data. SET requires digital signatures to verify that the customer, the merchant, and the bank are legitimate. It also uses multiparty messages that prevent credit card numbers from ending up in the wrong hands. Electronic money schemes include electronic cash, wallets, and cybercash (Sheldon 1998).

Electronic cash is a scheme that makes purchasing easier over the Internet. One method is for electronic cash companies to sell "Web dollars" that purchasers can use when visiting Web sites. Web sites would accept these Web dollars as if they were coupons.   *electronic cash*

A **wallet** is an electronic cash component that resides on the user's computer or another device such as a smart card. It stores personal security information such as private keys, credit card numbers, and certificates. It can be moved around, so users can work at different computers, and have its access controlled by departmental policies. The Personal Information Exchange is a protocol that enables users to transfer sensitive information from one environment or platform to another. With this protocol, a user can securely export personal information from one computer system to another.   *wallet*

Cybercash uses the wallet concept, where a consumer selects items to purchase and the merchant presents an invoice and a request for payment. The consumer can then use the cybercash wallet to pay for the purchase electronically. Encrypted messages are exchanged between the merchant server, the consumer, the cybercash server, and the conventional credit card networks to transfer the appropriate funds.

Smart Card

A **smart card** is about the size of a credit card and contains a microprocessor for performing a number of functions. A phone card is an example of a simple smart card. Unlike stripe cards, which have magnetic stripes on the outside, smart cards hold information internally and are more secure. They are often used as token authentication devices that generate access codes for secure systems. They have gained industry acceptance for a number of applications.

Smart card manufacturers are hopeful that the SET standard will help build a market for the cards in the United States. VISA has led the way, however universal adoption of the SET standard has been slow. Adoption of the VISA offering within the U.S. could boost the SET standard, which would benefit e-commerce.

13.6 IMPLEMENTATION ISSUES

Implementing a thorough security system is not as simple as installing technologies and forgetting them. There are a number of challenges that must be addressed for a security plan to work effectively. These include the following issues:

- Ease of use—Requiring a user to jump through hoops to access a system or complete a transaction will negate the value of security.
- Cost versus risk—Consideration must be given to the dollar amount of the transaction that is being protected.
- Lack of standards—Standards have not yet been developed that ensure technologies work together seamlessly.
- Internal security risks—There is a possibility that an internal breach will occur.
- System defaults—System defaults are usually set at the low end of security coverage.

Additionally, regulations such as the **Gramm-Leach-Bliley Financial Services Modernization Act** make implementing security systems difficult. State regulators can further complicate security issues for financial services firms.

The implementation strategy for financial security systems can be subdivided into four separate stages:

- Security audit—The audit will examine processes, infrastructure, and applications to determine if security weaknesses exist.
- Technology audit—This audit measures the available technologies and applications against those that are needed.
- Planning and implementation—Planning and implementation should consider the organization's business processes and those employees who use them.
- Training and monitoring—Users must be thoroughly trained and monitored for acceptance of the security provisions.

13.7 E-SECURITY VERSUS E-THIEVES

E-merchants must evaluate the steps being taken to prevent online fraud, security and privacy violations, and intruder attacks. Although the cost of implementing these **e-security** measures can be significant, the alternative is even more costly. There are actions that can be taken to prevent theft, protect sensitive information, and foil destructive attacks (Vreede 2001).

e-merchants

e-security

Online fraud can occur in a number of ways, however identification fraud and electronic price tag alteration are the most predominant forms. Identification fraud occurs when **e-thieves** use a consumer's credit card to make online purchases or when consumers make an online purchase and then later deny that they made the transaction. E-merchants often have little or no recourse and are held liable when transactions conducted at their sites are later determined to be fraudulent. This means that the merchants must pay higher discount rates and fees and higher penalties based on their charge-back percentages. Electronic price tag alteration occurs when hackers manipulate the shopping cart software code and alter prices. It is estimated that one-third of all shopping cart applications have software holes.

e-thieves

Traditional password protection, which is used to control system and database access, is not sufficient for e-commerce security. Secure e-commerce requires both parties to a transaction be positively mutually identified. This mutual authentication process prevents intruders from acquiring valuable information or goods under false pretenses. E-commerce transaction information must be secured against theft. Encrypting the information before it is sent, then decrypting it when received, can protect transaction communications. This security technique is designed to foil intruders who may intercept the transaction somewhere in the transmission path.

E-commerce transaction information, which is stored on computer systems and servers, must be carefully protected. This is usually accomplished by defensive hardware and software, which can include firewall and intrusion detection mechanisms. The goal is to keep systems safe from external and internal theft attempts. Confidential data, such as credit card numbers and account information, should be encrypted. Encryption can make the information unusable in the event that e-thieves gain access to the system.

Attacks that use destructive techniques such as denial of service and viruses also pose a threat to e-commerce. Interruption of service caused by such attacks can be very costly to an e-commerce organization. Since access to the site may be blocked, both immediate customers and potential customers may be lost. A number of security measures can be effective in mitigating these types of attacks. Antivirus software and the same measures used to foil theft are useful against these threats.

13.8 SECURITY PROTOCOLS

Security protocols for e-commerce provide authentication by using public key methods and implementing encryption of communications between senders and receivers. Authentication and authorization access methods are discussed in Chapter 2, as are encryption protocols and standards. This section reviews public and private key methods, digital certificates, and SSL, IPSec, and AES protocols—as they can be applied to the e-commerce environment. CryptoSystem techniques are presented in Chapter 3.

public key infrastructure (PKI)

A **public key infrastructure (PKI)** enables users of a public network such as the Internet to securely and privately exchange data and funds through the use of cryptographic keys that are obtained and shared through a trusted authority. The PKI provides digital certificates that can identify individuals or organizations, and directory services that can store and revoke certificates. It uses public key cryptography both to authenticate message senders and recipients and to encrypt and decrypt messages.

A digital certificate includes the certificate's name, a serial number, expiration dates, a copy of the certificate holder's public key, and the digital signature of the certificate-issuing authority so that a recipient can verify that the certificate is legitimate. A private key can also be used to authenticate the source of the message. When a digital certificate is encrypted with a private key the certificate can only be decrypted by using the matching public key, which identifies the certificate's source.

Building and maintaining a PKI to distribute digital certificates is a complex and time-consuming effort. It may be more appropriate to contract with a third-party organization like VeriSign to provide for general Internet e-commerce authentication services.

Protocols that can be deployed to provide security services in the e-commerce environment include SSL, IPSec, and AES. SSL provides for privacy, authentication, and message integrity and is a standard for securing Web traffic. IPSec secures each packet of data by encrypting it inside another packet that is then sent over the Internet. At the receiving end the message is unpacked and decrypted. AES is in development and will provide a more robust encryption method using 128-bit keys instead of the 56-bit DES key.

IPSec and E-commerce
IPSec provides the capability to secure communications across a LAN, private and public WANs, and the Internet. Functions that allow e-commerce systems to operate with a high degree of integrity and security include the following (Stallings 2000):

- Enhanced e-commerce security provisions
- Secure remote access over the Internet
- Secure branch office connectivity over the Internet
- Extranet connectivity with partners over the Internet
- Intranet and extranet connectivity to the Internet

The principal feature of IPSec that enables it to support these varied applications is that it can encrypt and/or decrypt all traffic at the IP level. This means that all distributed applications that are used to make e-commerce work can be secured. These applications include remote logon, e-mail, file transfer, Web access, and client/server access.

13.9 E-COMMERCE SYSTEM DESIGN CONCERNS

The primary concern of many information system and audit professionals is the profitability of e-commerce systems. E-commerce installations are often rushed into production for competitive reasons, without much thought given to security systems. Security measures must be a part of the basic systems' design and implementation plan, which should be a requirement of any automated system. Designing e-commerce systems that afford low vulnerability to attacks requires a rigorous approach based on the advice of security experts.

Often, organizations engaged in e-commerce overestimate the effectiveness of the security measures that have been deployed. Studies of e-commerce systems show little correlation between satisfaction with security by systems managers and the actual existence of effective security policies. It requires an actual breach of security before many systems managers become concerned enough to spend significant resources on security measures.

Many companies engaged in e-commerce are not aware that they are e-consumers of other e-companies, and they fail to consider security issues that occur when engaging in e-commerce with these vendors. The bottom line is simple—security must be part of the initial development cycle of an e-commerce system and not an afterthought.

E-commerce System Operation
E-commerce systems that are well organized, documented, and managed are most likely to maintain an acceptable security level. Security must be a part of the overall e-commerce strategy and not a reactive response after a breach has occurred.

Figure 13–5
Firewall Barrier Between
Buyers and Suppliers

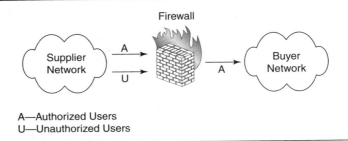

A—Authorized Users
U—Unauthorized Users

Firewall Mechanism

E-commerce servers should be behind a firewall, which will isolate the internal network from the untrusted outside world, and allow only legitimate exchanges between e-consumers and e-commerce services (Figure 13–5). Although these firewall mechanisms are effective for external threats, remember that most fraud occurs from internal tampering.

System Capacity

The computer resources and servers must be configured to handle volume spikes. There are busy hours where the transaction load might exceed the capacity of the system. Customers can be lost if access to the system is not available, however idle resources will not be productive to the organization. Market research on buying patterns would assist in determining the best mix of resources to deploy. A bonus can be realized with excessive capacity, which could make the system less vulnerable to hackers who could try to flood the system and shut it down.

Servers

Commercial Web servers support authentication and authorization methods to establish legitimate users. These basic methods coupled with SSL and IPSec protocols should provide a sound security foundation. Chapter 2 provides an in-depth analysis of access methods and Chapter 6 addresses the use of secure servers. Servers connected to the Internet and other accessible networks are vulnerable to many different types of attacks. Prices can be changed on items added to e-commerce shopping carts, bogus overflow character inputs can be used to penetrate password barriers, and Web cookies can be altered to access unauthorized accounts.

Credit Cards

If **credit cards** can be used on the e-commerce site, and usually they can, safeguards must be in place to prevent unauthorized use. A number of actions can be taken to minimize unauthorized credit card use:

- Obtain real-time authorization from a credit card company.
- Use credit card verification codes from the card's face.

- Use address verification systems for customers' billing addresses.
- Implement rule-based detection software for transaction criteria.
- Use predictive statistical models for fraudulent transactions.

Physical Access

Steps should be taken to limit physical and administrative access to e-commerce computer, database, and server resources. The fewer people, both employees and otherwise, who can access the systems, the safer they will be. Each person with authorized access should possess a unique ID and there should be some method of tracking that ID. Chapter 9 discusses the steps necessary to provide a secure physical environment for computer and networking systems.

Maintenance and Upgrades

Regular system maintenance and upgrades should be performed on a scheduled basis. This means that server and computer operating systems have had security patches applied and that the systems are configured properly. This process must also apply to the e-commerce application software. Regular audits must be performed to verify the configuration and usage patterns. Audits must also monitor intrusion systems to ensure they are working properly. An administrator must be assigned the responsibility of assessing and correcting security deficiencies. This creates a centralized responsibility and ensures that someone is maintaining acceptable security levels.

Disaster Plan

A process missing from many organizations is a **disaster plan**. Chapter 5 discusses the overall process of systems management, which includes database backup and the steps necessary to protect the organizations' resources. System backups must be conducted on a scheduled basis, and there must be offsite database storage. Use of early detection and loss mitigation systems can help in this area. A program must be in place to detect problems early, which can help avoid a major crisis.

disaster plan

13.10 DISTRIBUTED SECURITY RECOMMENDATIONS

A well-devised and implemented security program needs to be in place to thwart the efforts of unwanted visitors. Security software alone cannot stop a security breach if these people have a valid ID and password. The following are recommendations to consider when creating a secure networking environment:

- Set and keep policies for maintaining network security.
- Determine the security levels required for each area of the organization.
- Work toward an acceptable level of security without inhibiting network access and use.

- Be vigilant and keep a lookout for security holes.
- Protect vital areas under several levels of security.
- Maintain software for recording activity and user identities.
- Develop good passwords and keep them secure.
- Do not allow the sharing of passwords.
- Use a virtual network with security features.

13.11 INTERNET AND WEB RESOURCES

CommerceNet	www.commerce.net
Electronic Commerce World Institute	www.ecworld.org
W3Consortium	www.w3.org
Electronic Frontier Foundation	www.eff.org
Open Market, Inc.	www.openmarket.com
DISA	www.disa.org
VISA	www.visa.com
MasterCard	www.mastercard.com
American Express	www.americanexpress.com
American Society for Industrial Security	www.asisonline.org
Computer Security Institute	www.gocsi.com
System Administration, Networking and Security Institute	www.sans.org
Security Management Online	www.securitymanagement.com
Security Portal	www.securityportal.com
IETF IPSec	www.ietf.org/rfc/
VeriSign	www.verisign.com

■ SUMMARY

There are many components that comprise the distributed network-computing environment. Organizations are expanding their operations to include customers and suppliers that are located throughout the world. A common network that is accessible to all parties is essential to conduct business on this scale.

Network elements that can be integrated to form such a network include intranets, extranets, private networks, the Internet, and the Web. Each of these autonomous networks can be designed to provide an interface to the common Internet.

E-commerce uses a combination of network resources, protocols, and applications to provide system access for users, customers, suppliers, and partners. The extranet is a major player in the e-commerce network. Techniques such as EDI, CTI, and IVR are used in the e-commerce network to provide features and functionality that supports the parent organization's objectives.

A major consideration for implementing an e-commerce system is the level of security required to provide security and integrity for all transactions carried on the network. All electronic transactions must be secure, including those involving credit cards. Requirements include the following:

- Identifying authentic sources
- Preventing unauthorized monitoring
- Providing integrity for all transactions
- Preventing nonrepudiation

Security services for e-commerce include SSL, SET, PCT, and S-HTTP standards and specifications. Financial transactions can be protected by the following technologies:

- Digital signatures
- Firewalls
- Passwords
- Authentication certificates
- Biometrics

A number of payment protocols and methods are used in the e-commerce environment. These include electronic cash, smartcard, wallet, and cybercash.

Security must be addressed both during design and continuously during operation of an e-commerce system. E-merchants must evaluate the steps taken to prevent online fraud, security and privacy violations, and intruder attacks. Recommendations for maintaining a secure environment include the following:

- Set and enforce security policies
- Be vigilant in searching for security breaches and holes
- Keep an audit trail on everything
- Develop a good password scheme and procedure
- Use a virtual network with security features
- Match security levels with requirements
- Develop a disaster recovery plan

Key Terms

Automatic Voice Recognition (AVR)
CommerceNet
Computer Telephone Integration (CTI)
Credit card
Cryptographic systems

Cybercash
Disaster plan
Distributed application
Distributed architecture
Distributed computing environment

Distributed data processing
Distributed environment
Distributed file system
Distributed management environment
Distributed name service
Distributed network
Distributed processing
E-commerce
E-merchants
E-security
E-thieves
Electronic cash
Electronic Data Interchange (EDI)
Electronic Frontier Foundation (EFF)
eTRUST
Extranet
Firewall
Gramm-Leach-Bliley Financial Services Modernization Act
Interactive Voice Response (IVR)
Internet
Intranet
IPSec
Private Communication Technology (PCT)
Public key infrastructure (PKI)
Screen pops
Secure Electronic Transaction (SET)
Secure Hypertext Transfer Protocol (S-HTTP)
Secure Sockets Layer (SSL)
Security services
Smart card
Wallet
World Wide Web (WWW)

REVIEW EXERCISES

Questions and Analysis

Provide a short answer or in-depth analysis as required.

1. Provide an overview of the distributed environment.
2. Describe e-commerce.
3. What is EDI and how is it used?
4. Comment on the e-commerce infrastructure. Identify the components.
5. What is the function of an intranet?
6. What is the function of an extranet?
7. Describe CTI.
8. Describe IVR and AVR.
9. What are the four cornerstones of secure electronic transactions?
10. Comment on the requirement to secure electronic transactions.
11. Describe the secure sockets layer.
12. Describe the SET standard.
13. What major business requirements are addressed by the SET standard?
14. Describe the PCT standard.
15. What is eTrust?
16. Identify the components used to provide security to financial transactions.
17. Identify and explain the various electronic payment protocols.
18. Provide a list of challenges that must be addressed when implementing an e-commerce system.
19. What are the four strategies for implementing financial systems?
20. Comment on the issue of e-merchants versus e-thieves.
21. What are examples of e-commerce system design concerns.
22. Comment on the operation of an e-commerce system.
23. Produce a list of recommendations that can help provide a secure environment for an e-commerce system.

Definitions and Terms

Provide the most correct answer from the list of key terms.

1. The interchange of business documents via an electronic medium, such as the Internet.
2. A standard for communicating business transactions between organizations.
3. A set of protocols for protecting IP packets.
4. A standard by Microsoft Corporation for establishing a secure communication link using a public key system.
5. A trusted and efficient key and certificate system.
6. A message-oriented communications protocol that extends the HTTP protocol.
7. A cryptographic protocol, designed by Netscape, which provides data security at the socket level.
8. A credit-card-sized device with an embedded chip that can store digital certificates to establish one's identity.
9. A system used to provide a link between the voice systems and databases on the computer systems.
10. This protocol was designed to replace the SSL standard, and provide electronic customers and merchants with the assurance that e-commerce can be conducted in a secure manner.

True/False Statements

Indicate whether the statement is true or false.

1. A distributed application consists of a bus or ring LAN that uses shared communication media and shared access methods.
2. E-commerce can be used for the credit card purchase of airline tickets and books.
3. EDI can be used in conjunction with e-commerce to accomplish part of the transactions.
4. An intranet, extranet, the Web, and the Internet can all be part of an e-commerce network.
5. CTI is an alternative network software system to EDI and performs similar functions.
6. Four cornerstones of e-commerce security include preventing unauthorized monitoring of traffic, preventing the alteration of messages, ensuring an authentic message source, and ensuring repudiation.
7. SSL is a major component of e-commerce as it provides authorization features for a user.
8. SSL, SET, and PCT all provide similar security functions, using different techniques.
9. An initiative to establish more consumer trust and confidence in e-commerce transactions is called TrustMe.
10. Components that may be used to provide an adequate level of security support include digital signatures, authentication certificates, firewalls, and passwords.
11. Electronic money schemes may include electronic cash, wallets, cybercash, and quick money.
12. A smart card contains a microprocessor that can perform a number of security-related functions.
13. The implementation strategy for financial security systems may involve a total of three stages: a security audit, training, and planning.
14. A PKI enables users of the Internet to securely and privately exchange data and funds.
15. A firewall should be implemented in an e-commerce network that allows only legitimate users and activities.

Activities and Research

1. Research the Electronic Frontier Foundation (EFF) and CommerceNet organizations and prepare a document that outlines their initiatives.
2. Identify software products that can be used in an e-commerce system. Show the features of each.
3. Identify the standards and specifications that apply to e-commerce systems. Prepare an overview of each.
4. Identify organizations that use e-commerce systems. Provide the organization's name and the products offered. Identify payment options.
5. Research the product offering for EDI and CTI systems. Identify computer and telephone system requirements.
6. Compare the advantages and features of SSL, SET, and PCT.
7. Identify the latest security violations of e-commerce systems. Show the organization, the type of attack, and the cost of the attack.

CHAPTER 14

Wireless Networking Security

OBJECTIVES

Material presented in this chapter should enable the reader to

- Understand wireless technology and its impact on the current telecommunications environment.

- Become familiar with the wireless terms and definitions that are relevant in the Enterprise networking environment. Understand the history of mobile services and how they impact current wireless services.

- Understand the function and interaction of wireless technology in the Enterprise network environment with the wired WAN. Look at the technical aspects of wireless transmission and identify the various wireless network configurations that are possible.

- Understand how wireless security applies to the different applications of the technology, such as a packet data network, the cellular telephone network, and cellular digital packet data technology. Look at radio, mobile telephony, microwave, and satellite transmission capabilities.

- Identify components that comprise the wireless WAN and mobile computing environments and see how these environments can be compromised.

- Identify issues and considerations that are important when deciding to use a wireless networking solution. Look at the advantages and disadvantages for each of the different technologies.

- Become familiar with the security and integrity implications when implementing and managing a wireless WAN system.

INTRODUCTION

The use of wireless systems and devices is permeating all aspects of the telecommunications environment. In the past, both communication devices and end users were fixed in one location. Wireless communications breaks that barrier by allowing communication to and from a mobile communications device. Wireless, however, requires an infrastructure of wires in conjunction with the wireless system. It is becoming increasingly popular to interconnect computer equipment and to transmit data over wireless links rather than over conventional telecommunications circuits. Wireless transmission can be used to implement either long-distance WAN links or short-distance LAN links. There must be a way of interconnecting physically disjointed LAN and other groups of mobile users with stationary servers and resources. Networks that include both wired and wireless components are called hybrid networks. The ongoing development of a wireless network will allow its users to communicate with anyone, anywhere, and at any time.

Wireless technologies continue to play an increasing role in all types of networks. Since 1990, the number of wireless options has been increasing, while the cost of these technologies continues to decrease. As wireless networking becomes more affordable, demand will increase, and economies of scale will come into play. Most experts anticipate that wireless networking of all kinds will become more prevalent in the future.

The growth of wireless communications in North America and Europe has been astronomical. Wireless communications' technology has evolved from simple first-generation analog products designed for business use to second-generation digital wireless telecommunications systems for residential and business environments. Today, a next generation of wireless information networks is emerging. Complete Personal Communications Services (PCSs) will enable all users to economically transfer any form of information between any desired location. The new network will be built on and interface with the separate first- and second-generation cordless and cellular services and will also encompass other means of wireline and wireless access to Local Area Networks (LANs) and Specialized Mobile Radio (SMR) systems. Virtually any electronic communications device imaginable can become a part of the new wireless frontier.

The major problems faced with mobile telephone and data services are a limited number of users within the allowable frequency band, limited data rates within the allowable bandwidth, security, and signal fading. User numbers and data rates are limited by the very limited radio spectrum available to mobile telephone services. It is possible to monitor the transmissions at will. Efforts are underway to reduce these problems.

14.1 WIRELESS TRANSMISSION

The wireless network gives the impression of allowing the exchange of messages without plugging into a wire-based phone line. This is, however, a misconception, since most wireless systems interface with some type of wire-based network.

Unauthorized persons can easily receive wireless transmissions, including both terrestrial radio and satellite signals. The carrier-switched network does not have this problem because many different signals are multiplexed into large transmission pipes and are not accessible or readable. Although it is illegal to intercept messages intended for others, enforcement is almost nonexistent and scanners are readily available to enable eavesdroppers to monitor conversations. This section presents an overview of wireless topologies and technologies, along with security suggestions and proposals that can provide some level of secure transmission in this insecure environment.

Wireless Basics

Wireless transmissions involve electromagnetic waves, which are oscillating electromagnetic radiation caused by inducing a current in a transmitting antenna. The waves then travel through the air, or free space, where they are sensed by a receiving antenna. Free radio and television transmit signals in this manner. Figure 14–1 shows the electromagnetic spectrum.

This spectrum is measured in terms of the frequency of the waveforms used for communication. It is measured in cycles per second (cps), which are usually expressed in hertz (Hz). This name was given in honor of Robert Hertz, who was one of the inventors of the radio. The spectrum

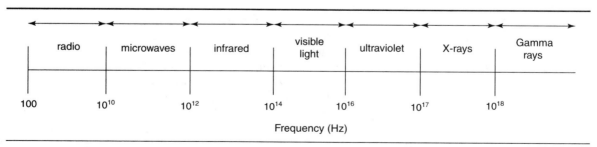

Figure 14–1
Electromagnetic Waves

starts with low-frequency waves, such as those for electrical power (60 Hz) and telephone systems (0 to 3 kHz), and goes all the way through the spectra associated with visible light. The Federal Communications Commission (FCC) assigns these frequency bands.

In wireless communications, the frequency affects the amount of data and the speed at which data may be transmitted. The strength of power of the transmission determines the distance that broadcast data can travel and still remain intelligible. In general, the principles that govern wireless transmissions dictate that lower frequency transmission can carry less data more slowly over longer distances, whereas higher frequency transmissions can carry more data faster over short distances.

The middle part of the electromagnetic spectrum is commonly divided into several named-frequency ranges, or bands. The most commonly used frequencies for wireless data communications are as follows:

- Radio—10 kHz to 1 GHz
- Microwave—1 GHz to 500 GHz
- Infrared—500 GHz to 1 THz

The important principles to remember about a broadcast medium focus on the inverse relationship between the frequency and distance and the direct relationship between frequency, data transfer rate, and bandwidth. It is also important to understand that higher frequency technologies will often use tight-beam broadcasts and require a line-of-sight between the sender and the receiver to ensure correct data delivery.

There are two types of media, **guided** and **unguided**. For unguided media, transmission and reception are achieved by means of an **antenna**. The antenna radiates electromagnetic energy into the medium (air), and for reception, the antenna picks up electromagnetic waves from the surrounding medium. The two basic types of configurations for wireless transmission are **directional** and **omnidirectional**. For the directional configuration, the transmitting antenna emits a focused electromagnetic beam, which requires the transmitting and receiving antennas be carefully aligned. In the omnidirectional case, the transmitted signal spreads out in all directions and can be received by many antennas. In general, the higher the frequency of a signal, the more it is possible to focus it into a directional beam.

guided
unguided
antenna

directional
omnidirectional

The three general ranges of frequencies that are discussed in this chapter are associated with **terrestrial microwave**, **satellite microwave**, and **broadcast radio**. Table 14–1 provides a list of the characteristics of unguided communications bands.

terrestrial microwave
satellite microwave
broadcast radio

14.2 WIRELESS NETWORK CONFIGURATIONS

Depending on the role that wireless components play in a network, wireless networks can be subdivided into three primary categories: Local Area Networks (LANs), extended LANs, and mobile computing. This chapter is concerned with mobile computing in the WAN.

Table 14–1
Unguided Communications

Frequency Band	Name	Analog Data Bandwidth	Digital Data Rate	Principal Applications
300–3,000 kHz	LF (Low Frequency)	To 4 kHz	10 to 1,000 bps	Commercial AM Radio
3–30 MHz	MF (Medium Frequency)	To 4 kHz	10 to 3,000 bps	Shortwave Radio
30–300 MHz	HF (High Frequency)	5 kHz to 5 MHz	To 100 kbps	VHF Television, FM radio
300–3,000 MHz	VHF (Very High Frequency)	To 20 MHz	To 10 Mbps	UHF Television, Terrestrial microwave
3–30 GHz	UHF (Ultra High Frequency)	To 500 MHz	To 100 Mbps	Terrestrial microwave, Satellite microwave
30–300 GHz	EHF (Extremely High Frequency)	To 1 GHz	To 750 Mbps	Experimental short Point-to-point

An easy way to differentiate among these uses is to distinguish in-house from carrier-based facilities. Both LAN and extended LAN uses of wireless networking involve equipment that a user owns and controls. Mobile computing typically involves a third party that supplies the necessary transmission and reception facilities to link the mobile part of a network with the wired part. Table 14–2 lists a number of wireless technologies.

Table 14–2
Wireless Technologies

Service	Usage	Network
Cellular	Circuit switched roaming modem and telephone connections	WAN
Microwave	Long-range data transmission	WAN
Mobile Satellite	Highly mobile messaging and location services	WAN
Packet radio	Packet switched roaming data transmission	WAN, messaging services
Infrared	Line-of-sight data transmission	LAN, peripheral sharing device
Spread spectrum	Short-range data transmission	LAN

Radio Technologies

In addition to its uses for the public broadcast of radio and television programs and for private communications with devices such as portable phones, electromagnetic radiation can be used to transmit computer data. A network that uses electromagnetic radio waves is said to operate at radio frequency, and the transmissions are referred to as **radio frequency (RF)** transmissions. Unlike networks that use wires or optical fibers, networks using RF transmissions do not require a direct physical connection between computers. Instead, each participating computer attaches to an antenna, which can both transmit and receive RF.

radio frequency (RF)

Network signals are transmitted over radio waves similar to the way your radio station broadcasts, but network applications use much higher frequencies. AM radio stations transmit in a frequency range of 535 kHz to 1605 kHz, and FM radio stations transmit in a frequency range of 88 MHz to 108 MHz. In the United States, network signals are transmitted at much higher frequencies of 902 MHz to 5.85 GHz.

A radio network transmits a signal in one direction or multiple directions, depending on the type of antenna that is used. The wavelength is very short with a low transmission strength, which means that it is best suited for line-of-sight transmissions. A line-of-sight transmission is one in which the signal goes from point to point, rather than bouncing off the atmosphere to skip across the country or across continents. A limitation of line-of-sight transmissions is that they are interrupted by tall landmasses, such as mountains.

Most wireless network equipment employs spread spectrum technology for packet transmissions. This technology uses one or more adjoining frequencies to transmit the signal across greater bandwidth. Spread spectrum frequency ranges are in the 902 MHz to 928 MHz range and higher. Spread spectrum transmissions typically send data at a rate of 2 MHz to 6 Mbps.

14.3 MOBILE TELEPHONY

Mobile telephony provides the luxury and convenience of placing calls through the telephone company facilities directly from one's own vehicle. The conventional mobile configuration depicted in Figure 14–2 includes a single base station connected to a local CO and the vehicle in which the mobile telephone set is installed. The base station is capable of transmitting and receiving on several UHF channels in succession. A high-power transmitter delivers 200 to 250 watts (W) to the base station's antenna, which is typically located on a tower or tall building. The mobile unit can travel within a 30-mile radius of the base station and reliably communicate with a transmission power output of up to 25 watts. Such large amounts of signal power have caused interference between adjacent

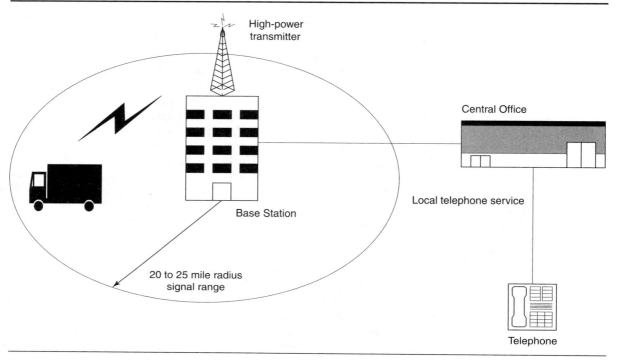

Figure 14–2
Conventional Mobile telephone service

channels when mobile units are in proximity to each other or the base. Only one conversation can be held at a time on the limited number of frequency channels available within a given service area. Under these circumstances, mobile telephone use has been limited to public safety services such as fire and police, forestry services, construction companies, and other private organizations.

14.4 MICROWAVE TECHNOLOGIES

Electromagnetic radiation beyond the frequency range used for radio and television can also be used to transport information. Microwave is a form of radio transmission using ultra high frequencies in the gigahertz (GHz) range to complete a link from one point to another. These signals are propagated from the transmitter to the antenna by using **waveguides**, which are hollow rectangular tubes that form a resonant cavity for passing a selected frequency signal. These cavities are tuned by inserting or extracting

waveguides

a "tuning slug" into the cavity, causing its resonance to change. Frequencies other than the resonant frequency are highly attenuated. A figure of merit for waveguides is the standing wave ratio, which is a measure of the ratio of the standing wave power to the main signal power. The lower the ratio, the less attenuated is the transmitted main signal. Transmitted signals are often degenerated by weather and other factors. Temperature layers can cause the signal to be refracted and portions of the signal can be reflected from the earth's surface. These effects are called multipath fading.

Although microwaves are merely a higher frequency version of radio waves, they behave differently. Microwave is considered to be a line-of-sight transmission method because the transmission and reception facilities must be within a line of sight. Distances are usually limited to less than 50 miles because of the curvature of the earth, and are actually placed within 26 miles of each other, as depicted in Figure 14–3. Each tower receives the signal, amplifies and otherwise regenerates it, and sends it to the next tower.

Microwave systems work in one of two ways. Terrestrial microwave transmits the signal between two directional antennas, shaped like dishes. These transmissions are performed in the frequency ranges of 4 GHz to 6 GHz and 21 GHz to 23 GHz and require the operator to obtain a FCC license. As with other wireless media, microwave solutions are applied where transmission cable costs are too high or where cabling is impossible. Terrestrial microwave may be a good solution for communications between two tall buildings in a metropolitan area. Satellite microwave, discussed next, is another solution for joining networks across a country or across continents.

Both types of microwave media transmit at speeds from 1 Mbps to 10 Mbps, which is a limitation where higher network speeds are desired.

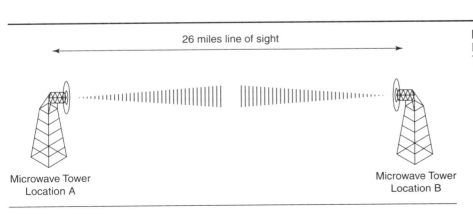

Figure 14–3
Line-of-Sight Microwave Transmission

14.5 SATELLITE TRANSMISSION

Much of today's communications between distant localities involves the use of communications satellites. Worldwide area networks such as the Internet are being established and maintained through an interface of satellites and ground communications systems. Figure 14–4 depicts such an example of wireless Internet access. The user has access to the Internet Service Provider (ISP) through a wireless company that uses both satellite and terrestrial communications links.

Although radio transmissions do not bend around the surface of the earth, RF technology can be combined with satellites to provide communication across longer distances. Figure 14–5 shows how a communications satellite in orbit around the earth can provide a network connection across an ocean.

Communications satellites are essentially electronic repeaters located many miles above the earth's surface. A satellite link is a channel connection using radio frequency waves between an earth station transmitter and a device on a satellite called a **transponder**. The transponder is a receiver and amplifier coupled to a transmitter that receives the incoming signal, amplifies it, and then retransmits it to another receiving earth station or satellite. The satellite then receives the transmission at one carrier frequency, amplifies it, and retransmits the information at a different carrier frequency. Satellites receive their operating power from solar cells that are attached to the satellite body.

transponder

Figure 14–4
Wireless Internet Access

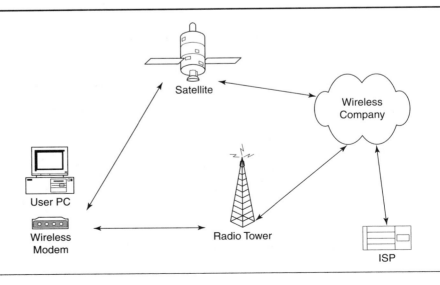

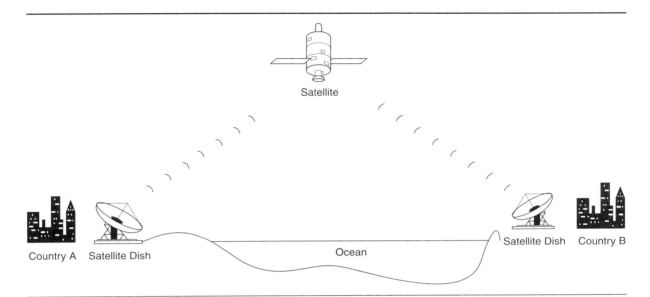

Figure 14-5
A Satellite with Two Ground Stations

Satellites act as relay stations for very-high-speed communications from one point to another. Typically, a satellite provides for multiple channels, each having a capacity of 1.544 Mbps. Satellite microwave transmits the signal between three directional antennas, one of which is on a satellite in space. These transmissions are in the 11 GHz to 14 GHz frequency range.

Data communications connections using a satellite link require ground stations for transmission and reception of the high-speed signals. Like microwave, satellite transmission is considered a line-of-sight transmission method. Because of the distance involved and satellite location and placement, however, its signal can cover a wide area. This wide area is referred to as its footprint and it is determined by the altitude and position of the satellite.

Communications satellites can be grouped into categories according to the height at which they orbit. The easiest types to understand are the **geosynchronous** or **geostationary** satellites, so named because a geosynchronous satellite is placed in an orbit that is exactly synchronized with the rotation of the earth. Such an orbit is classified as a **Geostationary Earth Orbit (GEO)** because when viewed from the ground, the satellite appears to remain at exactly the same point in the sky at all times. The distance from the earth required for geosynchronous orbit is approximately 36,000 km (20,000 miles) and is defined as a high earth orbit.

geosynchronous
geostationary
Geostationary Earth Orbit (GEO)

458   CHAPTER 14

Low Earth Orbit (LEO)

The second category of communications satellites operates in **Low Earth Orbit (LEO)**, which means that they orbit several hundred miles above the earth. The chief disadvantage of a LEO is the speed at which a satellite must travel. Because their period of rotation is faster than the rotation of the earth, satellites in lower earth orbits do not stay above a single point on the earth' surface. The satellite can only be used during the time that its orbit passes between two ground stations and the ground stations must track the satellite for signal alignment.

Middle Earth Orbiting (MEO)

The most recent category of communications satellites is the **Middle Earth Orbiting (MEO)** satellite. They operate much like the LEOs, although in slightly higher orbits. They are capable of supporting both voice and data services.

propagation delay

Propagation delay is the delay caused by the finite speed at which electronic signals can travel through a transmission medium. The round trip time for satellite transmission is about 0.25 seconds. This makes using satellite communications for voice transmissions very aggravating.

Satellite and terrestrial microwave systems can be implemented for sending and receiving multiplexed channel signals on their high-frequency carriers, which provides a way of handling large volumes of data transfers without using wires. They are also used as an extension of the cellular mobile communications telephone system.

14.6 CELLULAR TELEPHONY

The cellular mobile telephone system was developed to allow dial-up telephone service using mobile telephone handsets and radio transmissions. The cellular system is characterized by intermittent connections between users and flexible communications times. Cellular technology's most common use is to provide cellular telephone services, although the cellular network is also being used more and more by data communications devices such as pagers and personal digital assistants (PDAs). Cellular technology is a form of high-frequency radio in which antennas are spaced strategically throughout a metropolitan area. A service area or city is divided into many coverage areas, each with its own antenna. This arrangement generally provides subscribers with reliable mobile service of a quality that almost equals the hardwired telephone system.

cell

Cellular radio systems, commonly known as cellular telephone systems, use a large number of FM radio links, frequency division multiplexed (FDM), to accommodate a large number of simultaneous calls on the same frequency. This is accomplished by splitting up the region, typically a city, into a number of roughly hexagonal cells (Figure 14–6) that resemble a honeycomb. In each **cell** there is one base station with a transmitter capable of spanning the entire area of the cell, but not much more. Thus, each FM radio link, operating in one of the four groups of

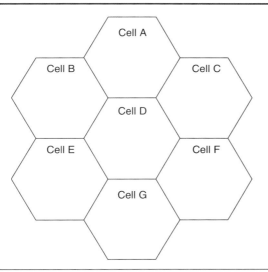

Figure 14–6
Cellular Honeycomb Structure

possible frequencies, is limited to a single cell and its adjacent partners. Since the power output of the base station is low, those frequencies can be used by another cell's base station to communicate with another user just several miles away in another nonadjacent cell. Therefore, the same four groups of frequencies are used many times over in any cellular system. As long as no two sides of the honeycomb cells touch, any frequency that has been used previously can be used again.

The central concept of cellular lies in the reuse of frequency channels in the same geographic area. This is necessary because the allocation of bandwidth is expensive for an industry to purchase. It follows that the carriers are looking for ways to accommodate more users in the same frequency bands.

Cellular Frequencies

The cellular frequencies in the United States are shown in Table 14–3. Note that the frequencies are from 824 MHz to 896 MHz. Transmit and receive

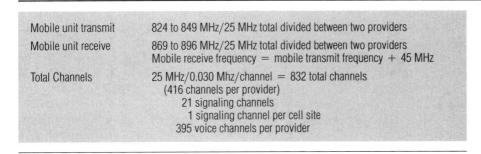

Table 14–3
Cellular Frequencies in the United States

Mobile unit transmit	824 to 849 MHz/25 MHz total divided between two providers
Mobile unit receive	869 to 896 MHz/25 MHz total divided between two providers Mobile receive frequency = mobile transmit frequency + 45 MHz
Total Channels	25 MHz/0.030 Mhz/channel = 832 total channels (416 channels per provider) 21 signaling channels 1 signaling channel per cell site 395 voice channels per provider

frequency channels occupy 25 MHz each. The mobile unit's frequency (downlink or forward channel) is always 45 MHz above transmit (up-link or reverse channel) frequency.

The up-link always occupies the lower band, because radio propagation is slightly better for lower frequencies and system designers want to give the maximum advantage to the mobile transmitter. This saves batteries and allows the use of smaller, less efficient antennas. The base station usually operates from the electric power grid and has higher gain antennas than the mobile units.

Advanced Mobile Phone Service

Advanced Mobile Phone Service (AMPS) was developed by AT&T and has become known as cellular telephone. The basic concept behind this wireless technology is to divide heavily populated areas into many small regions called cells. Each cell is linked to a central location called the **Mobile Telephone Switching Office (MTSO)**. Each MTSO is connected to a central office and other MTSOs. This connection is made by coax, twisted pair, optical fiber, or microwave. The MTSOs are connected to the Public Switched Telephone Network (PSTN) via a central office. The MTSO coordinates all mobile calls between an area comprised of several cell sites and the local central office. In this way, calls from fixed telephone locations can be routed to a mobile subscriber, or vice versa. Time and billing information for each mobile unit is accounted for by the MTSO. Figure 14–7 provides an example of the cell environment.

At the cell site, a base station is equipped to transmit, receive, and switch calls to and from any mobile unit operating within the cell to the MTSO. The cell itself encompasses only a few square miles, thus reducing the power requirements necessary to communicate with cellular telephones. This structure permits the same frequencies to be used by other cells since the power levels emitted diminish to a level that does not interfere with other cells. Heavily populated areas can be serviced by several transmission towers, rather than one, the tower used by conventional mobile techniques.

Cellular systems are full duplex. Different frequencies are used for base-to-mobile (downlink) and mobile-to-base (up-link) transmissions. Full duplex allows the MTSO to send control information to the mobile unit at any time. This control information is piggybacked onto the voice channel, and the user is often not aware of this transfer. The MTSO controls the transmitter power and frequency channel of the mobile unit. The MTSO finds an unused frequency channel to use and controls the mobile transmitter power to keep all received signals at the base station at the same level.

When a cellular phone is turned on, its microprocessor samples dedicated setup channels and tunes to the channel with the strongest signal. A closed loop is effectively established between the mobile unit and the cell

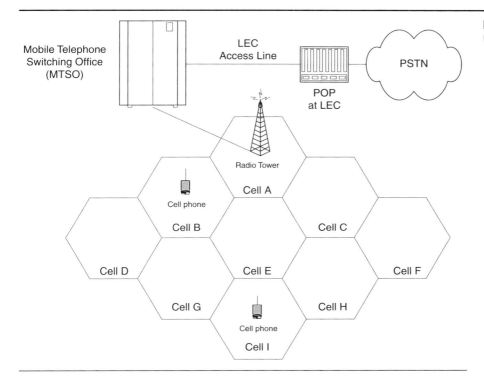

Figure 14-7
Cellular Telephone System

site at all times. If the cellular phone's signal strength significantly diminishes as a result of traveling outside one cell and entering another, the MTSO polls through all of its cell sites to determine the cellular phone's new cell location. This will be the cell in which the maximum signal strength is received.

Handoff occurs when a mobile unit leaves one cell and enters another, and this process is designed to be transparent to the user. Handoff includes the following activities:

- The MTSO receives signal strength reports from both cells. When the mobile unit's signal strength is consistently stronger in the new cell, the MTSO begins the handoff procedure.
- The mobile unit is told to use a new frequency via a data burst on the voice channel (downlink). The new frequency is the one assigned to the new cell.
- The mobile unit responds with an acknowledgment (ACK) on the reverse (up-link) setup channel.
- The conversation on the mobile unit continues uninterrupted.

The handoff procedure when a mobile unit travels from one MTSO's region to another MTSO's region is as follows:

- MTSO A notes that the signal is getting weaker and asks the new MTSO (which will be designated as MTSO B) for a signal strength report from the mobile unit.
- When the mobile unit's signal strength is significantly greater in MTSO B's region, MTSO A will transfer control to MTSO B.
- MTSO B will assign a new frequency to the mobile unit and adjust its power via a data burst on the voice (downlink) channel.
- The mobile unit responds with an acknowledgment (ACK) on the reverse (up-link) channel.
- The conversation on the mobile unit continues uninterrupted.

When everything is working properly, the handoffs between cells and between MTSOs should be transparent to the mobile user, although it is possible to hear *clicks* from the up-link and downlink control commands.

Normally, the cellular phone is used only within the metropolitan area where it is registered, which may include several counties or cities. Frequently, a user needs to operate a cellular phone outside of the home area. This **roaming** capability is possible anywhere throughout the country, provided that cellular services are available and a prearranged agreement has been made between telephone companies and their users.

AMPS Security and Privacy

The AMPS system is not very private since voices are transmitted over ordinary FM frequencies and are subject to monitoring by any FM receiver. Base stations often repeat mobile transmissions, so both sides of the conversation can be monitored with one receiver. The change in channels as a mobile is handed off does make it harder to follow conversations when the cell phone user is talking from a moving vehicle.

In 1988, the United States government banned the import or sale of scanners or other receivers that can tune to cellular frequencies, however there are many of these receivers in the public sector. It should be assumed that AMPS voice transmissions are public, so confidential information should not be used over the air.

Stolen cell phones work only until the owner cancels the service. There are no encryption capabilities for creating the Electronic Serial Number in the cell phone and the data fields are publicly available. Once the numbers are available, they can be cloned, and the calls billed to the legitimate subscriber. It is possible for the service provider to detect two calls at the same time from the same cell phone number.

Another fraud is to use a cloned or stolen cell phone on another network as a roamer. If the foreign network is not capable of checking the cell phone's home network in real-time, it may accept the call (Blake 2001).

Personal Communication Services

Personal Communication Services (PCS) is a multibillion-dollar industry representing the latest advancements in digital wireless telecommunications. A broad goal of PCS is for calls to follow a user as the user travels. A single cordless handset that could be used at home, on a mobile basis, and in the office is being widely discussed. At home, the handset would operate as a cordless phone. Outside or in a vehicle, it would automatically switch to a cellular service, and at work it would talk to the company's PBX. Users could also direct their calls to traditional wired phones when they reach their destination. Although most of the technology for providing this type of service exists, there are still topics of debate for choosing the methods to make the service work. The premise of PCS is that a user has a personal phone number that would be the user's interface to PCS and the vast array of transparently available telecommunications services.

The cellular system uses many different types of media to complete calls, including microwave and satellite transmissions. This is all transparent to the user who has the ability to call anywhere, anytime. Figure 14–8 depicts this PCS environment.

Many products and services marketed under the name of "PCS" are now available through long-distance and local-exchange carriers. The broad range of PCS services include digital cellular telephony with voice mail, e-mail, caller ID, and alphanumeric two-way paging systems, all of which are available on a handheld wireless PCS telephone or two-way alphanumeric pager. In many cases, these products are just a combination

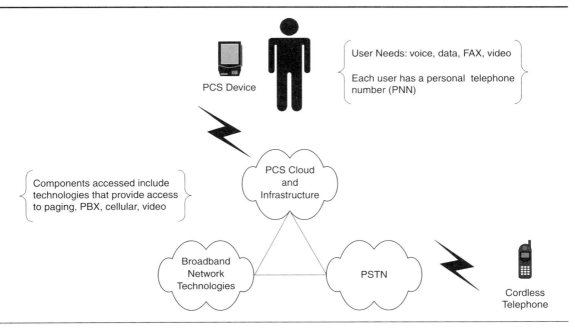

Figure 14–8
Personal Communication Environment

of existing services that use the conventional digital cellular technology, like TDMA and CDMA systems.

Global System for Mobile

Global System for Mobile (GSM)

The **Global System for Mobile (GSM)** standards offer the most promise as a global infrastructure for carrying the PCS service. Wireless data communications schemes will probably be greatly affected by the evolution of PCS standards in the future. Although widespread deployment of PCS is envisioned, it is not yet clear if demand will be sufficient to support this deployment in less-developed areas of the world.

Subscriber ID Module (SIM)
smart card

The **Subscriber ID Module (SIM)** is unique to the GSM system. It is a **smart card** with eight kilobytes of memory that can be plugged into any GSM phone. The SIM also offers some protection against fraudulent use since the phone is useless without it. The GSM SIM is only part of the effort for securing the system. Both the data used in authorizing calls and the digitized voice signal are usually encrypted. The security in GSM is better than in IS-136 and much better than in analog AMPS.

Although GSM and Cellular Digital Packet Data (CDPD) will offer many improvements, they still use large cells and have a limited bandwidth. The most important gains in digital systems are their use of much smaller cells called microcells. Instead of being a mile or more in diameter, a PCS microcell may only be a quarter of a mile in diameter or even smaller. Figure 14–9 provides an example of cell splitting. The number of cells increases as the inverse square of the cell size. Although the gains in capacity are not completely proportional, microcells can support about 10 times as many subscribers as large cells. Having many microcells and the capacity for massive amounts of reuse is the real key to capacity increases

Figure 14–9
Cellular Cell Splitting

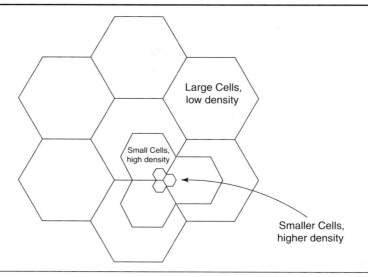

in PCS. Power requirements fall by the cube of the cell size. This means that reduced power requirements will make PCS cell phones very small, light, and inexpensive. It should also reduce concerns about radiation. Radiation is a health concern with many cell-phone users and this innovation should address this issue.

The next generation of wireless services in the communications revolution is called **Personal Communications Satellite Services (PCSS)**. Advances in the satellite systems and wireless technology will put the cellular phone system in the sky. Some satellite-based wireless telephone systems have been launched that allow practically universal coverage with a single network.

Personal Communications Satellite Services (PCSS)

The FCC has allocated frequency bands for nationwide PCS coverage in the 902 to 928 MHz and 1.8 to 2.2 GHz frequency bands. Narrowband FM is used in the 902 to 928 MHz band. Broadband PCS services operating in the 1.8 to 2.2 GHz band employ 100% digital technology, which is much more efficient in terms of bandwidth than conventional narrowband FM. Modulation techniques such as time-division multiple access (TDMA), code-division multiple access (CDMA), frequency-division duplex (FDD), and frequency-division multiple access (FDMA) are employed.

FDMA

Frequency-Division Multiple Access (FDMA) is a technology that is used to separate multiple transmissions over a finite frequency allocation. FDMA refers to the method of allocating a discrete amount of frequency bandwidth to each user to permit many simultaneous conversations. In cellular telephony, each caller occupies approximately 25 kHz of frequency spectrum. The cellular telephone frequency band consists of 416 total channels, or frequency slots, available for conversations. Within each cell, approximately 48 channels are available for mobile users. Different channels are allocated for neighboring cell sites, allowing for reuse of frequencies with a minimum of interference. This technique of assigning individual frequency slots, and reusing these frequency slots throughout the system, is known as FDMA.

Frequency-Division Multiple Access (FDMA)

TDMA

In North America, PCS 1900 **Time Division Multiple Access (TDMA)** has been chosen by a number of service providers as the second generation technology for mobile-wireless networks. TDMA is similar to GSM 900/DCS 1800, which operates in the 1900 MHz spectrum, and uses the same protocol. TDMA, as with FDMA, is a technology that is used to separate multiple conversation transmissions over a finite frequency allocation of through-the-air bandwidth. As with Frequency-Division Multiple Access (FDMA), TDMA is used to allocate a discrete amount of frequency bandwidth to each user to permit many simultaneous conversations. Each caller is assigned a specific time slot for transmission. A digital

Time Division Multiple Access (TDMA)

cellular telephone system using TDMA assigns 10 time slots for each frequency channel, and cellular telephones send bursts, or packets, of information during each time slot. The packets of information are reassembled into the original voice components by the receiving equipment. TDMA promises to significantly increase the efficiency of cellular telephone systems, allowing a greater number of simultaneous conversations.

Digital Cellular Radio Security

Privacy is considerably improved in digital cellular radio compared with the analog system. Ordinary analog scanners cannot monitor the digitized voice signals. Even decoding the signal from digital to analog is not straightforward, due to the need for a voice coder (**vocoder**). A vocoder includes electronics for digitizing voice at a low data rate by using knowledge of the way in which voice sounds are produced. A vocoder is present in all digital cell phones, so a modified cell phone could be used as a scanner.

There is some encryption of the authorization information in the TDMA system, which makes cloning and impersonation of cell phones difficult. In general, the level of security of the TDMA system is considerably better than with AMPS and is probably adequate for general use (Blake 2001).

CDMA

Code Division Multiple Access (CDMA) is the newest and most advanced technique for maximizing the number of calls transmitted within a limited bandwidth by using a spread-spectrum transmission technique. Rather than allocate specific frequency channels within the allocated bandwidth to specific conversations as in the case of TDMA, CDMA transmits digitized packets from numerous calls at different frequencies spread all over the entire allocated bandwidth spectrum.

The "code" part of CDMA lies in the fact that to keep track of these various digitized voice packets from various conversations spread over the entire spectrum of allocated bandwidth, a code is appended to each packet indicating which voice conversation it owns. This technique is similar to the datagram connectionless service used by packet-switched networks to send data over numerous switched virtual circuits within the packet-switched network. By identifying the source and sequence of each packet, the original message integrity is maintained while maximizing the overall performance of the network. Additional CDMA features include the following:

- Transmission of data is up to 14.4 kbps.
- Hand off is soft because two cells share the call during hand off.
- A power control system adjusts transmit power to enhance the quality of the signal.
- Multipath receptions of signals are coherently combined at the receiver, which enhances the quality of the signal.

CDMA Security

CDMA offers excellent security; a casual listener with a scanner will hear only noise on a CDMA channel (Blake 2001). Decoding a call requires a spread-spectrum receiver and the correct despreading code. Decoding bits are regenerated for each call, so the chances of eavesdropping are slim. Identification is accomplished using private-key encryption, which is presented in Chapter 2.

14.7 TWO-WAY MESSAGING

Two-way messaging, sometimes referred to as enhanced paging, allows short text messages to be transmitted between devices such as **personal digital assistants (PDA)** and **alphanumeric pagers**. Two distinct architectures and associated protocols have the potential to deliver these services: cellular digital packet data (CDPD) and telocator data protocol (TDP). CDPD uses idle capacity in the circuit-switched cellular network to transmit IP-based data packets. TDP is a suite of protocols that define an end-to-end system for two-way messaging to and from paging devices. Figure 14–10 depicts two-way messaging components using telocator data protocol. The activities are as follows:

personal digital assistants (PDA)
alphanumeric pagers

- A message is entered according to telocator message entry (TME) format.
- The paging service's message processor breaks the message into smaller packets, and
- Transmits it to the destination via telocator radio transport (TRT).
- The remote user's alphanumeric pager receives packets via TRT.
- Reassembles the packets into the full message, and
- Outputs the full message via telocator mobile computer (TMC).

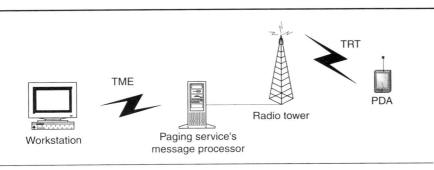

Figure 14–10
Two-way Messaging Using TDP

Pagers

pager

The **pager** best represents wireless messaging. A "beeper" device that was once worn primarily by the medical community, the pager has grown into one of the most visible and widely used message retrieval systems today. Today's pagers are available with numerous display and alert features as well as PCS options.

Pagers are available in a variety of shapes, colors, features, and sizes. Pagers can be classified into several categories:

- Tone-only pagers
- Numeric pagers
- Alphanumeric pagers
- Two-way pagers

The most basic function of the pager is to alert its user of an incoming message. To do this, the pager must receive the transmitted message from the calling source. Figure 14–11 illustrates a paging system. Transmitters throughout the service area are linked to a device called a paging terminal, which is also linked to the public switched telephone network

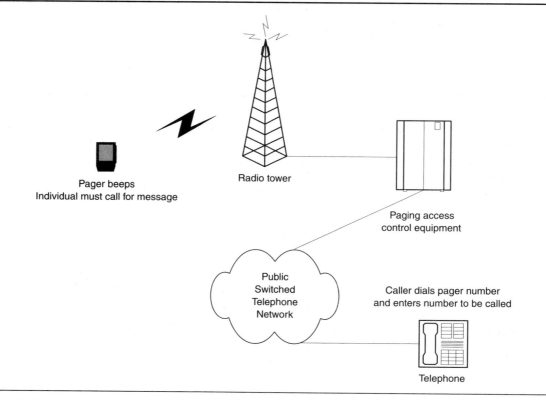

Figure 14–11
Radio Paging a "Roaming Individual"

14.10 SECURITY AND PRIVACY IN WIRELESS SYSTEMS

On shared media, any user of the media can intercept communications and anyone with access to the media can receive or transmit on the media. When the media are shared, privacy and authentication are lost unless some method is established to regain it. Cryptography, discussed in Chapter 2, is a method that provides the means to regain control over privacy and authentication. Some of the cryptographic requirements are in the wireless interface between the personal station (PS) and the radio system (RS). Other requirements are on databases stored in the network and on information shared between systems in the process of handoffs or giving service for roaming units.

Generally speaking, a secure wireless system can be realized if the following functions are addressed (Nichols 2002):

- The identity of the user can be validated through an authentication technique.
- Eavesdroppers on a data conversation can be blocked via an encryption process.
- Information is restricted to only authorized participants through some type of access control.
- A stolen or lost wireless device can be disabled from some central location.

Cell Phone Security

The Wireless Application Protocol (WAP) and Wireless Transport Layer Security (WTLS) can be attacked using various methods. Poor protocol design and implementation usually cause these problems. The attacker must be familiar with the WTLS protocol and its cryptography features to successfully launch an attack. The purpose of WTLS is to protect the data transferred between a cellular phone and the WAP gateway. These attacks can include the following:

- Key-search shortcut
- Message forgery
- Datagram truncation
- Chosen plaintext data recovery

It is essential that the wireless network be configured properly and security provisions deployed at other layers of the Enterprise network.

Security Levels

There are four levels of voice privacy, which are used to identify security and privacy requirements. These four privacy levels are as follows:

- Level 0—No privacy
- Level 1—Equivalent to wireline
- Level 2—Commercially secure
- Level 3—Military and government secure

The security and privacy requirements for these four levels include the following categories:

- Radio system performance
- Theft resistance
- Physical requirements
- System lifetime
- Privacy requirements
- Law enforcement needs

Radio System Performance

When a cryptographic system is designed for wireless systems, it must function in a hostile radio environment that is characterized by a number of impediments, including interference, thermal noise, multipath fading, and jamming. Performance can also be impacted by handoff activities.

Theft Resistance

The system operator may or may not care if a call is placed from a stolen personal station as long as the call is billed to the correct account. The owner of a personal station, however, will care if the unit is stolen. The following requirements are needed to accomplish the reduction of theft:

- Unique user ID
- Unique personal station ID
- Clone-resistant design
- Eliminate repair and installation fraud

Physical Requirements

Any cryptographic system used in a personal station must work in a mass-produced consumer product. The cryptographic system must meet basic handset and import/export requirements. It must also be low-cost wireline compliant have the capability of being mass-produced.

System Lifetime

It has been estimated that computing power doubles every two years, which means that a cryptographic algorithm that is secure today may be breakable in 5 to 10 years. A reasonable security requirement is that procedures must be viable for many years. It is necessary for the security designers to consider the best cracking algorithms today and have provisions for field upgradeability in the future.

Privacy Requirements

A user of a PCS personal station needs privacy in the following areas:

- Call setup information
- Speech
- Data

- User location
- User identification
- Calling patterns

All of the different uses of PCS communications need to be private so that the user can send information on any channel, whether it is voice, data, or control, and be assured that the transmission is secure.

Law Enforcement Needs

When a valid court order is obtained, analog telephones that are currently in use are relatively easy to tap by the law enforcement community. There are several methods that a PCS system operator can use to meet the requirements of the order. It is essential that the method used does not compromise the security of the system.

14.11 INTERNET AND WEB RESOURCES

Wireless research	www.cs.umd.edu/~waa/wireless.html
Security of the WEP algorithm	www.isaac.cs.berkeley.edu/isaac/wep-faq.html
Bay Area Wireless Group	www.bawug.org/
Security in the WTLS	www.hut.fi/~jtlaine2/wtls/

■ SUMMARY

Wireless networking is taking over an increasing portion of the networking load. Wireless technologies work well to extend the span of LANs, to provide WAN links, and to support mobile computing requirements.

A typical wireless network acts like its wired counterpart. A network adapter facilitates communications transfer across the networking medium like on a wired network, except the wires are not needed to carry the signals involved. Otherwise, users communicate as they would on any other network.

Mobile computing involves using broadcast frequencies and communications carriers to transmit and receive signals using packet radio, cellular, or satellite communications techniques. It requires specialized software including mobile-aware operating systems, mobile-aware applications, and mobile middleware to interface between multiple applications and multiple wireless WAN services.

Wireless WAN services vary widely in terms of availability, bandwidth, reliability, and cost. No single wireless WAN service is appropriate for all mobile-computing applications. Limitations often associated with wireless WAN data links include lack of security, insufficient range of coverage, and lack of standards for interoperability. The three principal technologies used to implement wireless WAN data transmission are packet radio networks, cellular telephone networks, and Cellular Digital Packet Data.

Wireless networking appears poised to grab an increasing share of network installations as newer and more powerful technologies and standards come online. To gain the advantage of mobility, it will be necessary to update the routing protocols and to enhance the security of the Internet, so that the network identifies the mobile-computing applications in a trusted manner.

Key Terms

Advanced Mobile Phone Service (AMPS)
Alphanumeric pager
Antenna
Broadcast radio
Cell
Code Division Multiple Access (CDMA)
Directional
Frequency-Division Multiple Access (FDMA)
Geostationary
Geostationary Earth Orbit (GEO)
Geosynchronous
Global System for Mobile (GSM)
Guided
Low Earth Orbit (LEO)
Middle Earth Orbiting (MEO)
Mobile Telephone Switching Office (MTSO)
Omnidirectional
Pager
Personal Communication Services (PCS)
Personal Communications Satellite Services (PCSS)
Personal Digital Assistant (PDA)
Propagation delay
Radio Frequency (RF)
Roaming
Satellite microwave
Smart card
Subscriber ID Module (SIM)
Terrestrial microwave
Time Division Multiple Access (TDMA)
Transponder
Transport Layer Security (TLS)
Unguided
Vocoder
Waveguides
Wireless Application Protocol (WAP)
Wireless Transport Layer Security (WTLS)

REVIEW EXERCISES

Questions and Analysis

Provide a short answer or in-depth analysis as required.

1. What are the security issues that permeate the WAN wireless environments?
2. What is a hybrid network?
3. Define wireless transmission.
4. What are the two types of configurations for wireless transmission?

5. What are the three subdivisions or categories of wireless networks? How do you distinguish among the uses of these networks?
6. List the different wireless technologies. How are each of these technologies used?
7. What are the most commonly used frequencies for wireless data communications?
8. Give an overview of mobile telephony.
9. What is microwave technology? How are the signals propagated?
10. Describe terrestrial microwave.
11. Describe satellite microwave. What is a transponder?
12. What is a geosynchronous satellite?
13. What is a LEO satellite? How is it different from a geosynchronous satellite?
14. What is a MEO satellite?
15. Comment on satellite wireless services. What phenomenon causes satellite microwave signals as long as 5 seconds to travel from sender to receiver? Why is this so?
16. What is cellular telephony? Describe the cellular radio systems.
17. What is a cell? How do they interact with each other? Why is this arrangement necessary?
18. What is a handoff process?
19. How is the handoff process different when the user is in the same MTSO versus when the user travels into another MTSO region?
20. Describe AMPS security and privacy.
21. What is PCS? What is a personal phone number?
22. What is GSM? What is an SIM?
23. What is TDMA? How is it different from FDMA?
24. Describe CDMA. What are the CDMA features?
25. What is two-way messaging?
26. Comment on the wireless advances in technology.
27. Describe the WAP standard.
28. Describe the WTLS protocol.
29. What are the four levels of voice privacy in wireless systems?
30. Identify some issues and considerations when deploying a wireless network.

Terms and Definitions

Provide the most correct answer from the list of key terms.

1. There are two types of media: guided and unguided. For unguided media, transmission and reception are achieved by means of this device.
2. A satellite that is placed in an orbit that is exactly synchronized with the rotation of the earth.
3. A multibillion-dollar industry representing the latest advancements in digital wireless telecommunications.
4. A technology that is used to separate multiple transmissions over a finite frequency allocation.
5. This device includes electronics for digitizing voice at a low data rate by using knowledge of the way in which voice sounds are produced.
6. This technique transmits digitized packets from numerous calls at different frequencies spread all over the entire allocated bandwidth spectrum.
7. This protocol is a worldwide standard for providing Internet communications and advanced telephony services on digital mobile phones, pagers, personal digital assistants (PDAs), and other wireless devices.
8. This protocol is intended for use over narrowband communication channels, and ensures authentication, data integrity, privacy, and denial-of-service protection.
9. The EIA/TIA Interim Standard that succeeded IS-54, which addresses digital cellular systems employing TDMA.
10. There are four levels of voice privacy, which are used to identify security and privacy requirements. Identify the level associated with no privacy.

True/False Statements

Indicate whether the statement is true or false.

1. Guided wireless communication is made possible with an antenna, whereas unguided transmissions use copper wire media.

2. The three ranges of frequencies that apply to the wireless environment include terrestrial microwave, satellite microwave, and cellular radio.

3. A transponder is a device that receives, amplifies, and retransmits signals between an earth station and a satellite.

4. The only difference between a LEO, MEO, and GEO is the size of the antenna.

5. AMPS requires the use of an MTSO.

6. AMPS is the most secure of all cellular communications.

7. The primary goal of PCS is for calls to follow the user as the user travels.

8. A SIM is the security component of the GSM system.

9. TDMA, CDMA, FDD, and FDMA are all modulation techniques that are used in narrowband cellular services.

10. Two-way messaging allows for usage of PDAs and pagers to transmit very long messages.

11. WAP is a worldwide standard for Internet communications on digital mobile phones and other wireless devices.

12. Wireless standards include IS-136, IS-54, IS-95, and IS-300.

13. The purpose of WTLS is to protect the data transferred between a cellular phone and the WAP gateway.

14. The four voice privacy levels are no privacy, commercial secure, government secure, and business secure.

15. Common problems with security in cryptographic systems include thermal noise, handoffs, and jamming.

Activities and Research

1. Identify the standards and specifications relating to wireless WAN systems. Prepare an overview of the features of each.

2. Identify the wireless devices available for today's users. Produce a matrix showing features and benefits of each.

3. Identify the wireless network's common equipment that must reside in an Enterprise organization's facilities. Describe its usage.

4. Identify software and middleware that supports the wireless market. Develop a matrix showing features, benefits, and cost.

5. Identify satellite network service providers. Provide an argument for selecting a provider.

6. Arrange for a mobile carrier to speak to the class. Issues covered can include products, services, and cost.

7. Arrange for a site visit to a satellite ground station. Produce a logical drawing of the site and components.

8. Design a wireless WAN system that would include all types of wireless technologies.

APPENDICES

APPENDIX A
OSI Model

The seven layers of the OSI model are described in this Appendix. Addressed are issues relating to security topics.

APPENDIX B
Enterprise Systems Security Review

A sample audit is included in this appendix. It includes sample questions that can be asked to determine the level of security and integrity of the Enterprise computer and network assets and resources.

APPENDIX C
Request for Comments

RFCs and other standards references relating to security are listed by category in this appendix.

APPENDIX D
ISO 17799 Security Standard

The ten components of the ISO 17799 model are presented.

APPENDIX E
Certification Programs

Security certification programs are presented.

Appendix A

OSI Model

Computer software and hardware has been developed and introduced by many separate and independent computer and network vendors worldwide over the past decades. Many user application problems were introduced, most based on the complexity of emerging computer and network technologies, which were lacking in compatibility. To overcome these incompatibilities, the International Standards Organization (ISO) adopted a layered functional approach for data communications and computer networking.

In the context of data communications, a protocol, or set of rules, is necessary for two devices to communicate. In this context, these rules do expedite communication, but they must be followed exactly for any communication to take place. Because there are so many functions that a communications architecture must provide, it was necessary to develop an architecture called a protocol stack.

A widely accepted structuring technique, and the one chosen by the ISO, is layering. The communications functions are partitioned into a hierarchical set of layers. Each layer performs a related subset of the functions required to communicate with another system. It relies on the next lower layer to perform more primitive functions and to conceal the details of the functions. It provides services to the next higher layer. Communication equipment manufacturers could then develop software implementations of protocols for one layer that are independent of implementations for any other layer. Standardized interfaces are then used to integrate these protocols into a working architecture. Standards are needed to promote interoperability among vendor equipment and to encourage economies of scale.

The task of ISO was to define a set of layers and the services performed by each layer. The partitioning should group functions logically and should have enough layers to make each manageably small, but should not have so many layers that the processing overhead imposed by the collection of layers is burdensome. Figure A–1 provides a general view of the OSI architecture.

The seven-layer stack of protocols is called the Open Systems Interconnection (OSI) Reference Model. This reference model was established by the ISO in 1983 to provide a target for the future development of interoperable communications standards. The ISO charter is to encourage, facilitate, and document the cooperative development of standards that meet with international approval.

Protocols have been developed that do not conform exactly with the OSI model, and the long-term goal of the standards organizations has been to have all developers convert to this model. As an example, the TCP/IP stack consists of four layers, rather than the seven layers of the OSI model.

An overview of each of the seven layers will be discussed in the following sections. Included in the

Figure A–1
OSI Model

```
User A                           User A
Device 1                         Device 2

Layer 7                          Layer 7
Application     ⇅                Application     ⇅

Layer 6                          Layer 6
Presentation    ⇅                Presentation    ⇅

Layer 5                          Layer 5
Session         ⇅                Session         ⇅

Layer 4                          Layer 4
Transport       ⇅                Transport       ⇅

Layer 3                          Layer 3
Network         ⇅                Network         ⇅

Layer 2                          Layer 2
Data-Link       ⇅                Data-Link       ⇅

Layer 1      ← Communications Path →   Layer 1
Physical                         Physical
```

discussions of each layer are excerpt from ISO 7498-2, the Security Architecture document (Goldman 1998).

PHYSICAL LAYER—LAYER 1

The Physical Layer covers the physical interface between devices and the rules by which bits are passed from one to another. The Physical Layer has four important characteristics:

- Mechanical
- Electrical
- Functional
- Procedural

The Physical Layer provides the procedural characteristics to activate, maintain, and deactivate the physical link through the transmission media. This layer is concerned with both the physical and electrical interface between the user equipment and the network terminating equipment (NTE). The Physical Layer also defines the electrical and mechanical characteristics such as voltage levels, pin connection design, cable lengths, and other functions associated with the physical connection of data communications equipment. Devices that operate at this layer are repeaters, hubs, and multistation access units (MAUs).

The Electronics Industries Association/Telecommunications Industry Association (EIA/TIA) standard 233-E is a good example of a Physical Layer specification. It defines precisely the connectors and cable that connect a computer (DTE) and a modem (DCE).

The Physical Layer provides the Data Link layer with a means of transmitting a serial bit stream between two corresponding systems. It can also be used to provide for traffic flow confidentiality.

DATA LINK LAYER—LAYER 2

The purpose of the Data Link Layer is to ensure the orderly and reliable exchange of information across the physical link. The principal service that is provided by the Data Link Layer to higher layers is error detection and control and retransmission of messages. To accomplish this, the data units carry synchronization, sequence number, error-detection fields, other control fields, and data. This layer is also responsible for the rate of data flow on the link.

The ISO/IEC standard 8802-3 specifies an implementation of the OSI Data Link Layer protocols. In this implementation, the Data Link layer is divided into two sublayers: the Logical Link Control (LLC) and the Medium Access Control (MAC) sublayers. Examples of devices that operate at this layer are bridges and LAN switches. These devices are used basically for segmentation of a LAN.

Security at this layer can provide for connection and connectionless confidentiality to a peer layer elsewhere. It can also provide for selective field confidentiality, such as a PIN for an automated teller machine.

NETWORK LAYER—LAYER 3

The Network Layer provides for the transfer of information between end systems across a communications network. It is responsible for establishing, maintaining, and terminating the connection between the communicating end systems. The principal example of a device that operates at this layer is a router. Network functions include:

- Network connection—The establishment of the path between the communication end systems (call setup)
- Data transfer—The exchange of data between the connected end systems via the established path (Data multiplexing and error control are employed, with a means to monitor and ensure the sequential delivery (flow control) of data units.)
- Connection release—The termination of the connection on completion of the communication (call disconnect)

The Network Layer relieves the higher layers of the need to know about the underlying data transmission and switching techniques used to connect systems. At this layer, the computer system engages in a dialogue with the network to specify the destination address and to request certain network facilities such as a priority.

There are a number of provisions for security at the Network Layer. These include peer entity and data origin authentication, access control, connection and connectionless confidentiality, and traffic confidentiality. Also included are functions for connection and connectionless integrity.

TRANSPORT LAYER—LAYER 4

The Transport Layer provides a mechanism for the exchange of data between end systems. The connection-oriented transport service ensures that the data are delivered error free, in sequence, with no losses or duplications. The Transport Layer and the three lower layers are collectively referred to as the Network or Transport service.

This layer provides both reliable and unreliable transport protocols. The reliable protocol is the Transport Control Protocol (TCP) in the TCP/IP stack. Also in the TCP/IP stack is the unreliable transport protocol, the User Datagram Protocol (UDP).

The Transport Layer provides end-to-end service. Transport functions, which concern such issues as quality of service and cost optimization, may include:

- Connection management
- Data verification
- Error control
- Flow control
- Sequencing
- Multiplexing

The size and complexity of a transport protocol depends on the extent of the reliability of the underlying network and network layer services. Just as there are various POTS customer classes of service, there are different classes of network transport services.

The Transport Layer provides for a number of security features. These include connection and connectionless integrity and connection and connectionless confidentiality. It also provides for peer entity and data authentication and access control services.

SESSION LAYER—LAYER 5

The Session Layer provides a means to transfer data and control information in an organized and synchronized manner. The Session Layer provides the following mechanisms:

- Dialogue discipline—Includes either half duplex or full duplex.
- Grouping—The flow of data can be marked to define groups of data.
- Recovery—The session layer can provide a checkpointing mechanism for a retransmission process.

The Session Layer is responsible for establishment, management, and release of each "session" between the user and the network. Network logon and user identification is an example of a Session Layer function.

PRESENTATION LAYER—LAYER 6

The Presentation Layer provides an interface between the Application Layer and the layers below the Presentation Layer. The layers below the Presentation Layer use information in a format that is useful for transmission across the network. This layer provides for translation between these two information formats. The functions performed include authentication, compression, encryption/decryption, and translation.

The Presentation Layer provides the services to allow the application processes to interpret the meaning of the information exchanged. This layer is responsible for the "presentation" of information to the user. The Presentation Layer defines the syntax used between application entities and provides for the selection and subsequent modification of the representation used. Examples of services performed at this layer include data compression, format conversion, and encryption.

Encryption is the process that scrambles data so it cannot be read if intercepted by an unauthorized person. This layer allows for password encryption on a LAN or credit card number encryption through the Secure Sockets Layer (SSL) technique on a WAN (Palmer 1999).

APPLICATION LAYER—LAYER 7

The Application Layer provides the end user with a transparent communications "window" to the network. It allows an application process to access the OSI environment and to communicate with another application process. This layer contains management functions and generally useful mechanisms to support distributed applications. In addition, general-purpose applications such as file transfer, electronic mail, and terminal access to remote computers resides at this layer.

Numerous security services reside at the Application Layer. These are as follows:

- Nonrepudiation origin and delivery
- Connection and connectionless integrity
- Connection and connectionless confidentiality
- Connection and connectionless field confidentiality
- Peer entity and data authentication
- Access control service
- Traffic flow confidentiality

Table A–1 provides an example of the communications services as provided by the OSI protocol stack.

Table A–1
OSI Protocol Stack Services

Personal Computer Protocol Stack	Services Provided
Application Layer	Application support
Presentation Layer	Translation
Session Layer	Logical connections
Transport Layer	Network addresses
Network Layer	Application ports Error recovery
Data Link Layer	CSMA/CD protocol Hardware address Error recovery
Physical Layer	Controller Transceiver Connectors Cables Signals

Appendix B

Enterprise Systems Security Review

This security review provides an audit device for determining the level of security in an Enterprise organization. Elements and components that could impact the security and integrity of the organization include (Green 2001): For each component enter a check mark for compliance or leave it blank for noncompliance. Additional elements can be added to the miscellaneous section of the form.

- Policies and procedures
- Training
- Data integrity and security
- Computer and network access
- Building equipment rooms, raised floors, and closets
- Computer and networking equipment
- Wiring and cabling plant
- Trouble reporting and maintenance
- System Administration
- Operational performance
- Disaster prevention and recovery
- Fraud prevention
- Voice systems
- Personnel

Check each block for compliance.

POLICIES AND PROCEDURES
❏ Documented security policy
❏ Enforced security policy and procedures
❏ Documentation sufficient for system restoration
❏ Documentation sufficient for data and database restoration
❏ Paperwork and sensitive documents shredded or destroyed
❏ Scheduled changing of passwords
❏ Background checks for sensitive area and document access
❏ Audits conducted for compliance
❏ Maintaining current CERT publications
❏ Distribution of current CERT alerts

TRAINING
❏ Security awareness for new employees
❏ Annual security training for all personnel
❏ Educating users to consequences of fraud and negligence
❏ Educating users on dial-up issues
❏ Train users on client/server security issues
❏ Train system administrator on security tools

DATA INTEGRITY AND SECURITY
❏ Media security methods and procedures in force
❏ Off-site backup database maintained

- ❏ Virus checking software in-place and up-to-date
- ❏ Password protection enabled
- ❏ Reviews of accounting and log information conducted
- ❏ Restriction of uploads to databases and hard drives
- ❏ Maintaining authorized file permissions database
- ❏ Protection of centralized database servers
- ❏ Regular backups of system and application software

COMPUTER AND NETWORK ACCESS
- ❏ Equipment and cable secure from monitoring and tapping
- ❏ Access control on all sensitive areas
- ❏ Terminal passwords guarded
- ❏ Are there different security levels for sensitive systems and databases
- ❏ Protection of dial-up ports
- ❏ Firewall protection with ACLs
- ❏ Router and gateway optioning in-band or out-of-band
- ❏ Enforcement of idle time-outs for dial-up connections

BUILDING EQUIPMENT ROOMS, RASIED FLOORS, AND CLOSETS
- ❏ Humidity and temperature controls monitored
- ❏ Perimeter and access control procedures
- ❏ Equipment cabinets and rooms locked
- ❏ Building and equipment free of fire hazards
- ❏ Equipment room free of flammable materials
- ❏ No water pipes in electronic areas
- ❏ Equipment rooms properly ventilated
- ❏ Backup air conditioning
- ❏ Entrance access secured

COMPUTER AND NETWORKING EQUIPMENT
- ❏ Facility earthquake proofed
- ❏ Fire suppression system operational
- ❏ Backup power and UPS tested routinely
- ❏ Power spike and surge suppression active
- ❏ Fire extinguishers full
- ❏ Smoke and fire detectors operational
- ❏ Redundant device configuration

WIRING AND CABLE PLANT
- ❏ Wiring closets secure
- ❏ Entrance cable protected from damage
- ❏ Duplicate facilities
- ❏ Records of all wiring and cable plant
- ❏ Wiring diagrams and cable assignments
- ❏ Cables labeled or tagged
- ❏ Provide covers for all copper cable
- ❏ Uses a cable scanner to detect faults

TROUBLE REPORTING AND MAINTENANCE
- ❏ Detection alarms operational
- ❏ Management review of trouble reports and alarm reports
- ❏ Database record of trouble reports and maintenance
- ❏ Formal trouble report process
- ❏ MTTR and MTBF documents for all devices

SYSTEM ADMINISTRATION
- ❏ Equipment inventory
- ❏ System baseline
- ❏ Scheduled audits
- ❏ Scheduled system upgrades
- ❏ Protect configuration files
- ❏ Maintain full backup image

OPERATIONAL PERFORMANCE
- ❏ Call accounting equipment secured
- ❏ Trouble reports documented and retained
- ❏ Monitor network for illegal traffic

DISASTER PREVENTION AND RECOVERY
- ❏ Fire prevention system
- ❏ Backup power systems
- ❏ Off-site backup systems

- ❑ Servers and databases backed up regularly
- ❑ Critical services redundant
- ❑ Backup circuits
- ❑ Power-fail transfer for PBX
- ❑ Test of disaster recovery procedures

FRAUD PREVENTION

- ❑ Trouble reports analyzed
- ❑ Unusual trouble patterns identified and investigated
- ❑ Evidence of repeated troubles and incidents documented
- ❑ Remote access DISA deactivated
- ❑ Consoles secure
- ❑ Calls blocked to high-fraud areas
- ❑ International calls blocked
- ❑ Barriers in place to counter toll fraud
- ❑ Voice-mail secure
- ❑ Wireless security

VOICE SYSTEMS

- ❑ Remote access (DISA) disabled or controlled
- ❑ Password for system administration console secure
- ❑ International calls to unneeded countries blocked
- ❑ Barriers to toll fraud in place
- ❑ Paperwork such as bills and company directories shredded
- ❑ Voice-mail programmed to prohibit dialing to unassigned extensions

PERSONNEL

- ❑ Incidents of personal use of telecommunications documented
- ❑ User passwords changed on a regular basis
- ❑ Passwords changed when employees are no longer employed by the organization
- ❑ Unused e-mail boxes deactivated
- ❑ Audits on system administrators
- ❑ Multiple system administrators
- ❑ Background checks for sensitive positions

MISCELLANEOUS

- ❑
- ❑
- ❑
- ❑
- ❑
- ❑
- ❑
- ❑

Appendix C

Request for Comments

Request for Comments (RFCs) are document series, begun in 1969, which describe the Internet suite of protocols, related experiments, and other Internet-related information and ideas. They are proposals, which when reviewed by the Internet Task Force, can be formalized into official protocols called Internet standards. There are number of search engines available for locating these documents. The IETF Web page has a link (www.rfc-editor.org/) that provides the ability to search the Web based on a number of criteria. The status of these documents changes frequently and can include both revisions and replacements.

The RFCs listed here are related in some way to security, privacy, and integrity of computer and networking resources.

RFC1157—Simple Network Management Protocol (SNMP)
RFC1213—Management Information Base for Network Management of TCP/IP-based Internets
RFC1321—The MD5 Message-Digest Algorithm
RFC1352—SNMP Security Protocols
RFC1421—Privacy Enhancement for Internet Mail: Part I: Message Encryption and Authentication Procedures
RFC1446—Security Protocols for Version 2 of the SNMPv2
RFC1457—Security Label Framework for the Internet
RFC1507—DASS-Distributed Authentication Security Service
RFC1510—The Kerberos Network Authentication Service
RFC1511—Common Authentication Technology Overview
RFC1675—Security Concerns for IPng
RFC1704—Internet Authentication
RFC1731—IMAP4 Authentication Mechanisms
RFC1760—The S/KEY One-time Password System
RFC1636—Security in the Internet Architecture
RFC1828—IP Authentication using Keyed MD5
RFC1847—Security Multiparts for MIME: Multipart/Signed and Multipart/Encrypted
RFC1848—MIME Object Security Services
RFC1858—Security Considerations for IP Fragment Filtering
RFC1910—User-based Security Model for SNMPv2
RFC1929—Username/Password Authorization for SOCKS V5
RFC1948—Defending Against Sequence Number Attacks
RFC1961—GSS-API Authentication Method for SOCKS V5
RFC1984—IAB and IESG Statement on Cryptographic Technology and the Internet
RFC1994—PPP Challenge Handshake Authentication Protocol (CHAP)

RFC2015—MIME Security with PGP
RFC2040—The RC5, RC5-CBC, RC5-CBC-Pad, and RC5-CTS Algorithms
RFC2045—Multipurpose Internet Mail Extensions (MIME)
RFC2046—Multipurpose Internet Mail Extensions (MIME)
RFC2078—Generic Security Service Application Program Interface
RFC2082—RIP-2 MD5 Authentication
RFC2084—Considerations for Web Transaction Security
RFC2085—HMAC-MD5 IP Authentication with Replay Prevention
RFC2104—HMAC: Keyed-Hashing for Message Authentication
RFC2144—The CAST-128 Encryption Algorithm
RFC2196—Site Security Handbook
RFC2222—Simple Authentication and Security Layer (SASL)
RFC2228—FTP Security Extensions
RFC2289—A One-time Password System
RFC2316—Report of the IAB Security Architecture Workshop
RFC2323—IETF Identification and Security Guidelines
RFC2350—Expectations for Computer Security Incident Response
RFC2401—Security Architecture for the Internet Protocol
RFC2402—IP Authentication Header
RFC2406—IP Encapsulating Security Payload (ESP)
RFC2407—The Internet IP Security Domain of Interpretation for ISAKMP
RFC2408—Internet Security Association and Key Management Protocol (ISAKMP)
RFC2411—IP Security Document Roadmap
RFC2444—The One-time Password SASL Mechanism
RFC2474—Physical Link Security Type of Service
RFC2479—Independent Data Unit Protection Generic Security Service Application Program Interface
RFC2480—Gateways and MIME Security Multiparts
RFC2484—PPP Challenge Handshake Authentication Protocol (CHAP)
RFC2504—User's Security Handbook
RFC2521—ICMP Security Failures Messages
RFC2535—Domain Name System Security Extensions
RFC2541—DNS Security Operational Considerations
RFC2574—User-Based Security Model for SNMPv3
RFC2577—FTP Security Considerations
RFC2617—An Extension to HTTP: Digest Access Authentication
RFC2623—NFS Version 2 and Version 3 Security Issues and the NFS Protocol's Use of RPCSEC_GSS and Kerberos V5
RFC2634—Enhanced Security Services for S/MIME
RFC2659—Security Extensions for HTML
RFC2709—Security Model with Tunnel-mode IPSec for NAT Domains
RFC2712—Addition of Kerberos Cipher Suites to Transport Layer Security (TLS)
RFC2725—Routing Policy System Security
RFC2726—PGP Authentication for RIPE Database Updates
RFC2743—Generic Security Service Application Program Interface Version 2, Update 1
RFC2744—Generic Security Service API Version 2: C-bindings
RFC2747—RSVP Cryptographic Authentication
RFC2755—Security Negotiation for WebNFS
RFC2827—Network Ingress Filtering: Defeating Denial of Service Attacks Which Employ IP Source Address Spoofing
RFC2828—Internet Security Glossary
RFC2829—Authentication Methods for LDPA
RFC2830—Lightweight Directory Access Protocol (v3): Extension for Transport Layer Security
RFC2841—IP Authentication Using Keyed SHA
RFC2853—Generic Security Service API Version 2: Java Bindings
RFC2868—Remote Authentication Dial-In User Service (RADIUS)
RFC2941—Telnet Authentication Option
RFC2942—Telnet Authentication: Kerberos Version 5
RFC2977—Mobile IP Authentication, Authorization, and Accounting Requirements

RFC3008—Domain Name System Security Signing Authority
RFC3013—Recommended Internet Service Provider Security Services and Procedures
RFC3062—LDPA Password Modify Extended Operation
RFC3067—TERENA's Incident Object Description and Exchange Format Requirements
RFC3090—DNS Security Extension Clarification on Zone Status
RFC3097—RSVP Cryptographic Authentication- Updated Message Type Value
RFC3112—LDPA Authentication Password Schema
RFC3118—Authentication for DHCP Messages
RFC3127—Authentication, Authorization, and Accounting: Protocol Evaluation
RFC3156—MIME Security with OpenPGP
RFC3163—ISO/IEC 9798-3 SASL Mechanism

Appendix D

ISO 17799 Security Standard

ISO 17799 is a detailed security standard. It is organized into ten major sections, each covering a different topic or area. A brief description of each section follows:

1. **Business Continuity Planning**
   The objectives of this section are to counteract interruptions to business activities and critical business processes from the effects of major failures or disasters.

2. **System Access Control**
   The objectives of this section are to
   1. control access to information;
   2. prevent unauthorized access to information systems;
   3. ensure the protection of networked services;
   4. prevent unauthorized computer access;
   5. detect unauthorized activities; and
   6. ensure information security when using mobile computing and telenetworking facilities.

3. **System Development and Maintenance**
   The objectives of this section are to
   1. ensure security is built into operational systems;
   2. prevent loss, modification, or misuse of user data in application systems;
   3. protect the confidentiality, authenticity, and integrity of information;
   4. ensure IT projects and support activities are conducted in a secure manner; and
   5. maintain the security of application system software and data.

4. **Physical and Environmental Security**
   The objectives of this section are to
   1. prevent unauthorized access, damage, and interference to business premises and information;
   2. prevent loss, damage, or compromise of assets and interruption to business activities; and
   3. prevent compromise or theft of information and information processing facilities.

5. **Compliance**
   The objectives of this section are to
   1. avoid breaches of any criminal or civil law; statutory, regulatory, or contractual obligations; and of any security requirements;
   2. ensure compliance of systems with organizational security policies and standards; and
   3. maximize the effectiveness of and to minimize interference to/from the system audit process.

6. **Personnel Security**
   The objectives of this section are to
   1. reduce risks of human error, theft, fraud, or misuse of facilities;

2. ensure that users are aware of information security threats and concerns, and are equipped to support the corporate security policy in the course of their normal work; and
3. minimize the damage from security incidents and malfunctions and learn from such incidents.

7. **Security Organization**
   The objectives of this section are to
   1. manage information security within the Company;
   2. maintain the security of organizational information processing facilities and information assets accessed by third parties; and
   3. maintain the security of information when the responsibility for information processing has been out-sourced to another organization.

8. **Computer and Operations Management**
   The objectives of this section are to
   1. ensure the correct and secure operation of information processing facilities;
   2. minimize the risk of systems failures;
   3. protect the integrity of software and information;
   4. maintain the integrity and availability of information processing and communication;
   5. ensure the safeguarding of information in networks and the protection of the supporting infrastructure;
   6. prevent damage to assets and interruptions to business activities; and
   7. prevent loss, modification, or misuse of information exchanged between organizations.

9. **Asset Classification and Control**
   The objectives of this section are to maintain appropriate protection of corporate assets and to ensure that information assets receive an appropriate level of protection.

10. **Security Policy**
    The objectives of this section are to provide management direction and support for information security. Within each section are the detailed statements that comprise the standard.

Appendix E

Certification Programs

Certification programs for security professionals are in the development stage. A search of the Web produces a number of education programs that are directed toward a particular vendor or product. Examples include Check Point, Linux/UNIX, Prosoft, Novell, Sun Microsystems, and other training programs. Security issues and provisions are embedded in other technical certifications such as the Sniffer University series.

CHECK POINT
Check Point has a program entitled the Check Point Certified Professional that provides students with the necessary skills to implement their security-based products. Information is available at www.checkpoint.com/services/education/certification/index.html.

LINUX/UNIX
A Security Bootcamp for Linux provides administrators with practical experience with obscure and difficult security tasks. This certification provides a tangible way to show prowess in the operating system. Information is available at www.linuxcertified.com/certification.html.

PROSOFT
There are Internet/Web certifications provided by Austin, Texas-based Prosoft. The Certified Internet Webmaster (CIW) includes certifications for Security Professional and e-Commerce Professional. Testing is provided by Prometric.

The CIW Security Professional is responsible for the successful deployment of e-business transaction and payment security systems, as well as implementation of security policy and development of countermeasures using attack-recognition and firewall technologies.

The CIW e-Commerce Professional uses Secure Electronic Transactions (SET), cryptography standards, and certificate authorities to monitor relationships between cardholders, issuers, merchants, and other elements.

NOVELL
Novell has introduced an Internet training and certification program that focuses on vendor-independent, cross-platform skills covering a number of job functions. The Certified Internet Architect teaches skills to connect an organization's existing infrastructure to the Internet and provide the necessary security skills to protect the Web-enabled infrastructure from unauthorized access.

I-NET+ COMPTIA
i-Net+ CompTIA certification provides for vendor-neutral, entry-level Internet certification programs that establish technical knowledge of Internet, intranet, and extranet topics, including security and business concepts.

SUN MICROSYSTEMS

Sun Microsystems offers several certification programs. The Sun Certified Network Administrator program covers network configuration and remote installation procedures, fundamental security, and troubleshooting techniques.

WEB REFERENCES

A number of URLs point to certification and training programs where security and integrity programs may be available. These include the following:

- AS/400: www-1.ibm.com/certify/certs/a4_index.shtml
- Cisco: www.cisco.com/public/training_cert.shtml
- CIW: www.ciwcertified.com/
- Cyberstate University: www.cyberstateU.com/
- CTIA: www.comptia.org/
- Global Knowledge: www.globalknowledge.com/
- Novel: http://education.novell.com/
- ITAA: www.itaa.org/
- Microsoft: www.microsoft.com/train_cert/
- Red Hat: www.redhat.com/training
- Sun: http://suned.sun.com/usa/certification/index.html
- Prometric: www.prometric.com/
- VUE: www.vue.com/

A brief look at the current available training programs listed as "security certification training" reflects the following URLs:

- UMBC: www.continuinged.umbc.edu/cctc
- Sun: http://suned.sun.com/US/
- Check Point: www.checkpoint.com/services/education/certification/
- IEEE Computer Society: www.computer.org/education/
- NSTISSI: www.ncisse.org
- CISSP: www.scmagazine.com/scmagazine/1998_04/lastword/lastword.html
- ISACA: www.isaca.org/cert1.htm
- Symantec: www.symantec.com/education/certification/

The administrator can expect a multitude of education providers to jump on the security training and certification bandwagon.

References and Other Resources

Amoroso, Edward. 1994. *Fundamentals of Computer Security Technology.* Upper Saddle River, NJ: Prentice Hall.

Andrews, Jean. 2001. *Enhanced A+ Guide to Managing and Maintaining Your PC.* Enhanced 3rd ed. Cambridge, MA: Thomson Learning.

Awwad, Mike Mutasem. 2000. *IntranetWare/NetWare 4.11: Administration, Troubleshooting, and TCP/IP.* Upper Saddle River, NJ: Prentice Hall.

Badgett, Tom. 1999. *A Guide to Operating Systems: Trouble Shooting and Problem Solving.* Cambridge, MA: Thomson Learning.

Barr, James G. 2001. *Distributed Denial of Service Attacks: What the Enterprise Should Know.* World Wide Web: http://www.faulkner.com/products/faccts/00018254.htm, October 4, 2001.

Beyda, William J. 2000. *Data Communications: From Basics to Broadband.* 3rd ed. Upper Saddle River, NJ: Prentice Hall.

Blacharski, Dan. 1998. *Network Security in a Mixed Environment.* Foster City, CA: IDG Books Worldwide.

Black, Uyless. 2000. *Internet Security Protocols.* Upper Saddle River, NJ: Prentice Hall.

Blake, Roy. 2001. *Wireless Communication Technology.* Albany, NY: Thomson Learning.

Burke, Richard. 1999. *A Guide to TCP/IP on Microsoft Windows NT. 4.0. Course Technology.* Cambridge, Mass.

Caudle, Kelly. 2001. *MCSC Guide to Microsoft Windows 2000 Networking.* Albany, NY: Thomson Learning.

CCDA. 2000. *Cisco Certified Design Associate Study Guide.* Osborne McGraw-Hill. Berkeley, CA: Cisco Press.

CCNA. 2000. *Cisco Certified Network Associate Study Guide.* Osborne McGraw-Hill. Berkeley, CA: Cisco Press.

Ciampa, Mark. 2001. *Guide to Designing and Implementing Wireless LANs.* Boston, MA: Course Technology.

Cisco. 2001. Internetworking Technologies Handbook. 3rd ed. Indianapolis, IN: Cisco Press.

Coleman, David. 1997. *Groupware: Collaborative Strategies for Corporate LANs and Intranets.* Upper Saddle River, NJ: Prentice Hall.

Comer, Douglas E. 2001. *Internetworking with TCP/IP:* Volume 3. Upper Saddle River, NJ: Prentice Hall.

Comer, Douglas E. 2001. *Computer Networks and Internets with Internet Applications.* 3rd ed. Upper Saddle River, NJ: Prentice Hall.

Coulouris, George. 2001. *Distributed Systems: Concepts and Design.* 3rd ed. Addison-Wesley.

Dempsey, Rob. 1997. *Security in Distributed Computing.* Upper Saddle River, NJ: Prentice Hall.

Denning, Dorothy E. 2000. "Cyberterrorism." May 23, 2000. World Wide Web: http://www.terrrorism.com/documents/denning-testimony.shtml, November 21, 2001.

Dent, Jack. 2000. *Guide to UNIX Using Linux*. Cambridge, MA: Course Technology.

Dooley, Bian J. 2001. *Carrier Managed VPN Services*. World Wide Web: http://www.faulkner.com/products/faccts/00016988.htm, October 4, 2001.

Elahi, Ata. 2001. *Network Communications Technology*. Albany, NY: Thomson Learning.

Falk, Howard. 2001. *Network Security*. World Wide Web: http://www.faulkner.com/products/faccts/00003467.htm, October 4, 2001.

Falk, Howard. 2001. *Biometrics Technology*. World Wide Web: http://www.faulkner.com/products/faccts/00016761.htm, October 4, 2001.

Faulkner Information Services. 2001. "Secure Server Environments." World Wide Web: http://www.faulkner.com/products/faccts/00018098.htm, October 4, 2001.

Ferraiola, Diane. 2001. *Conducting a Gap Analysis*. World Wide Web: http://www.faulkner.com/products/faccts/00018422.htm, November 7, 2001.

FitzGerald, Jerry. 1999. *Business Data Communications and Networking*. 6th ed. New York: John Wiley & Sons.

Flynn, Ida M. 2001. Understanding Operating Systems. 3rd ed. Albany, NY: Thomson Learning.

Foose, Bob. 2001. *Wireless Security Tutorial*. World Wide Web: http://www.faulkner.com/products/faccts/00017873.htm, October 4, 2001.

Forouzan, Behrouz A. 2000. *TCP/IP Protocol Suite*. New York: McGraw-Hill.

Forouzan, Behrouz A. 2001. *Data Communications and Networking*. 2nd ed. New York: McGraw-Hill.

Gallo, Michael A. 2002. *Computer Communications and Networking Technologies*. Albany, NY: Thomson Learning.

Garrett, Paul. 2001. *Making, Breaking Codes: An Introduction to Cryptology*. Upper Saddle River, NJ: Prentice Hall.

Gelber, Stan. 1997. *Data Communications Today*. 2nd ed. Upper Saddle River, NJ: Prentice Hall.

Goldman, James E. 1997. *Local Area Networks: A Client/Server Approach*. New York: John Wiley & Sons.

Goldman, James E. 1998. *Applied Data Communications: A Business-Oriented Approach*. 2nd ed. New York: John Wiley & Sons.

Goldman, James E. 1999. *Client/Server Information Systems: A Business-Oriented Approach*. New York: John Wiley & Sons.

Gollman, Dieter. 2000. *Computer Security*. New York: John Wiley & Sons.

Goncalves, Marcus. 2000. *Firewalls: A Complete Guide*. New York: McGraw-Hill.

Green, James Harry. 1996. *The Irwin Handbook of Telecommunications Management*. 2nd ed. New York: McGraw-Hill.

Green, James Harry. 2001. *The Irwin Handbook of Telecommunications Management*. 3rd ed. New York: McGraw-Hill.

Held, Gilbert. 1996. *Dictionary of Communications Technology*. New York: John Wiley & Sons.

Held, Gilbert. 1997a. *Virtual LANs: Construction, Implementation, and Management*. New York: John Wiley & Sons.

Held, Gilbert. 1997b. *High-speed Networking with LAN Switches*. New York: John Wiley & Sons.

Held, Gilbert. 2001. *Data Over Wireless Networks: Bluetooth, WAP, & Wireless LANs*. New York: McGraw-Hill.

Herrick, Clyde. 2001. *Telecommunications Wiring*. Upper Saddle River, NJ: Prentice Hall.

Hioki, Warren. 2001. *Telecommunications*. 4th ed. Upper Saddle River, NJ: Prentice Hall.

Hudson, Kurt. 2000. *CCNA Guide to Cisco Networking Fundamentals*. Cambridge, MA: Course Technology.

Hudson, Kurt. 2000. *CCNA Guide to Cisco Routing*. Cambridge, MA: Course Technology.

Hudson, Richard O. 2002. *Windows NT Administration and Security*. Upper Saddle River, NJ: Prentice Hall.

Kaeo, Merike. 1999. *Designing Network Security*. Cisco Press.

Kosiur, Dave. 1998. *Building and Managing Virtual Private Networks*. New York: John Wiley & Sons.

Larson, Robert E. 2001. *CCNA All-in-one Exam Guide*. Berkeley, CA: Osborne McGraw-Hill.

Ledford, Jerri L. 2001. *Implementing the Government Information Security Reform Act*. World Wide Web: http://www.faulkner.com/products/faccts/00017873.htm, October 4, 2001.

Martin, James. 1996. *Enterprise Networking: Data Link Subnetworks*. Upper Saddle River, NJ: Prentice Hall.

McLaren, Tim. 1996. *MCSE Study Guide: TCP/IP and SMS*. Indianapolis, IN: New Riders Publishing.

McClure, Stuart. 2001. *Hacking Exposed*. 3rd ed. Berkeley, CA: Osborne/McGraw-Hill.

Miller, Michael A. 2000. *Data & Network Communications*. Albany, NY: Delmar Thomson Learning.

Muller, Nathan J. 2001. *"Extranet Applications."* World Wide Web: http://www.faulkner.com/products/faccts/00017870.htm, October 4, 2001.

Muller, Nathan J. 2001. *Public Key Encryption*. (4 October 2001). World Wide Web: http://www.faulkner.com/products/faccts/00018097.htm.

Newman, Robert C. 2002. *Broadband Communications*. Upper Saddle River, NJ: Prentice Hall.

Newton, Harry. 1999. *Newton's Telecom Dictionary*. New York: Miller Freeman.

Nichols, Randall K. 2002. *Wireless Security: Models, Threats, and Solutions*. New York: McGraw-Hill.

Oppenheimer, Priscilla. 1999. *Top-Down Network Design*. Cisco Press.

Palmer, Ian. 2001. *"Internet Explorer Security Options."* World Wide Web: http://www.faulkner.com/products/faccts/00018265.htm, October 4, 2001.

Palmer, Michael. 1999. *A Guide to Designing and Implementing Local and Wide Area Networks*. Cambridge, MA: Course Technology.

Panko, Raymond R. 1999. *Business Data Communications and Networking*. 2nd ed. Upper Saddle River, NJ: Prentice Hall.

Panko, Raymond R. 2000. *Business Data Communications and Networking*. 3rd ed. Upper Saddle River, NJ: Prentice Hall.

Perry, James T. 1999. *The Internet*. Cambridge, MA: Course Technology.

Pfleeger, Charles P. 1997. *Security in Computing*. Upper Saddle River, NJ: Prentice Hall.

Proctor, Paul E. 2001. *The Practical Intrusion Detection Handbook*. Upper Saddle River, NJ: Prentice Hall.

Quinn-Andry, Terri. 1998. *Designing Campus Networks*. Cisco Press.

Ramteke, Timothy S. 2001. Networks. 2nd ed. Upper Saddle River, NJ: Prentice Hall.

Ratliff, Randy L. 1999. *Network + Certification Training Guide*. MARCRAFT.

Regan, Patrick. 2000. *Networking with Windows 2000*. Upper Saddle River, NJ: Prentice Hall.

Rowe, Stanford H. 1999. *Telecommunications for Managers*. Upper Saddle River, NJ: Prentice Hall.

Sheldon, Tom. 1998. *Encyclopedia of Networking*. New York: McGraw-Hill.

Sheldon, Tom. 2001. *Encyclopedia of Networking and Telecommunications*. New York: McGraw-Hill.

Shelly, Gary B. 1998. *Business Data Communications: Introductory Concepts and Techniques*. 2nd ed. Cambridge, MA: Course Technology.

Skoudis, ed. 2002. *Counter Hack*. Upper Saddle River, NJ: Prentice Hall.

Stallings, William. 2000a. *Network Security Essentials*. Upper Saddle River, NJ: Prentice Hall.

Stallings, William. 2000b. *Computer Organization and Architecture*. Upper Saddle River, NJ: Prentice Hall.

Stallings, William. 2000c. *Local & Metropolitan Area Networks*. Upper Saddle River, NJ: Prentice Hall.

Stallings, William. 2000. *Data and Computer Communications*. 6th ed. Upper Saddle River, NJ: Prentice Hall.

Stallings, William. 2001. *Business Data Communications*. 4th ed. Upper Saddle River, NJ: Prentice Hall.

Stamper, David A. 1999. *Business Data Communications*. 5th ed. Addison Wesley Longman.

Stamper, David A. 2001. *Local Area Networks*. 3rd ed. Upper Saddle River, NJ: Prentice Hall.

Teare, Diane, Ed. 1999. *Designing Cisco Networks*. Cisco Press.

Thompson, Arthur. 2000. *Understanding Local Area Networks: A Practical Approach.* Upper Saddle River, NJ: Prentice Hall.

Tittel, Ed. 1998. *A Guide to Networking Essentials.* Cambridge, MA: Course Technology.

Tittel, Ed. 2000. *Guide to Microsoft Windows 2000 Core Technologies.* Cambridge, MA: Course Technology.

Traeger, Cynthia. 2001a. *Government Encryption Policies.* World Wide Web: http://www.faulkner.com/products/faccts/00017844.htm, October 4, 2001.

Traeger, Cynthia. 2001b. *Common Data Security Architecture.* World Wide Web: http://www.faulkner.com/products/faccts/00017479.htm, November 7, 2001.

Velasco, Nancy B. 2001. *"Introduction to Networking Using NetWare (4.1x)."* 2nd ed. Upper Saddle River, NJ: Prentice Hall.

Vreede, Sheree Van. 2001. *E-Commerce Security Solutions.* World Wide Web: http://www.faulkner.com/products/faccts/00016949.htm, October 4, 2001.

ACRONYMS

AAA	Accounting, authentication, and authorization
ACID	Atomicity, consistency, isolation, and durability
ACL	Access Control List
AES	Advanced Encryption Standard
AFCA	Air Force Communications Agency
AFCERT	Air Force Computer Emergency Response Team
AFIWC	Air Force Information Warfare Center
AH	Authentication Header
AIS	Automated Information Systems
API	Application Programming Interface
APPN	Advanced Peer-to-Peer Networking
B2B	Business-to-Business
BXA	Bureau of Export Administration
C2	Command and Control
C2W	Command and Control Warfare
C4	Command, Control, Communications, and Computers
C4I	Command, Control, Communications, Computers, and Intelligence
CA	Certificate Authority
CAPI	Cryptography Application Program Interface
CBC	Cipher Block Chaining
CDSA	Common Data Security Architecture
CERN	European Laboratory for Particle Physics
CERT	Computer Emergency Response Team
CHAP	Challenge Handshake Authentication Protocol
CMIP	Common Management Information Protocol
CO	Central Office
COCOM	Coordinating Committee for Multilateral Export Controls
COMSEC	Communications Security
CPE	Customer Provided Equipment
CRC	Cyclic Redundancy Check
CRL	Certificate Revocation List
CSCI	Commercial Satellite Communications Initiative
CTI	Computer Telephone Integration
DARPA	Defense Advanced Research Project Agency
DBMS	Database Management System
DCE	Data Communications Equipment
DDoS	Distributed DoS
DES	Data Encryption Standard
DISA	Defense Information Security Administration
DISN	Defense Information System Network
DMI	Desktop Management Interface
DNS	Domain Name Service
DoS	Denial of Service
DSS	Digital Signature Standard
DSU	Data Service Unit
DTE	Data Termination Equipment
EAP	Extensible Authentication Protocol
E-commerce	Electronic Commerce
EDI	Electronic Data Interchange
EIA	Electronic Industries Standard
EKMS	Electronic Key Management System
ELINT	Electronic Intelligence
E-mail	Electronic Mail
EMI	Electromagnetic interference
EMP	Electromagnetic pulse
EMSEC	Emissions Security
EPS	Electronic Protection System
ESP	Encapsulating Security Payload
EUID	Effective User ID
EW	Electronic Warfare
FIPS	Federal Information Processing Standard
FTP	File Transfer Protocol
GCCS	Global Command and Control System
GCSS	Global Combat Support System
GSSAPI	Generic Security Service API
HERF	High Energy Radio Frequency
HTTP	Hypertext Transfer Protocol
IANA	Internet Assigned Numbers Authority
IASE	Information Assurance Support Environment
ICMP	Internet Control Message Protocol
IDEA	International Data Encryption Algorithm
INFOSEC	Information Security
IPMO	INFOSEC Program Management Office
IPSec	Internet Protocol Security
ISAKMP	Internet Security Association and Key Management Protocol
ISO	International Standards Organization
ISP	Internet Service Provider
ISSO	NSA Information Systems Security Organization
ITSEC	Information Technology Security Evaluation Criteria
IW/C2W	Information Warfare/Command and Control Warfare
J6	Joint Staff director for C4
KDC	Key Distribution Center
L2F	Layer 2 Forwarding Protocol
LAN	Local Area Network
LATA	Local Access Transport Area
LDAP	Lightweight Directory Access Protocol
LEAF	Law Enforcement Access Field

LSA	Local Security Authority	RPC	Remote Procedure Call
MAC	Message Authentication Code	RSA	Rivest, Shamir, Adelman
MD5	Message Digest 5	SATAN	Security Administrators Tool for Auditing Networks
MIB	Management Information Base		
MIME	Multipurpose Internet Mail Extension	SET	Secure Electronic Transaction
Modem	Modulator/Demodulator	SHA1	Secure Hash Algorithm 1
NAIC	National Air Intelligence Center	S-HTTP	Secure Hypertext Transfer Protocol
NAT	Network Address Translation	SMI	Structure of Management Information
NIPC	National Infrastructure Protection Center	S/MIME	Secure Multipurpose Internet Mail Extension
NIST	National Institute of Standards and Technology	SMLI	Stateful Multilayer Inspection
NMS	Network Management System	SMTP	Simple Mail Transport Protocol
NOS	Network Operating System	SNMP	Simple Network Management Protocol
NSA	National Security Agency	SSL	Secure Sockets Layer
NT-1	Network Termination 1	TCB	Trusted Computing Base
OPSEC	Open Platform for Secure Enterprise Connectivity	TCP/IP	Transmission Control Protocol/Internet Protocol
OSF	Open Systems Foundation	TDEA	Triple DEA
OSI	Open Systems Interconnection	TDR	Time Domain Reflectometer
PAP	Password Authentication Protocol	TFTP	Trivial File Transfer Protocol
PBX	Private Branch Exchange	TGS	Ticket-Granting Server
PCT	Private Communication Technology	TGT	Ticket-Granting Ticket
PGP	Pretty Good Privacy	TLS	Transport Layer Security
PIN	Personal Identification Number	TNI	Trusted Network Interpretation
PKI	Public Key Infrastructure	UPS	Uninterruptible Power Supply
POP	Point of Presence	URL	Uniform Resource Locator
POTS	Plain Old Telephone Service	VLAN	Virtual Local Area Network
PPP	Point-to-Point Protocol	VPN	Virtual Private Network
PPTP	Point-to-Point Tunneling Protocol	WAN	Wide Area Network
PSTN	Public Switched Telephone Network	WBEM	Web-based Enterprise Management
RC2	Rivest Cipher 2	WLAN	Wireless LAN
RC4	Rivest Cipher 4	WWW	World Wide Web
RFC	Request for Comment	XBSS	X/Open Baseline Security Services
RIPE	RACE Integrity Primitives Evaluation	XOR	Exclusive OR
RMON	Remote Monitoring		

GLOSSARY

The terms in this glossary have been assembled from a number of sources, including *Newton's Telecom Dictionary* (Newton 1999) and the *Dictionary of Communications Technology* (Held 1996). Two useful Web sites include whatis.com and pcwebopedia.com.

Abstract Syntax Notation One (ASN-1). The OSI language for describing abstract syntax.
access .The ability of information to flow between a subject and an object.
access control. Limiting the flow of information from the resources of a system only to authorized persons, programs, processes, or other systems in a network.
Access control list (ACL). A listing of users and their associated access rights, that is examined when determining whether or not access to a resource should be granted.
access rights. The "keys" that define a user's ability to access resources on a network.
accountability. Holding users responsible for their actions. The ability to reconstruct events on a computer system or network.
accreditation. The evaluation of security controls against a predefined set of specified security requirements.
ACID. Atomicity, consistency, isolation, durability.
address spoofing. See spoofing.
Address Resolution Protocol (ARP). The Internet protocol used to dynamically map Internet addresses to physical addresses on LANs.
administrator account. A default account on the Windows NT system that has high-level privileges.
Advanced Peer to Peer Networking (APPN). An IBM defined protocol used in the IBM system network architecture.
agent. Part of the SNMP structure that is loaded into each managed device to be monitored.
algorithm. A prescribed finite set of well defined rules or processes for a solution of a problem in a finite number of steps.
Anonymous FTP. A special FTP implementation that permits users to login under the user ID "anonymous."
API. See Application Programming Interface.
application gateway. A type of firewall that bases access decisions on the nature of the application's communication.
Application layer. The top-most layer of the OSI model providing such communication services as e-mail and file transfer.
Application Programming Interface (API). Software that an application program uses to request and carry out lower-level services performed by the computer's or telephone system's operating system.

asymmetric encryption. A form of cryptosystem in which encryption and decryption are formed using two different keys. Also known as public-key encryption.
attack. A method of breaking the integrity of a cipher.
audit. The review of a computer system or network by an independent, knowledgeable individual.
audit ID. An ID, used in constructing an audit trail, that is associated with each user.
auditing. The process of tracking activities of users by record and selected types of event in the security log of a server or a workstation.
audit trail. A set of records that list in chronological order selected events on a computing system or network device. Used to enable reconstruction and examination of a given sequence of events.
authenticate. The process of establishing whether an individual claiming an identity is really that individual—you are who you say you are.
authentication. The process of validating the claimed identity of an end user or a device such as a host, router, server, switch, and so on.
Authentication Header (AH). A header included in IPSec, which provides authentication and integrity of the IP packet.
authenticator. Additional information appended to a message to enable the receiver to verify that the message should be accepted as authentic.
authorization. The act of granting access rights to a user, groups of users, system, or program.
autokey. A cipher whose key is generated by message data.
availability. The amount of time that a computing system is available for use by the user.
avalanche effect. A characteristic of an encryption algorithm in which a small change in the plaintext or key gives rise to a large change in the ciphertext.
awareness. Awareness refers to the general knowledge of the user community about the importance of the security environment and requirements.
B1 security. A security level for operating systems that requires a number of security measures, including mandatory access controls.
backbone. The primary connectivity mechanism of a hierarchical distributed system.
backup and data archiving. The process of backups, including copying data to magnetic tape or optical disks, or by copying or replicating information to other systems.
bastion host. A computer system that interacts directly with an untrusted network.
Berkeley services. These services extend host equivalency to other trusted systems. Included are rlogin, remote shell, and remote copy.

biometric. The process of using hard-to-forge physical characteristics of individuals, such as fingerprints, to authenticate users.

biometric-based authentication. Authentication that is based on a specific personal trait such as a fingerprint.

biometric device. A device used in authenticating access to a system. It authenticates a user by measuring some hard-to-forge physical characteristic, such as a fingerprint.

block chaining. A procedure used during symmetric block encryption that makes an output block dependent not only on the current plaintext input block and key, but also on earlier input and/or output.

block cipher. An encryption method in which data are encrypted and decrypted in fixed-size blocks.

break. The result of an attack that causes the integrity of a cipher to be compromised.

broadcast. A packet delivery system where a copy of a particular packet is provided to all hosts attached to the network.

Brown book. The Department of Defense Guide to Understanding Trusted Facility Management.

browser. Software that looks at various types of Internet resources, also called a WWW client.

brute force attack. A type of attack under which every possible combination is attempted.

burden of proof. The duty of proving a disputed assertion.

Bureau of Export Administration (BXA). Part of the United States Department of Commerce.

business code of conduct. A formal set of regulations, usually endorsed by senior management, covering the business activities of employees.

C2 security. A security level for operating systems that dictates a number of security measures including protection of encrypted password strings.

catenet. A network in which hosts are connected to networks that possess varying characteristics and are interconnected by routers.

Common Data Security Architecture (CDSA). A security framework for developing security and authentication application programs.

Computer Emergency Response Team (CERT). A group of computer experts at Carnegie-Mellon University. CERT exists to facilitate Internet-wide response to computer security events.

CERN. A particle physics laboratory in Geneva, Switzerland.

certificate. A unique collection of information, transformed into unforgeable form, used for the authentication of users.

Certificate authority (CA). An entity that distributes public and private key pairs.

Certificate revocation list (CRL). A digitally signed list of all certificates created by a given certificate authority that have not yet expired but are no longer valid.

certification systems. Vendors offer systems such as Sentry CA, e-Lock, and Certificate Server.

challenge-response. A process where a unique value or identification is generated and presented as a challenge that is then subjected to a defined algorithmic process to produce a response.

change management. A defined process for the orderly management of change from one state to another.

chroot. A UNIX system call that allows the system administrator to change a particular directory to the root directory for a user.

ciphertext. Encrypted text that must first be decrypted to produce readable text.

cipher. A security method designed to hide data content by acting on individual bits, without regard to the semantics of the actual content.

Cipher Block Chaining (CBC). A method of using a block cipher to encrypt an arbitrarily sized message in which the encryption of each block depends on the previous blocks.

ciphertext. The output of an encryption algorithm; the encrypted form of a message or data.

cleartext. A message that is not encrypted; also called plaintext.

client. A software application that works on the user's behalf to extract some service from a server somewhere on the network.

clipper chip. A hardware implementation of the Skipjack encryption algorithm.

code. An unvarying rule for replacing a piece of information with another object.

Common Management Information Protocol (CMIP). The OSI network management protocol.

compromise. To invade something by getting around its security procedures.

computationally secure. Secure because the time and/or cost of defeating the security is too high to be feasible.

confidentiality. Ensuring that information is not disclosed to people who are not explicitly intended to receive it.

confusion. A cryptographic technique that seeks to make the relationship between the statistics of the ciphertext and the value of the encryption key as complex as possible.

conventional cipher. A secret key cipher.

conventional encryption. Symmetric encryption.

cookie. An Internet mechanism that lets site developers place information on the client's computer for later use.

cooperative processing. The ability to split application processing and logic between a client and server.

Coordinating Committee for Multilateral Export Controls (COCOM). A treaty among many leading western nations that coordinated export control regulations.

covert channel. A communications channel that enables the transfer of information in a way unintended by the designers of the communications facility.

Cyclic Redundancy Check (CRC). An error-checking hash mechanism.

credential. A security token used by security environments such as Kerberos to store security information.

cron. A process scheduler that is a standard feature on UNIX operating systems.

cryptanalysis. The branch of cryptology dealing with the breaking of a cipher to recover information; forging encrypted information that will be accepted as authentic.

cryptographer. A creator of ciphers.

cryptographic checksum. An authenticator that is a cryptographic function of both the data to be authenticated and a secret key.

cryptographic key. A digital code that can be used to encrypt, decrypt, and sign information.

cryptographic mechanism. The process of enciphering and deciphering.

cryptography. The branch of cryptology dealing with the design of algorithms for encryption and decryption.

Cryptography Application Program Interface (CAPI). The first API developed for encryption programs.

cryptology. The study of secure communications, which encompasses both cryptography and cryptanalysis.

daemon. A UNIX program that runs in the background and waits for requests to perform a tast.

data classification. The labeling of information according to its sensitivity.

data confidentiality. The process of ensuring that only the entries allowed to see the data can see it in a usable format.

data custodian. A trustee who has been given responsibility for the protection of a corporate information source.

data diddling. An attack in which the attacker changes the data while en route from source to destination.

data encryption. A method for protecting information in which the actual data is scrambled and can only be understood with the use of a special key.

Data encryption Standard (DES). A secret key cryptographic scheme standardized by NIST.

data guardian. The individual responsible for the protection of a corporate information resource.

data integrity. The state of data when it has not been compromised or altered in any way.

data owner. The owner who has created or been given ownership of an informational resource.

data remanence. The residual data that may be left over on storage media after it has been erased.

data warehouse. A subject-oriented, integrated, time-variant, nonvolatile collection of data used primarily to support organizational decisions.

DB groups. Collections of users who have the same set of privileges for database access.

debit card. A card used for identification to aid the direct access and manipulation of financial transactions, usually money transfers.

decipher. The process of uncovering the data hidden with ciphertext.

decryption. The process of extracting data that has been hidden by encryption.

degaussing. The process of demagnetizing electronic storage media to erase its contents.

delegation of authority. The act of transferring, for a limited time or function, high-level privileges.

Demilitarized Zone. A perimeter network that adds an extra layer of protection by establishing a server network that exists between the protected network and the external network.

demon dialer. A program used to automatically dial a given telephone number and guess passwords to gain access.

Denial of Service (DoS). An attack that attempts to deny corporate computing resources to legitimate users. Service is denied to users by either clogging the system with a deluge of irrelevant messages or sending disruptive commands to the system.

DES. See Data Encryption Standard.

Desktop Management Interface (DMI). A standard that establishes a common, standard interface for different management applications.

detection. The ability of a computer system, application, or organization to discover unauthorized use of its resources.

diagnostic attack. A hacking technique that uses network diagnostic tools and utilities to gain illicit access.

dialback. An authentication technique for modem users that disconnects the line and dials the user back at a predetermined telephone number.

differential cryptanalysis. A technique in which chosen plaintexts with particular XOR difference patterns are encrypted.

Diffie-Hellman key exchange. An algorithm that provides a way for two parties to establish a secret key that only they know, even though they are communicating over an insecure channel.

diffusion. A cryptographic technique that seeks to obscure the statistical structure of the plaintext by spreading out the influence of each individual plaintext digit over many ciphertext digits.

digital certificate. A password protected and encrypted file that contains identification about its holder. It includes a public key and a private key.

digital signature. A block of data that is appended to a message such that the recipient of the message can verify the contents and originator of the message.

Digital Signature Standard (DSS). A digital signature algorithm developed by the NSA.

digram. A two letter sequence. Also called a digraph.

directed broadcast. A packet sent to the broadcast IP address of a network.

discretionary access control. A means of restricting access to objects based on the identity of subjects and/or groups to which they belong. A subject with a certain access permission is capable of passing on that permission to any other subject.

Distributed Denial of Service (DDoS). A type of attack that uses a large number of computers to simultaneously send packets to a target.

DMZ. See Demilitarized Zone.

documentary evidence. Court of law requirements for a printout of computer-produced evidence.

domain. The objects to which a subject has access; the part of a computer network in which the data processing resources are under common control.

domain name. A name used to locate a computer, such as *www.weather.com*.

Domain Name System (DNS). A network utility used to resolve system names and IP addresses.

DSS. See Digital Signature Standard.

dual homed bastion host. A bastion host is equipped with two LAN cards and is used to filter network traffic. One LAN card is connected to an untrusted network and the other to a trusted network.

dumpster diving. The process of searching for useful information by looking through an organization's trash.

durability. The ability of a system to perform its functions over an extended period of time without any negative effects of its performance.

EDI. See Electronic Data Interchange.

Effective User ID (EUID). The user ID under whose authority a user is currently operating.

electronic commerce (e-commerce). The interchange of business documents via an electronic medium, such as the Internet.

Electronic Data Interchange (EDI). A standard for communicating business transactions between organizations.

Encapsulating Security Payload (ESP). AN IPSec header used when encrypting the contents of an IP package.

encapsulation. A technique where data and the procedures for operations on the data are packaged together to form a single identifiable entity.

encryption. A method of scrambling information in such a way that it is not readable by anyone except the intended recipient, who must decrypt it to read it.

encryption algorithm. The algorithm used to convert information into unintelligible form using a key.

Ethernet. 802.3 LAN, operating at 10, 100, and 1000 Mbps.

EUID. See Effective User ID.

Exploit. A program or technique designed to take advantage of a vulnerability on a target computer.

Extensible Authentication Protocol (EAP). A general protocol for PPP authentication that supports multiple authentication mechanisms.

Federal Information Processing Standard (FIPS). Standards published by NIST with which all U.S. government computer systems should comply.

file access permissions. A set of flags that are used to determine whether or not to grant access to a file or directory.

File Transfer Protocol (FTP). A TCP/IP service that permits the transfer of files between systems.

filtering router. A router capable of performing packet filtering.

firewall. A system, based on either hardware or software, used to regulate traffic between two networks.

finger. Two programs that comprise a database of e-mail addresses. The finger server stores the addresses and processes inquiries; the finger client is used to send the inquiries.

FORTEZZA card. A cryptographic peripheral that provides encryption/decryption and digital signature functions.

FTP. See File Transfer Protocol.

gateway. In data networks, it is a node on the network that connects two otherwise incompatible networks. A gateway operates at Layer 7 of the OSI model. It also can block the transmission of certain types of network traffic.

gecos. A field in the UNIX password files used to store user information.

Generic Security Services API (GSSAPI). A standard programming interface for security.

Gopher. This protocol and software program uses a menu structure to provide smooth access to Internet sites.

granularity. The relative degree of detail to which any mechanism or process can be adjusted.

group. A logical association of users whose access rights will be determined based on their membership in the group.

groupware. A general term used to describe software that is used to support the shared information needs of a department or organization.

GSSAPI. See Generic Security Services API.

guest account. An account on NOSs used by casual users.

guideline. Recommendations on best security practices for a computing environment.

hacker. A person who "hacks" away at a computer until access is successful.

hash. The process of creating a fixed value from arbitrary data, such that if the arbitrary data is altered, the hash value also changes.

hash function. A mathematical computation that results in a fixed-length string of bits from an arbitrary size input.

heterogeneous. A heterogeneous environment has a mixture of different technologies.

high level privileges. Rights bestowed to only the most trusted users of the system, such as the system or database administrator.

home directory. The default directory into which a user is placed after login.

honeypot. A system designed to deceive an attacker in a classic bait-and-switch scheme.

host equivalency. A relationship that occurs when one system trusts another system and does not require users of programs to authenticate themselves before gaining access to resources.

hub. A layer one device used to connect devices on an Ethernet LAN.

HyperText Transfer Protocol (HTTP). The protocol used to transmit data across the WWW.

ICMP. See Internet Control Message Protocol.

IDEA. See International Data Encryption Algorithm.

identification. A method of determining the identity of a computer user. It is generally the result of an authentication process.

IETF. The group responsible for creating and maintaining the Internet protocols.

IKE. See ISAKMP.

impersonation. The process of pretending to represent something you are not.

information warfare. A form of warfare whose aims are to nullify or destroy the computing resources of the opponent.

inheritance. An object that acquires the rights and properties of another object.

initialization vector. A random block of data that is used to begin the encryption of multiple blocks of plaintext; used with a block-chaining encryption.

integrity. The process of ensuring that the data has not been altered except by the people who are explicitly intended to do so.

integrity checks. A utility or task that reviews a computing environment for unauthorized modifications.

International Data Encryption Algorithm (IDEA). A secret key, or block cipher, used in Pretty Good Privacy (PGP).

Internet Control Message Protocol (ICMP). AN internet protocol used for the monitoring and control of an IP network.

Internet Security Association and Key Management Protocol (ISAKMP). A key management protocol for IPSec that is a required part of the complete IPSec implementation.

intranet. An organization's internal network.

intruder. An individual who gains, or attempts to gain, unauthorized access to a computer system or unauthorized privileges on that system.

IP address. A 32-bit number representing the address of an Internet device.

IP datagram. The fundamental unit of information passed across the Internet.

IP forwarding attack. A hacking technique used to attempt to pass TCP/IP packets to a protected network.

IPSec. See IP Security protocol.

IP Security protocol (IPSec). A set of protocols for protecting IP packets.

IPv4. The current version of the Internet Protocol widely deployed on today's Internet.

IPv6. A proposed version of the Internet Protocol that will feature major security improvements.

ISAKMP. See Internet Security Association and Key Management Protocol.

Java. Sun Microsystem's technology used to distribute active content across the WWW.

Kerberos. A secret-key network authentication protocol, developed at the Massachusetts Institute of Technology (MIT), that uses the DES cryptographic algorithm for encryption and a centralized key database for authentication.

kernal. The core of UNIX and Windows NT/2000 OS, which is responsible for sharing the processor among running processes.

key. A secret value used to protect data. Keys may be used to protect files and restrict access.

key distribution center. A system that is authorized to transmit temporary session keys to principals.

key escrow. The practice of storing cryptographic keys with one or more third parties.

key fingerprint. A human-readable code that is unique to a public key; used to verify ownership of the public key.

keyspace. The number of keys supported by a cipher.

Law Enforcement Access Field (LEAF). A special component of the Clipper Chip encryption system to provide access to encrypted data by law enforcement bodies.

LEAF. See Law Enforcement Access Field.

leakage. Occurs when a corporate firewall or router inadvertently releases information about a trusted network to an untrusted network.

listener. A process that listens for connections from the network, usually acting as a backdoor for an attacker to gain access to the system.

Local Area Network (LAN). A network that provides connectivity to computers and other devices in close proximity.

Local Security Authority (LSA). A security subsystem that runs on a Windows NT system.

Location-based authentication. Authentication that is based on an expected exact location of the device used for requesting authentication.

logic bomb. Logic embedded in a computer program that checks for a certain set of conditions to be present on the system. When these conditions are met, it executes some function resulting in authorized actions.

MAC. The data link layer address of a network interface.

macro virus. A small set of instructions carried out by a certain program. It implements destructive or mischievous tasks.

Management Information Base (MIB). A collection of objects that can be accessed via a network management protocol.

mandatory access control. A type of security controls, such as sensitivity labeling, which cannot be turned off or deleted by a user. It limits access to objects based on flags or labels contained in the objects.

Man-in-the-middle attack. A type of attack that takes advantage of the store-and-forward mechanism used by insecure networks such as the Internet. An attacker places himself between two parties and intercepts messages before transferring them on to their intended destination.

masquerading. An attack in which an attacker pretends to be someone else.

master key. A long-lasting key that is used between a key distribution center and principal for the purpose of encoding the transmission of session keys.

Message authentication code (MAC). Cryptographic checksum.

message digest. An algorithmic representation of a message that is encrypted and appended to a message to be used for integrity checking; a value or hash that is calculated based on arbitrary data.

Message Digest 5 (MD5). A one-way hash algorithm that generates a 128-bit output.

message key. An encrypted key that is transported with the original message. The message key, once deciphered, is then used to decipher the full message.

middleware. A software layer that provides a common interface and translation between the application, the data, and the operating system.

MIME. See Multipurpose Internet Mail Extensions.

MTU. The largest possible unit of data that can be sent on a given physical medium.

multilevel encryption. Repeated use of an encryption function, with different keys, to produce a more complex mapping from plaintext to ciphertext.

Multipurpose Internet Mail Extensions (MIME). A set of agreed-on formats enabling binary files to be sent as e-mail or attached to e-mail.

mutual authentication. A form of authentication where both parties are authenticated to one another, usually through a trusted third party.
Name resolution. The process of mapping a name into the corresponding address.
National Institute of Standards and Technology (NIST). An agency of the U.S. government that established national technical standards.
NAT. See Network Address Translation.
National Security Agency (NSA). An agency of the U.S. government responsible for listening in on and decoding all foreign communications of interest to the security of the United States.
Network Address Translation (NAT). The process of converting one IP address to another IP address; often used to connect networks with an illegal address space to the Internet.
NIST. See National Institute of Standards and Technology.
nonce. An identifier or number that is only used once; generated for a special occasion.
nonrepudiation. A property of a cryptographic system that prevents a sender from denying later that a message was sent or a certain action performed.
Novice. An inexperienced attacker who uses tools written by more experienced developers, with little knowledge of how the tools actually function.
NSA. See National Security Agency.
object. An entity that contains or receives information, such as a record, block, file, or directory.
one-time password. A password, generated by a special algorithm, that is never reused.
one-way function. A function that is easily computed, but the calculation of its inverse is infeasible.
Open Platform for Secure Enterprise Connectivity (OPSEC). A platform for integrating and managing Enterprise security through an open and extensible management framework.
OPSEC. See Open Platform for Secure Enterprise Connectivity.
Orange Book. The Department of Defense Trusted Computer System Evaluation Criteria. This document presents a method for evaluating the security controls of a data processing system.
packet filtering. The limiting of TCP/IP traffic, based on the type of traffic, source and destination IP addresses, ports, or other information.
packet monitoring. The active monitoring of traffic on a network. Hackers use this technique to discover passwords and other sensitive information.
packet snarfing. Also known as eavesdropping. An analyzer can set its interface on promiscuous mode and copy packets for later analysis.
padding. Meaningless data added to the beginning or end of messages.
password. A secret character string used for authentication.
password aging. A method used to enforce user password changes.
Password Authentication protocol (PAP). A simple authentication method used with PPP.

password cracking. A method that seeks to discover passwords by passing a dictionary of terms through a password algorithm, and then comparing the output to the encrypted password string.
password generator. A computing device or software program that constructs unique, and supposedly uncrackable, password strings.
password guessing. A technique used to gain unauthorized access to a computer or network device by supplying a wide range of passwords for a user account.
PBX. A telephone switch used by an organization, usually purchased.
PEM. See privacy enhanced mail.
performance. The ability of a computing environment to perform its required operations in a timely manner.
Personal Identification Number (PIN). An individual number used as part of an authentication process.
PGP. See pretty good privacy.
physical access. The ability to interact directly with a computer system.
PIN. See Personal Identification Number.
ping. A packet based on the ICMP that is used to determine if a computer is accessible on the network.
plaintext. The original message, in unencrypted, readable form.
Point-to-Point Protocol (PPP). A protocol that allows a computer to use a telephone line and a modem to connect directly to the Internet.
port number. A number between 1 and 65,535 used in the header of a TCP or UDP packet that identifies where on the source computer a packet originated and the intended destination.
Pretty Good Privacy (PGP). An encryption scheme, developed by Phillip Zimmerman, based on the RSA encryption algorithm.
Private Communication Technology (PCT). A standard by Microsoft Corporation for establishing a secure communication link using a public key system.
private key. One of the two keys used in an asymmetric encryption system.
private key encryption. Refers to an encryption method in which one or more identities share a password. Also called shared secret key.
procedure. In terms of computing security policy, a procedure is the actual method used to enact computing policy for a particular computing or network environment.
process. The environment in which a program executes.
promiscuous mode. When the interface is in this mode, it gathers all data from the network, regardless of the destination.
protocol. A formal description of messages to be exchanged and rules to be followed for two or more systems to exchange information.
Protocol Data Unit (PDU). OSI terminology to packet.
proxy. An application running on a gateway that relays packets between a trusted client and an untrusted host. See proxy server.
proxy server. A proxy server, like a firewall, is designed to protect internal resources on networks that are connected to other networks such as the Internet.

pseudorandom number generator. A function that deterministically produces a sequence of numbers that are apparently statistically random.

public account. An account on NOSs that can be accessed by any user on the network.

public key. One of the two keys used in an asymmetric encryption system; a digital code used to encrypt information and verify digital signatures.

public key cipher. An asymmetric cipher that uses one key to encipher a message and a second key to decipher the ciphertext.

public key encryption. Asymmetric encryption.

Public key infrastructure (PKI). A trusted and efficient key and certificate system.

RADIUS. A service for authenticating and authorizing dial-up users.

RAID. Defines techniques for combining disk drives into arrays. Data are written across all drives, which improves performance and protects data.

RAS. See remote access server.

read access. Permission to read information.

Red Book. The Department of Defense Trusted Network Interpretation Environments Guide. This document provides insights into maintaining a trusted network environment.

remote access server. A piece of computer hardware that sits on a corporate LAN. Employees use the PSTN to get access to their e-mail, software, and data.

remote execution facility. A TCP/IP network service that permits the remote execution of commands without requiring the user to login.

Remote procedure call (RPC). A protocol governing the method with which an application activates processes on other nodes and retrieves the results. It is a method of communication between a client and a server.

replay attack. An attack in which a service already authorized and completed is forged by another "duplicate request" in an attempt to repeat authorized commands.

Request for Comments (RFC). Documents that specify Internet standards; some documents contain informational overviews and introductory topics.

rights and permissions. A network operating system has rights and permissions that grant users access to file systems, directories, and other resources.

risk. The possibility that a particular vulnerability will be exploited.

risk analysis. The process of identifying security risks, determining their impact, and identifying areas requiring protection.

risk assessment. Also called risk analysis; involves the analysis of controls in terms of their appropriateness and effectiveness.

Rivest Cipher 2 (RC2). A variable-key-size block cipher designed by Ron Rivest for RSA Data Security, Inc.

Rivest Cipher 4 (RC4). A variable-key-size stream cipher designed by Ron Rivest for RSA Data Security, Inc.

Rivest, Shamir, Adelman (RSA). A public key algorithm.

rlogin. A service offered by Berkeley UNIX that allows users of one computer to log onto other UNIX systems and interact as if their devices were directly connected.

role. A feature of many databases that permits access to be assigned by job function.

role based security. A collection of security algorithms that are given to defined roles or positions in an organization or entity.

root. Used to refer to the highest-level directory on a UNIX system.

RootKit. A type of tool that allows an attacker to maintain super-user access on a computer.

router. An OSI Layer 3 intelligent device that connects like and unlike networks (LANs and WANs). Routers typically support multiple protocols.

routing attack. A hacking technique that uses the IP source routing option to route packets to a destination where they can be manipulated.

RSA. See Rivest, Shamir, Adelman.

RSA algorithm. A public-key encryption algorithm based on exponentiation in modular arithmetic.

rule based security. A collection of security attributes.

S/key. An authentication program that relies on a one-way function for its security.

S/MIME. An extension of the multipurpose Internet mail extension that adds security to protect against interception and e-mail forgery.

SATAN. Security administrators tool for auditing networks.

screened subnet. A network protected by a filtering router or other network device.

secret key. The key used in a symmetric encryption system. Both participants must share the same key, and this key must remain secret to protect the communication.

secret key cipher. A symmetric cipher, where the same key is used to encipher a message and to decipher the ciphertext.

secure channel. A technology that provides privacy, integrity, and authentication in point-to-point communications.

Secure hash algorithm 1 (SHA1). A one-way hash algorithm designed by NIST that has a 160-bit digest.

Secure hypertext transfer protocol (S-HTTP). A message-oriented communications protocol that extends the HTTP protocol.

Secure RPC. Developed by SUN Microsystems; has secure authentication and data protection mechanisms.

secure shell. A protocol for secure remote login and other secure network services over an insecure network.

Secure sockets layer (SSL). A cryptographic protocol, designed by Netscape, which provides data security at the socket level.

security. A way of ensuring data on a network is protected from unauthorized use.

Security access mechanism. The central security authority for a Windows NT domain.

security advisories. Security information that alerts the computing community to security problems and their associated fixes. This information is usually distributed by e-mail bulletins.

security criteria. Refers to a recognized collection of required security attributes that form a standard used for the measurement or analysis of security functionality.

security descriptor. A Windows NT descriptor that contains information about the owner and access control lists.
security perimeter. The boundary at which security controls are placed to protect network assets.
security policy. The set of rules and practices that regulate how an organization manages, protects, and distributes sensitive information.
security strategy. A strategic plan to move an organization to an appropriate level of security.
security testing. The process of determining whether a security feature or process is implemented as designed and adequate for its intended purpose.
security token. A device or file, usually protected by encryption, which contains security information.
self assessment. An audit performed by a system administrator, user department, or any nonindependent entity.
sendmail. An e-mail transport mechanism used in the UNIX operating system and on the Internet.
sensitive information. Any information where unauthorized disclosure would result in a significant financial loss, embarrassment, or have other negative effects.
sensitivity labels. A method of applying data classification titles to files, directories, and other computing resources.
sequencing attack. A hacking technique that uses knowledge of TCP/IP sequence numbers to intercept and insert bogus traffic in an established TCP/IP communication.
session hijacking. A technique whereby an attacker steals an existing session established between a source and destination.
session key. A temporary encryption key used between two principals.
Set user ID bit. An executable UNIX program that will execute under the authority of the owner of the program rather than the user that invokes it.
S-HTTP. See Secure hypertext transfer protocol.
Simple Mail Transport Protocol (SMTP). A protocol for electronic mail applications.
Simple Network Management Protocol (SNMP). A standard protocol used to manage network devices and computing systems. It provides a framework for systems to report problems, configuration information, and performance data.
single sign on. The ability of a user to perform the authentication process required to access all required network and computing resources once and only once.
single user mode. A special UNIX operating system level, used primarily for diagnostic purposes.
SLIP. An Internet protocol used to run IP over serial lines such as telephone circuits or RS-232 cables connecting two systems.
smart card. A credit-card-sized device with an embedded computer chip, which can store digital certificates that can establish one's identity.
SMTP. See Simple Mail Transport Protocol.
smurf attack. A directed broadcast attack. An attacker sends packets with a spoofed source IP address to the broadcast IP address of a network.

snarfing. See packet snarfing.
sniffer. A program used to capture traffic from the network.
SNMP. See Simple Network Management Protocol.
social engineering. The collective name for techniques that deceive users into revealing sensitive information.
socket. The combination of an IP address and port number.
spam. The act of spewing out large numbers of electronic messages via e-mail or newsgroups to users who do not want to receive them.
spoof. A program that masquerades as legitimate, but performs illicit actions.
spoofing. A type of attack in which one computer disguises itself as another to gain access to a system.
stateful packet filter. A packet filter that includes memory, so that decisions to transmit or drop a packet can be based on the packet's headers as well as previous packets encountered by the device.
steganography. A method of encryption that hides the existence of a message.
stream cipher. An encryption method that encrypts and decrypts arbitrarily sized messages one character at a time.
strength. The ability of a cipher to hide its message.
Structure of Management Information (SMI). The rules used to define the objects that can be accessed via a network management protocol.
subject. An entity, such as a person or process, that causes information to move between objects.
substitution. A basic type of encryption that replaces one symbol with a corresponding symbol.
superuser. The UNIX system administrator.
SUPERVISOR. A special account on NOSs that provides a comprehensive set of privileges.
switch. A layer 2 device used to implement a LAN that selectively transmits data only to specific destinations on the LAN.
symmetric algorithm. A cryptographic algorithm that uses the same key for encryption and decryption.
symmetric cipher. A secret key cipher.
symmetric encryption. A form of cryptosystem in which encryption and decryption are performed using the same key. Also called conventional encryption.
SYN. A bit in the TCP header used to indicate the packet should be used to synchronize sequence numbers.
system. An organized assembly of equipment, personnel, procedures, and other facilities designed to perform a specific function or a set of functions.
system classification. The act of labeling a system in terms of the sensitivity to the organization of the information processed or stored by the system.
TCP/IP. See Transmission Control Protocol/Internet Protocol.
TCSEC. See Trusted Computer System Evaluation Criteria.
technology envy. The desire to be on the leading edge of technology.
Telnet. A standard Internet service that allows a user terminal access to a remote computer over a network.

TFTP. See Trivial File Transfer Protocol.

threat. Any circumstance or event with the potential to cause harm to a networked system.

Tiger team. A term used to describe a group of individuals given the task of responding to security incidents.

Time to Live (TTL). A field in the header of IP packets that indicates the number of network hops that the packet can take before it is discarded.

timing service. A DCE service used to synchronize clocks among distributed systems.

token-based authentication. Authentication that is provided by the possession of a unique physical card or device that supplies one component of the authentication process.

Traceroute. A UNIX program that uses the TTL field to determine the path between a source and destination computer.

transaction. A collection of operations that comprise a complete unit of work.

Transmission Control Protocol (TCP). The major transport protocol in the Internet suite of protocols providing reliable, connection-oriented, full-duplex stream.

trap door. Secret undocumented entry point into a program, used to grant access without normal methods of access authentication.

trap door one-way function. A function that is easily computed, and the calculation of its inverse is infeasible unless certain privileged information is known.

trigger. A database stored procedure initiated by a predefined event.

Triple DES. An algorithm that uses DES and one, two, or three keys to encrypt/decrypt/encrypt the data.

tripwire. Contributed software, developed by Kim and Spafford, which checks for the unauthorized modification of operating system and application data files and programs.

Trojan horse. A type of computer program that performs an ostensibly useful function, but contains a hidden function that compromises the host system's security.

trust. The composite of security, availability, and performance. It is a firm belief or confidence in the honesty, integrity, reliability, and so on, of another person or thing.

trusted system. A computer and operating system that can be verified to implement a given security policy.

trustee. A user or group of users who have specific access rights to work with a particular directory, file, or object.

tunneling. Provide a secure, temporary path over the Internet or another network.

unauthorized access. The capability of reaching a certain area, either physical or logical, without permission.

unconditionally secure. Secure even against an opponent with unlimited time and unlimited computing resources.

Uniform Resource Locator (URL). The protocol for identifying documents on the Web.

Usenet. A worldwide system of thousands of discussion areas, called newsgroups.

User Datagram Protocol (UDP). A transport layer protocol that delivers packets unreliably across the network.

User ID. An assigned name or acronym used to identify computer users.

Virtual LAN (VLAN). A means by which LAN users on different physical LAN segments are afforded priority access privileges across the LAN backbone so they appear to be on the same physical segment of an Enterprise-level logical LAN.

Virtual Private Network (VPN). A software defined network offering the appearance, functionality, and usefulness of a dedicated private network.

virus. Code embedded within a program that causes a copy of itself to be inserted in one or more other programs. In addition to propagation, the virus usually performs some unwanted function.

VLAN. See Virtual LAN.

VPN. See Virtual Private Network.

vulnerability test. An active test of security controls, performed to discover security weaknesses and exposures.

WAIS. A search engine that indexes large quantities of information and makes the indices searchable.

war dialer. A tool used to locate modems and repeat dial tones within a range of list telephone numbers.

Web spider. A type of keyword search software.

World Wide Web (WWW). A wide-area information retrieval initiative to provide universal access to Internet documents.

worm. Program that can replicate itself and send copies from computer to computer across network connections. The program usually performs some unwanted function.

write access. Permission to write an object.

X.509 certificate. A certificate that conforms to the X.509 authentication framework standard.

X/Open Baseline Security Services (XBSS). A set of security requirements defined by the X/Open organization.

XBSS. See X/Open Baseline Security Services.

XOR. Exclusive OR. A Boolean logic function.

INDEX

3DES, 50
3G, 221

AAA, 36
Abstract Syntax Notation One (ASN.1), 228
Access control, 54, 401
Access control mechanisms, 25
Access control list (ACL), 130, 329
Access line, 351
Access list, 42, 176
Access Point, 320
Access rights, 205
Access VPN, 396
Accounting, 37
Accounting management, 140
ACD, 190
ACL, 130, 329
Active attack, 15
Advanced Encryption Standard (AES), 73
Advanced Mobile Phone Service (AMPS), 460
AES, 73
Agent, 227
Aggregation problem, 201
Algorithm, 49
Alphanumeric pager, 468
AMPS, 460
Anonymous, 44
Antenna, 451
API, 143, 206
Application gateway, 177
Application Program Interface (API), 143, 206
Archive, 199
ASN.1, 228
Asymmetric encryption, 50
Asymmetrical, 46, 78
Attack, 15
Attenuation, 158
Auditing, 294
Authentication, 36, 401
Authentication Header (AH), 233, 401
Authentication server (AS), 102
Authentication mechanisms, 25, 130
Authenticode, 209
Authorization, 40
Automatic call distributor (ACD), 190
Automatic Voice Recognition (AVR), 432
Availability, 17
AVR, 432

Backup, 199
Bacteria, 259
Baseline, 125
Bastion host, 180
Bell-LaPadula Model, 81
Biba Model, 82
Biometric, 57
Biometric-based authentication, 131
Biometrics, 91, 131
Bit-level cipher, 68
Block cipher, 68
Blowfish, 72
Bluetooth, 219, 325
Boot Sector virus, 249
Bridge, 310
Broadcast domain, 330
Broadcast radio, 451
Browser, 211, 372
Browsing, 260
Brute force, 78, 254
B2B, 23
Bus, 305
Business to business, 23

Cable tester, 158
Cache, 211
Caesar cipher, 67
CAPI, 219
Capstone, 73
CAST-128, 73
CCC, 250
CCIPS, 269
CDMA, 466
CDMA2000, 221
CDSA, 218
Cell, 458
Cellular telephony, 458
Central Office (CO), 350, 355
Centralized, 358
CERN, 239
CERT, 250
CERT Coordination Center (CCC), 250
Certificate, 44
Certificate authority (CA), 44
Certification programs, 492
CGI, 262
Challenge Handshake Authentication Protocol (CHAP), 104, 215
Challenge/response, 215
Channel banks, 355
CHAP, 104, 215
Check value, 67
Checksum, 45

Chinese Wall Model, 82
Cipher, 65
Ciphertext, 66
Circuit filtering, 177
Circuit switching, 365
Clark-Wilson Model, 82
Client, 102
Client/Server, 6, 166, 308
Clipper, 73
Closed User Groups (CUG), 43
CMIP, 138
CO, 350
Code Division Multiple Access (CDMA), 466
CommerceNet, 436
Common criteria, 115
Common Data Security Architecture (CDSA), 218
Common gateway interface, 262
Common Management Information Protocol (CMIP), 138
Compression, 74
Compulsory tunnel, 399
Computer Crime and Intellectual Property Section (CCIPS), 269
Computer Emergency Response Team (CERT), 250
Compuer Security Institute (CSI), 249
Computer system, 166
Computer Telephone Integration (CTI), 190, 338, 432
Confidential, 18
Confidentiality, 35, 401
Configuration management, 138
Content filter, 417
Content management, 336
Contingency, 295
Continuity plan, 27
Controller, 168
Cookies, 374
Corporate security policy, 11
Covert channels, 80
CPE, 308
Cracker, 129, 254
Credentials, 103
Credit card, 442
Cryptanalysis, 65
Cryptanalyst, 78
Crypto API (CAPI), 219
Cryptographic systems, 433
Cryptography, 65
Cryptology, 65
Cryptosystem, 78

509

CTI, 190, 338, 432
CUG, 43
Customer Premises Equipment (CPE), 308
Customer security, 291
Cybercash, 438
Cybercrime, 269
Cyberspace, 270
Cyberterrorism, 269

DACS, 356
Data, 22, 198
Data Communications Equipment (DCE), 308
Data Encryption Algorithm (DEA), 69
Data Encryption Standard (DES), 50, 69
Data security, 49, 69
Data Service Unit (DSU), 310, 354
Data Terminal Equipment (DTE), 309, 354
Data warehouse, 198
Data warehousing, 198
Database, 22, 198
Database Management System (DBMS), 201
Database security, 22
Datagram, 368
DBMS, 201
DCE, 308, 354
DDoS, 260
Decentralized, 358
Deception, 328
Default gateway, 183
Default router, 183
Demarcation, 26
Demilitarized zone (DMZ), 176
DES, 50
Denial of service (DoS), 16, 130, 259, 328
Desktop management interface, 142
Dial-back, 131
Dictionary attack, 129, 258
Differential cryptanalysis, 79
Diffie-Hellman, 75
Diffuse, 323
Digest, 67
Digital Access Cross-connect Switch (DACS), 356
Digital certificate, 44
Digital envelope, 49
Digital signature, 45
Digital Signature Standard (DSS), 51
Digital watermark, 81
Direct sequence, 324
Directional, 451
Directory, 204
Directory service, 204
Disaster plan, 295, 443
Disaster recovery, 293

Disaster recovery plan, 27
Disclosure, 17
Disk mirroring, 168
Disk storage, 168
Disk striping, 168
Distributed, 358
Distributed application, 428
Distributed architecture, 428
Distributed computer environment, 143, 428
Distributed data processing, 428
Distributed Denial of Service (DDoS), 260
Distributed environment, 142, 428
Distributed file system, 428
Distributed management environment, 142, 428
Distributed name service, 428
Distributed network service, 428
Distributed nodes,
Distributed processing, 428
DMZ, 176
DNS, 374
Domain, 375
Domain Name System (DNS), 374
DoS, 16, 130, 259, 328
DSS, 51
DSU, 310, 354
DTE, 309, 354
Duplexing, 168

EAP, 217
Earthquake, 286
Eavesdropping, 129, 255, 327
E-commerce, 23, 97, 429, 433
E-commerce security, 433
EDGE, 221
EDI, 337, 429
EIA, 360
Electronic cash, 437
Electronic commerce, 23, 97, 429
Electronic Data Interchange (EDI), 337, 429
Electronic Frontier Foundation (EFF), 436
Electronic Industries Association (EIA), 360
Electronic Mail, 236
Encapsulating Security Payload (ESP), 233, 401
Enciphering, 49
Encryption, 49, 394
Enhanced Data GSM Evolution (EDGE), 221
Enterprise System Review, 484
Enterprise threats, 248
Environmental safeguards, 284
ESP, 233, 401
Ethernet, 303
Ethics, 268

E-mail, 236
E-merchants, 439
E-security, 439
E-thieves, 439
eTRUST, 436
Extensible Authentication Protocol (EAP), 217
Extranet, 19, 334, 431
Extranet security, 335
Extranet VPN, 396

Face recognition, 93
Factoring, 78
Fault management, 138
FDMA, 465
Federal Information Processing Standard (FIPS), 69
FEP, 169
Fibre Channel, 169
File Transfer Protocol (FTP), 212
File-infecting virus, 249
Filter, 42, 177, 311
Finger geometry, 92
Finger scanning, 91
FIPS, 69
Fire hazards, 285
Fire and water damage, 282
Firewall, 42, 175, 310, 404, 431
Flooding, 286
Forensics, 291
Frequency Division Multiple Access (FDMA), 465
Frequency Hopping, 324
Front-end processor, 169
FTP, 212

Gap analysis, 267
Gateway, 184, 311
General Packet Radio Service (GPRS), 220
Generic Security Services API (GSS-API), 219
Geostationary, 457
Geostationary Earth Orbit (GEO), 457
Geosynchronous, 457
Global positioning, 132
Global System for Mobile (GSM), 464
Gopher, 213
Government, 80
GPRS, 220
GPS, 132
Gramm-Leach-Bliley Financial Services Modernization Act, 438
GSM, 464
GSS-API, 219
Guided, 451

Hacker, 129, 255
Hand geometry, 92
Hardware, 13

Harrison-Ruzzo-Ullman Model, 82
HRU, 82
Hash, 67
Hash algorithm, 66
Hash code, 51
Hash function, 50
Hashing, 51
Heat and humidity, 284
HTML, 211, 372
HTTP, 96, 374
Hub, 312
Human attackers, 281
Hybrid system, 190
HyperText Markup Language (HTML), 372
Hypertext Transfer Protocol (HTTP), 96, 374

IANA, 380
IDEA, 50, 71
Identification,
Inference problem, 201, 328
Information, 22
Information Flow Model, 83
Information leaks, 21
Information Technology Security Evaluation Criteria (ITSEC), 113
Information warfare, 269
Integrity, 16, 417
Intellectual property, 261
Interactive Voice Response (IVR), 432
Interfaces, 360
International Data Encryption Algorithm (IDEA), 50, 71
International Standards Organization,
Internet, 19, 370, 431
Internet Assigned Numbers Authority (IANA), 380
Internet Explorer, 214
Internet Protocol Security (IPSec), 233, 412, 441
Internet Protocol Version 4 (IPv4), 238
Internet Protocol Version 6 (IPv6), 238
Internet Service Provider (ISP), 372
Internet telephony, 418
Internet VPN, 397
Internetworks, 357
Intranet, 19, 332, 430
Intranet VPN, 396
Intranetware, 210
Intrusion detection, 26, 105, 292
IPSec, 233, 412, 441
IPv4, 238
IPv6, 238
Iris imaging, 93
ISO, 354
ISO 17799, 114
ISP, 372
ITSEC, 113
IVR, 432

Java, 211, 372
JavaScript, 372

KDC, 102
Kerberos, 102
Kernel, 203
Key, 46
Key distribution center (KDC), 102
Key escrow, 54
Key management, 53
Key recovery, 80
Key systems, 190

L2F, 232
L2TP, 231
LAN, 18, 304
LATA, 350
Layer 2 Forwarding Protocol, 232
Layer 2 Tunneling Protocol (L2TP), 231
Line of sight, 323
Local Access Transport Area (LATA), 350
Local Area Network (LAN), 18, 304
Local loop, 351
Logic bomb, 21, 258
Low Earth Orbit (LEO), 458

Macro virus, 249
Mainframe, 6, 166
MAN, 18, 331
Managed object, 135
Managed device, 135
Management Information Base (MIB), 135, 223
Man-in-the-middle, 257
Masquerade, 255, 327
MAU, 313
MD4, 52
MD5, 52, 71
Media, 317, 340, 355
Media access, 288
Message digest, 51
Message Digest Algorithm 4 (MD4), 52
Message Digest Algorithm 5 (MD5), 52, 71
Metropolitan Area Network (MAN), 18, 331
Microwave technologies, 454
MIB, 135, 223
MIB tree, 144, 224
Middle Earth Orbiting (MEO), 458
MIME, 101
Mobile Telephone Switching Office (MTSO), 460
Mobile telephony, 453
Modem, 313, 354
Modulator/Demodulator (Modem), 313, 354
Monitoring, 292
Mono-alphabetic cipher, 67

MTSO, 460
Multipart virus, 249
Multipurpose Internet Mail Extension (MIME), 101
Multistation Access Unit (MAU), 313

National Computer Security Association (NCSA), 249
National Institute of Standards and Technology (NIST), 71, 198
Natural disasters, 280
NCSA, 249
NDS, 210
Netcat, 210, 262
NetWare, 210
NetWare Directory Services (NDS), 210
Network, 18
Network access , 287
Network Access Server (NAS), 411
Network analyzer, 272
Network Interface Card (NIC), 314
Network Management, 132, 145, 222
Network Management System (NMS), 222
Network monitoring, 145
Network Operating system (NOS), 128
Network resource, 35
Network resource access, 35
Network security, 13
Network Service Provider (NSP), 375
Network Termination One, 314
Network to Network Interface (NNI), 361
Newsgroups, 213
NIC, 314
NIST, 71, 198
NMS, 222
NNI, 361
Non-repudiation, 417
NOS, 128
NSP, 375
NT-1, 314

Object,
Object identifier (OID), 228
OID, 228
Omnidirectional, 451
One-time password, 40
One-way function, 74
Open Group, 113
Open Software Foundation (OSF), 113
Open System, 113
Open Systems Foundation (OSF), 142
Open Systems Interconnection (OSI), 214
Operating System (OS), 167, 202
Orange Book, 108
OS, 167, 202
Oscilloscope, 156
OSI, 214, 354, 479

Packet Data Network (PDN), 367
Packet filtering, 177
Packet filters, 405
Packet sniffer, 256
Packet Switching, 365, 367
Packets, 372
Pager, 468
Palm scanning, 92
PAP, 104
Passive attack, 15
Password, 38
Password Authentication Protocol (PAP), 104
Password cracker, 129
PBX, 189
PCN, 322
PCS, 462
PCT, 94, 435
Performance management, 139
Permissions, 206
Personal Communication Network (PCN), 322
Personal Communication Services (PCS), 462
Personal Communications Satellite Services (PCSS), 465
Personal Digital Assistant (PDA), 467
Personal identification number (PIN), 43
Personnel security, 291
PGP, 96
Physical access security, 290
Physical device security, 289
Physical security, 26, 280
Piconet, 220
PIN, 43
PKE, 53
PKI, 46, 77, 440
Plain Old Telephone Service (POTS), 347
Plaintext, 66
Point of Presence (POP), 155, 359
Point to Point Protocol (PPP), 104, 215, 400
Point to Point Tunneling Protocol (PPTP), 230, 394
Poly-alphabetic cipher, 67
Polyinstantiation, 202
Polymorphic virus, 250
POP, 155, 359
Port scanning, 254
POTS, 347
Power loss, 282
PPP, 104, 215, 400
PPTP, 230, 394
Premptive, 147
Pretty Good Privacy (PGP), 96
Private Branch Exchange (PBX), 189
Private Communication Technology (PCT), 94, 435

Private key, 46, 78
Privilege, 44
Promiscuous mode, 272
Propagation delay, 458
Protocol, 215
Protocol analyzer, 156, 272
Prototyping, 266
Proxy, 180
Proxy server, 180, 405
Proxy services, 181
PSTN, 350
Public key, 46, 74
Public Key Encryption (PKE), 46
Public key infrastructure (PKI), 77, 440
Public Switched Telephone Network, 350
Push, 373

RACE Integrity Primitives Evaluation (RIPE), 72
Radio Frequency (RF), 453
Radio waves, 320
RADIS, 37
RAID, 168
Rainbow Series, 108
RAS, 36, 394
RC5, 72
Realm, 103
Red Book, 110
Redundant Array of Inexpensive Disks, 168
Remote Access Server (RAS), 36, 394
Remote Authentication Dial-in User Service, 37
Remote Monitoring (RMON), 237
Repeater, 314
Replay, 257
Replay attack, 129
Replication, 199
Repudiation, 45
Retinal scanning, 93
RFC, 487
Rights, 206
Ring, 306
RIPEMD-160, 72
Risk analysis, 266
Rivest, Shamir, and Adleman (RSA), 51
RMON, 237
RMON MIB-II, 144
Roaming, 462
Roles, 44
Rootkits, 210, 262
Route authentication, 42
Router, 183, 315, 406
RS232-D, 360
RSA (Rivest, Shamir, Adleman), 51, 76

S/Key, 40
S/MIME, 101

Saboteurs, 287
SATAN, 155, 252
Satellite microwave, 451
Satellite transmission, 456
S-Box, 72
Scatternet, 220
Schema, 198
Screen pops,
Screened subnet, 180
SCSI Enclosure Standard (SES), 169
Search engine, 374
Secret key , 49, 78
Secret key algorithm, 50
Secure card, 104
Secure Courier, 383
Secure Electronic Transaction (SET), 97, 435
Secure hash algorithm (SHA), 52
Secure Hypertext Transfer Protocol (S-HTTP), 96, 436
Secure Multipurpose Internet Mail Extension (S/MIME), 101
Secure Sockets Layer (SSL), 95, 435
Securing electronic transactions, 434
Security accreditation, 107
Security Act, 159
Security administration, 122
Security Analyzer Tool for Analyzing Networks (SATAN), 155, 252
Security Association (SA), 233
Security audit, 154
Security breach, 259
Security cards, 42
Security certification, 107
Security consulting , 28
Security evaluation, 107
Security labels, 239
Security management, 139
Security models, 81
Security policy, 12, 125
Security services, 434
Security system design, 89
Sensors, 292
Server, 102, 171
Server security, 172
Service control point (SCP), 395
Service switching point (SSP), 395
SES, 169
Session hijacking, 256
Session key, 103
SET, 97, 435
SHA, 52
SHA-1, 71
S-HTTP, 96, 383, 436
Signature scanning, 93
Signaling System 7 (SS7), 351
Simple Network Management Protocol (SNMP), 136, 169, 226
Site security policy, 12
Skipjack, 73

Index

SLIP, 400
Smart card, 43, 132, 438, 464
SMI, 228
SMLI, 187, 406
Smurf attack, 260
Smurfing, 260
Sniffer, 129, 312
SNMP, 136, 169, 226
SNMPv2, 226
SNMPv3, 226
Social engineering, 262
Socket, 254
SOCKS, 217
SOCKS5, 218
Software, 13
Spoofing, 179, 256
Spread spectrum, 323
SS7, 351
SSL, 95, 383, 435
Stack-based buffer overflow, 203
Standards, 339
Standards organizations, 340
Star, 306
Stateful Multi-layer Inspection (SMLI), 187, 406
Stealth virus, 250
Steganography, 66
Storage management, 170
Store and Forward, 368
Storm damage, 286
Stream cipher, 69
Structure of Management Information (SMI), 228
Subscriber ID Module (SIM), 464
Superuser, 129
Switch, 316
Symmetric algorithm, 50
Symmetric encryption, 50
Symmetrical, 78
SYN flood, 260

TACACS, 37
TCB, 110
TCP/IP, 234, 371
TCSEC, 108
TCSEC Class A1, 110
TCSEC Class B1, 110
TCSEC Class B2, 110
TCSEC Class C1, 109
TCSEC Class C2, 109
TDEA, 69
TDMA, 465
TDR, 156
Telnet, 213
Terminal Access Controller-Control Access, 37
Terrestrial microwave, 451
Test equipment, 156
TFTP, 212
TGS, 103

TGT, 103
Third generation, 221
Threat, 14
Threat assessment, 264
Three-way handshake, 234
Ticket-granting server (TGS), 103
Ticket-granting ticket (TGT), 103
Time Division Multiple Access (TDMA), 465
Time domain reflectometer (TDR), 156
TLS, 95
TNI, 110
TOKEN, 131
Token Ring, 303
Token-based authentication, 131
Topology, 306
Transceiver, 317
Transmission Control Protocol/Internet Protocol (TCP/IP), 371
Transponder, 456
Transport Layer Security (TLS), 95, 470
Transposition cipher, 68
Trap, 141
Trapdoor, 21
Triple Data Encryption Algorithm (TDEA), 69
Triple-A, 36
Triple-DES, 50, 69
Trivial File Transfer Protocol (TFTP), 212
Trojan horse, 21, 258
Troubleshooting, 154
Trunk, 351
Trust , 10, 41
Trust model, 41
Trust relationship, 41
Trusted Computer System Evaluation Criteria (TCSEC), 108
Trusted computing base (TCB), 110
Trusted network, 10
Trusted Network Interpretation (TNI), 110
Trusted system, 41
Trustees, 205
Tunneling, 399
Two-factor authentication, 132
Two-way messaging, 467

UMTS, 221
Unguided, 451
UNI, 361
Uniform Resource Locator (URL), 374
Uninterruptible power supply (UPS), 282
Universal Mobile Telecommunications System (UMTS), 221
UNIX, 209
Untrusted network, 10

UPS, 282
URL, 374
USENET, 213
User Network Interface (UNI), 361

V.35, 361
Vandals, 287
Vigenere cipher, 67
Virtual circuit, 368
Virtual LAN, 329
Virtual Local Area Network (VLAN), 329
Virtual Private Network (VPN), 391
Virus, 21, 249
VLAN, 329
VLAN security, 330
Vocoder, 466
Voicemail, 191
Voice over IP (VoIP), 418
Voice systems, 188
Voluntary tunnel, 399
VPN, 229, 391
VPN security issues, 415
Vulnerabilities, 13, 329

Wallet, 437
WAN, 18, 348
WAP, 469
War dialer, 253, 364
Water damage, 286
Waveguides, 454
WBEM, 147
WCDMA, 221
Web, 371
Web-Based Enterprise Management (WBEM), 147
Wide Area Network (WAN), 18, 348
Wideband Code Division Multiple Access (WCDMA) , 221
Windows NT Server, 206
Windows NT Workstation, 206
Wireless, 219
Wireless Application Protocol (WAP), 469
Wireless LAN (WLAN), 319
Wireless LAN connectivity, 320
Wireless LAN security, 326
Wireless standards, 470
Wireless LAN technologies, 322, 469
Wireless Transport Layer Security (WTLS), 470
WLAN, 319
World Wide Web (WWW), 18, 371, 381, 431
Worm, 259
WWW, 371, 431

X.21, 360
X.500, 205
X.509, 78